Quantitative Analysis for System Applications

DATA SCIENCE AND ANALYTICS TOOLS AND TECHNIQUES

Daniel A. McGrath, Ph.D.

Technics Publications
BASKING RIDGE, NEW JERSEY

2 Lindsley Road
Basking Ridge, NJ 07920 USA
https://www.TechnicsPub.com

Cover design by Lorena Molinari

Edited by Jennifer Nichols

First Edition

First Printing 2018

Copyright © 2018 Daniel A. McGrath, Ph.D.

ISBN, print ed.	9781634624237
ISBN, Kindle ed.	9781634624244
ISBN, PDF ed.	9781634624268

Library of Congress Control Number: 2018954668

Contents

List of Figures

List of Tables

Introduction

When you can measure what you are speaking about, and explain it in numbers, you know something about it. But when you cannot express it in numbers, your knowledge is of a meager and unsatisfactory kind.

William Thompson (Lord Kelvin), 1891

This book is for data scientists – those knowledge workers who have experienced "The Meditation of 10,000 Mouse Clicks" over and over again. This meditation is experienced during deep dives into a mess of new data where manual cleaning and arranging must be performed before analysis can begin. As data holdings get bigger and bigger, and questions get harder and harder, data scientists must focus on the *systems*, the tools and techniques, and the disciplined process to get the correct answer, quickly!

The word 'system' is used with two meanings: a physical one and a virtual one. The data scientist must have a deep understanding of the working physical system – the subsystems, components, materials, interfaces, interactions, transactions, humans-in-the-loop, measurements, and all related knowledge. To perform quantitative analysis, the data scientist must also interface with the virtual systems that hold the numbers in all the many forms from XML files in "the Cloud", to classic databases, to ubiquitous spreadsheets, and to the almost extinct clipboard with a tally list. Hence the title: *Quantitative Analysis for Systems Applications* (QASA).

It is assumed that the data scientist reading this book is supporting a mid-size operation. The large operations have already achieved mature quantitative analysis processes that work with big data systems. Small operations are typically data deserts,[1] but whose management can still greatly benefit from the fundamental analytical tools. Managers of mid-size operations want to see results generated from the full suite of analytical tools, like the large operations use, but do not have the budgets to build out robust databases using star schemas or to write customized applications. Examples presented in this book apply to manufacturing plants with 3,000 people, distributed operations that have almost 1,000,000 transactions per day, and summary reports for the 3,221 counties in the United States (US). This abundance presents the data scientist with many opportunities to shine! While most answers can be obtained in Excel®,

---

[1] A data desert is a term used when there are few or no digital holdings associated with a process or operation which prevents quantitative analysis or modeling from being performed.

some professional grade tools are needed. This book is software independent with the examples presented having been generated using the most common tools available. The intent of this book is to present the concept rather than to be a how-to tutorial.

This book is intended for the continuum of analysts and managers in both industry and government who have needs that range from simple situational awareness in support of proposal development, to personal development of data science skills while between contracts, and ultimately to performance of a task to process large amounts of quantitative data in the course of analysis and production activities. A secondary audience is graduate students who need to process data using the latest technologies and who "need to up their game" in data science. The primary aim is to provide a common language to be used between business analysts and data scientists for communicating needs and results. The secondary aim is along the lines of a necklace stringing along the pearls of wisdom I have acquired during 35+ years of analyzing data.

The Economist, in a special report in the spring of 2010, defined the data scientist as an emergent professional "who combines the skills of software programmer, statistician, and storyteller/artist to extract the nuggets of gold hidden under the mountains of data." This book focuses on approaches, techniques, and tools (collectively called *technology*) that can be applied from a simple spreadsheet upward to the extensive holdings in "the Cloud." This manual is not about software tools. There are many vendors that provide full, turn-key solutions and specialized applications that all have an appropriate place to be deployed. This manual is focused on what can be done using analytics rather than how to do it.

Analytics is the latest in disruptive technology, especially when coupled with big data, but not totally new to the 21st Century. IBM's Watson advertises the power of computer-aided decision making in TV commercials, but the managers of small- and mid-sized businesses cannot access that capability. MBA classes introduce the tools, but do not go deep enough to teach the art and science needed to put the software to work to generate defensible models. As global sales become a reality for all economies in the world, thanks to the internet, supply chain and supporting data also are globalized and presented using many classic and cutting-edge data visualization tools. Two examples herein use data for world trade and soccer, since those two examples cover the modern geo-spatial spread of concentrations of wealth and people. Learning these locations will help prepare the reader for the places that will be addressed and influential during the next 20 years of decision making.

The nine chapters in this book are arranged in three parts that address systems concepts in general, tools and techniques, and future trend topics. Some chapters are presented using a question and answer approach like used in blogs. It is not a textbook since practice problems and detailed methodologies are not presented. Many examples are included that were generated using common software, such as Excel, Minitab, Tableau, SAS, and Crystal Ball, that are easily reproduced by a knowledgeable user. Detailed programming for cloud-based applications is not presented, but the logic does apply for the full-scale implementations. While words are good, examples can sometimes be a better teaching tool. For each example included, data files can be found on the companion website at https://technicspub.com/qasa. Warning: some of the Excel files are quite large (approaching 100 megabytes) so be mindful of the data connection being used for downloading them. Many of the data sets are tied to the global economy because they use data from shipping ports, air freight hubs, largest cities, and soccer teams. Regarding references: material I quoted is attributed with a footnote. A bibliography is included at the end that includes all of the works I reviewed in conjunction with the writing process and is accessible using author's last name and year of publication. Some citations are messy as they are websites. I did not cite vendor brochures and websites.

I hope reading QASA (sorry, being a Federal contractor for 35 years wired my brain for acronyms) will increase your skills for performing the role of a data scientist. The lessons I share are the result of countless projects using a never-ending stream of changing technologies to generate answers perfectly, cheaply, and instantaneously. Many of my assignments mimicked the chaos of a rugby game in which I ended up "muddy and bloody" by the end … every time. The concepts shared herein were developed for survival and I hope they help you as they helped me. And as we say in the southwest US, *mi QASA es su QASA.* ☺

Daniel A. McGrath
Lubbock, TX
daniel.mcgrath.lemsco@gmail.com

PART I

Foundation

He who is content with what he has,
would not be content with what he would like to have.

Socrates

CHAPTER 1

What Does Quantitative Analysis of a System Really Mean?

The 21ˢᵗ Century is full of computers that accumulate data about every phase of our lives and work.[2] Many tools exist that allow us to mine and analyze the data. Armed with nearly infinite computer power, informative graphs and charts can be generated nearly instantaneously. This plethora of information quickly becomes noise unless squelching filters are used to enhance the signal that delivers the needed knowledge. There is only one leverage point where the cycle of unmitigated information proliferation can be broken – that is the critical role data scientists play in analyzing systems. The analyst needs to ask the following questions before the noise suppression function is ever touched:

- Where did the data come from?
- What does the data really mean?
- Why am I looking at these data?
- Who needs the information these data provide?
- When will these data support a decision?
- How will I distill all these data into meaningful information?

There are many other questions that have been asked and explored by philosophers ever since stylus was applied to tablet. The classical Greeks coined the concept of science to describe this line of inquiry. Science is defined as "the state of knowing: knowledge as distinguished from ignorance or misunderstanding."[3] So what good is this knowledge? Sometimes it just describes nature. Most times it involves how humans live since critical decisions are made based on the knowledge at hand that is derived from the data that is gathered. In the workplace, decisions are made by managers. In everyday life, we all frequently operate in managerial mode, so we can thrive in an ever-changing world. Management is defined as "judicious use of means to accomplish an end."[3] So, in the modern world, this means using data to help make decisions.

---

[2] Steve Lohr, The Age of Big Data, New York Times, February 11, 2012.

[3] All definitions are from https://www.merriam-webster.com/dictionary/.

Formally, this approach of using data is called *management science*, which Wikipedia describes as:

> *the broad interdisciplinary study of problem solving and decision making in human organizations...It uses various scientific research-based principles, strategies, and analytical methods including mathematical modeling, statistics and numerical algorithms to improve an organization's ability to enact rational and accurate management decisions by arriving at optimal or near optimal solutions to complex decision problems. In short, management sciences help businesses to achieve goals using various scientific methods.*

Management science is a very 20th Century discipline that has been slightly re-branded for the 21st Century. It sets the foundation for using data and making decisions and benefits from advances in databases, accessibility, visualization, and interfaces from modern computer technology. It must be judiciously used for the betterment of the people that interface with the system in a manner that is socially responsible. For, if compassion is not shown or if human dignity is not honored, the system will not be sustained. Simulations and models do not address people, but rather avatars, and make decisions in a coldly logical manner for optimal results. Real life is messy, so all solutions will be suboptimal when real people are in the loop.

What is all the fuss about analytics these days? Table 1-1 list the terms currently in vogue that define the subject areas this book addresses. These terms all fall into the realm of management science and represent new tools and techniques. The fundamental questions still apply, with the first step always to gain a detailed understanding of the system itself.

What is a system?

A system is defined as "a regularly interacting or interdependent group of items forming a unified whole."[3] In common usage, it can be almost anything including ecosystem, computer system, entertainment system, surgical system, weapons system, business system, and even System of Systems. A challenge has always been how to describe the system. My car has a nice owner's manual, but when I need more detail, I find a copy of the shop manual that lists all of the components of all of the subsystems along with blue print views that decompose all assemblies into their piece-parts. Then there is still the electrical system with the on-board computer that I have never investigated. Now with self-driving cars, there are visual subsystems, radar subsystems, and real-time navigation systems that recalculate for traffic jams.

Table 1-1. Buzzword Bingo.

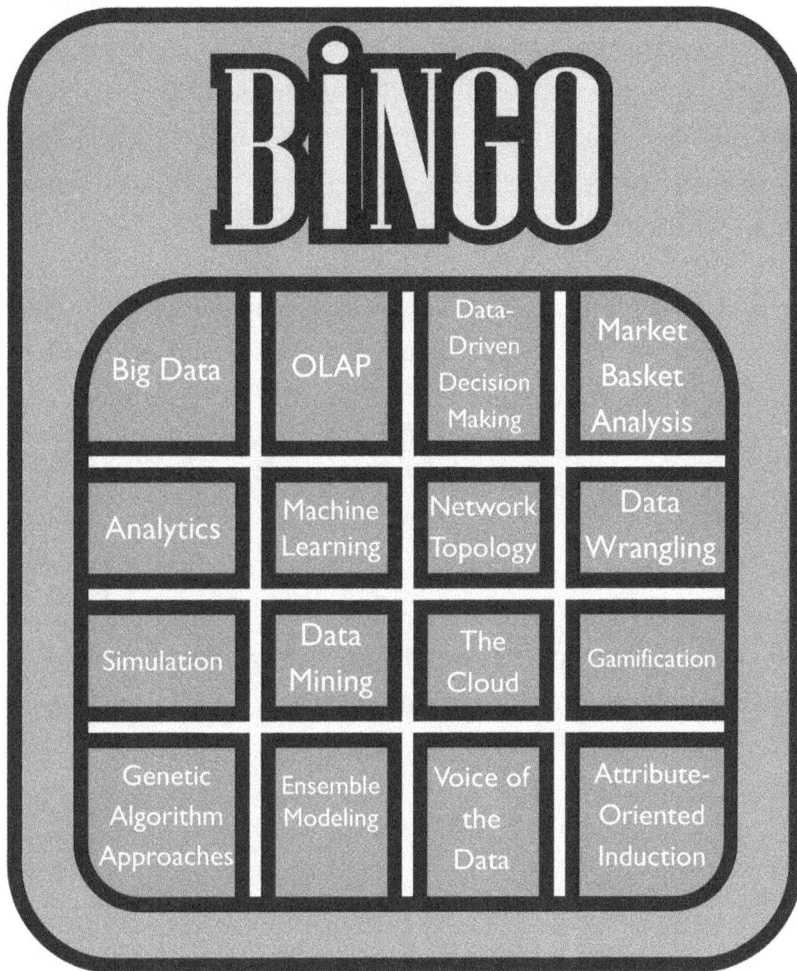

Each manufacturer has their own spin on how expected capability is defined – in systems speak, one can say that this class of objects follows the rule of isomorphism. Over time, centralization is achieved when one vendor makes the best solution and it is adopted by all manufacturers with massive economy of scale benefits realized by all manufacturers and owners.

The systems science terms most recognized are *open* and *closed*. In the physical world, an open system can receive or disperse matter and energy. This cannot occur in a closed system. In the digital world, there are more distinctions. Is the system connected to the web? Are the inputs pre-programmed or can they be dynamic signals? Is it memory resident or saved to a disk? Closed systems obtain a state of equilibrium following Newton's Second Law of Physics. Open systems are known to obtain a steady state condition where inputs are balanced against outputs irrespective of what the initial conditions were (equifinality). This becomes more challenging when feedback loops exist and when advanced processes are self-regulating (homeostasis).

Other concepts for describing systems include decomposition, force field analysis, and hierarchical structure. These descriptions are necessary when building a digital representation of a system – commonly known as a *model*. Model building is an example of progressive differentiation, or what programmers call spiral development. The initial builds are always simple so that they can be verified (Does the model work as designed?) and validated (Does the model give the right answer?). If utility is proven, the model is expanded to accept more inputs, perform more functions, and be used for more purposes.

This growth, like that of an organism, is governed by self-organization, results in higher complexity, and represents another instance of an open system – measured by the expansion of capabilities and utility. In thermodynamic terms, entropy is decreased, and degradation is inevitable unless energy is supplied to maintain the steady state condition.

There are a few pitfalls in not addressing the system, such as traditional business intelligence weaknesses because of flawed or incomplete information, strategies based on backwards looking data (historical financial performance), and insular decision making by top executives using tribal knowledge. With respect to modeling, failure to address all inputs and their interactions will yield results that do not mimic reality and will consequently fail validation efforts. The fuel that allows all of this to happen is data of good quality.

What are data?

Datum (singular form of data) is defined as "something given or admitted especially as a basis for reasoning or inference."[3] So, just about anything can be data, depending on the need. Most commonly it is a measurement of a property inherent to the system. It can simply be the flag position of a specified condition. Defining data is not a new problem.

Aristotle's categories are a good place to start to narrow down the aspects that data can take:

- Substance
- Quantity
- Quality
- Relation
- Place
- Time
- Situation
- Condition

- Action
- Passion

While these categories do not always directly apply to every data set – I doubt if passion can be applied to rainfall data – they are a starting point. The categories that are used relate to the context of the data, or the provenance to use an IT term. The point is that the data scientist needs to understand how the data was collected and what it represents before starting to process the numbers. Leaping blindly into the sea of data usually results in multiple sessions of reworking the analysis, which is A-ok when you get paid by the hour!

Then there is text and the tools specific to textual analysis. This book does not address that realm. Fundamentally these tools take categorization to the nth degree by looking at themes, contexts, centrality, density, and even individual words when context analysis is applied. The quantitative analysis consists of counting, discussed in detail below.

Social media is a special case of textual analysis that also has an additional component of "connections", which leads to network analysis. Logical processing is similar to performing quantitative analysis and becomes complicated when hundreds of entities become involved.

Figure 1-1 is a handy visualization tool that uses font size to indicate the frequency of a term in a specific body of knowledge. JFK, the large airport serving New York City, has the most visitors and is shown with the largest font. Miami and Los Angeles are second and third with the color being slightly redder than blue. Note that colors fade to purple, then red, reflecting the decrease in magnitude. (Apologies to those reading this in grayscale.)

Figure 1-1. Word cloud of top 50 US airports with the most international arrivals.

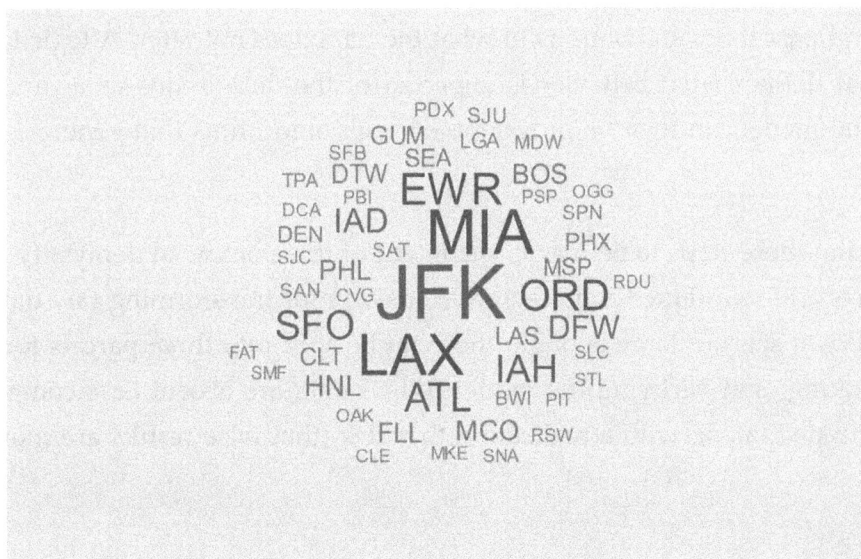

I could have customized this display, but left the defaults to show the groupings of sizes. The Word Cloud is an efficient way to show 50 data points, given that codes are used. If the full airport names were used, the graphic would not be readable. Yes, all 50 names are displayed…it is not that I don't trust the system, I have learned to verify outputs from every software package I ever use. Finally, before I get any emails, I want to explain why DCA is in this Word Cloud. DCA is the code for Reagan National Airport serving Washington, D.C. with Dulles (IAD) being the international airport for the National Capital Region. So, how does a national airport have international arrivals? The flights are from Canada and the US government has implemented a preclearance program at the major airports where travelers go through customs before boarding the flight.

For building a model of a system, data are the key ingredients. Before the model can be built, there are fundamental questions concerning the data that must be answered:

- What is the provenance of the data?
- How was it obtained?
- Was it ever transformed? If so, by whom, when, and using what tools?
- How accurate is it?
- How frequently is it generated?
- How often is it updated?
- Where is it stored?
- How is it stored?
- Who has read/write access?

When armed with these answers, construction can begin on a conceptual representation of the system that demonstrates to what the data refers; what processes generate the data, accept it, transform it, and pass it on; and bounds to what the data does not refer. A logic layer can then be overlain that defines what behavior is expected of the data trends as a function of time. Finally, a digital model can then be programmed with algorithms that generate the expected results.

While some claim these steps to be magic, the intent of this book is to demystify this belief by sharing the tools and techniques of quantitative analysis for transforming raw data into useful models. When data sets are large enough, they can be split into three parcels for purposes of training, calibrating, and verifying the model. When data are absent or incomplete, creative solutions are needed, along with a realization that risk (that false results are more likely) has been accepted.

Fundamental system characteristics: open and closed systems

An example of an open system is England – the island. During the last Ice Age, when sea level was lower, it was not an island, but was connected to France and Ireland via land bridges. This allowed many animal species, plants, and humans free access to a large land area as an open system. When sea levels rose, plants and animals became stranded, but man was able to sail back and forth to Europe, preserving an open system. The most profound aspect of openness was during Roman times in the first century AD, when Rome established colonies for the mining of Cornish tin for shipment to Rome and the making of bronze. This resulted in a complete change to the economic system with the establishment of cities, importation of luxury items, influx of gold and coins, and an attempt to close the social systems of the island with the construction of Hadrian's wall. The most recent openness development was the opening of the Chunnel railroad tunnel in 1994 that allows for the passage of high-speed trains carrying a wide variety of cargos.

An example of a closed system is air travel around the world as it is all inclusive. An example of an open system is air travel in the US with domestic flights being a closed aspect, international flights being an open aspect, and the US border being the boundaries for differentiation. The US Department of Transportation requires monthly forms to be submitted by all international carriers. These data for a single month are shown in Figure 1-2.

Figure 1-2. Venn diagram of international air travel to the US for April, 2014, for total number of passengers and total number of flights.

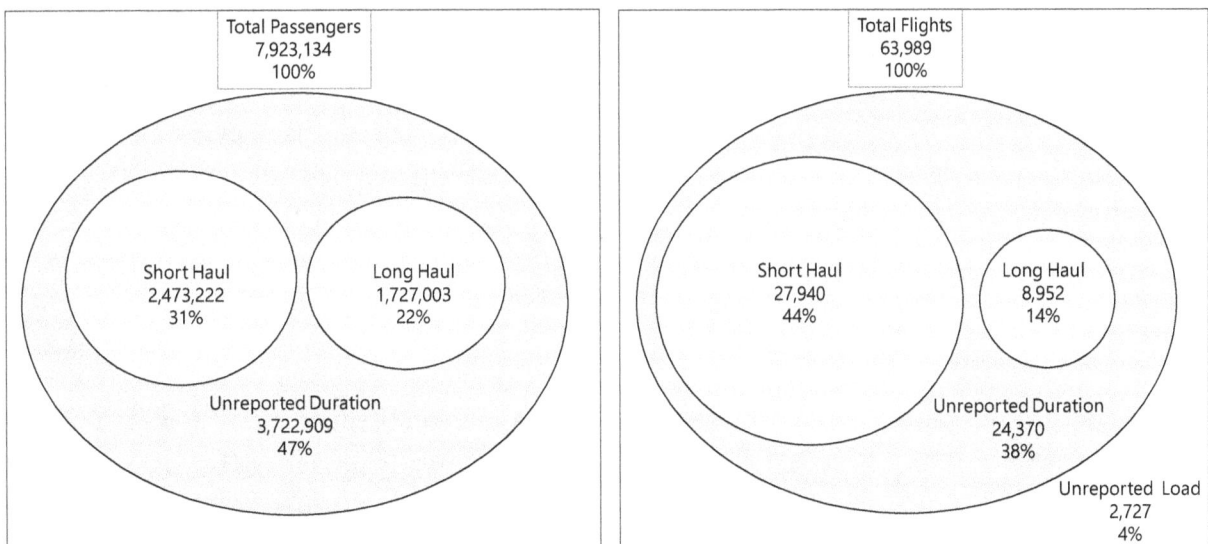

Source: US Department of Transportation Form 41: Schedule T-100(f).

Venn diagrams are very powerful logic diagrams to introduce a data set since it shows the order of magnitude of the data, the degree of independence of the categories, and the proportional relationship between categories. These diagrams can also be prepared for two or more time periods to show the degree of dynamic change to the system. For this example, the Venn diagram highlights two nuances that may limit the utility of the information that can be obtained. This example is conceptual since the proportionalities of the sets were approximated rather than rigorously calculated. It allows the reader to tell the differences from a glance.

The data in Figure 1-2 are categorized by the duration of the flight (ramp-to-ramp field in the database) with short haul international flights being less than five hours and long-haul flights being up to 16.7 hours. This is of interest to a vendor wanting to sell blankets for long haul flights to gauge the potential market. However, a substantial portion of the data did not have an entry for the duration. An extrapolation could be made based on the existing ratio between short and long haul flights. Typically, long haul flights use large, wide-body planes and some of the short haul flights, such as those in the Caribbean, use small, single engine planes. A degree of uncertainty for extrapolating the data is demonstrated by the different ratios for passenger data and flight data. The long haul to short haul ratio for passengers is 22% / 31% = 71%. The long haul to short haul ratio for flights is 14% / 44% = 32%. Using an extrapolated value will introduce a substantial amount of uncertainty. But even more fundamentally, the two data sets do not represent the same thing. There were 2,727 flights (4.3%) that did not report passenger data and are shown as being outside the circle for flights. This implies that the total passenger count is short by about 352,688 people. Again, this is an assumption that is clouded by the uncertainty of the comparative ratio and the frequency of duration not being reported, but is useable due to the Law of Large Numbers that is discussed in Chapter 5.

While it is ideal to discover these nuances at the start of an analysis, the reality is that the subtle nuances are typically found after a substantial amount of analysis has been performed and the results do not correspond to the *a priori* expectations. These discoveries require all the analyses to be re-performed and reinterpreted.

Case Study: Life on the Big Island

The Big Island of Hawaii is quite the case study for a system, because it has aspects that are both open and closed. The natural systems have been greatly impacted by man, but the geologic systems always will dominate all other systems. Short introductions are given below to the geologic, biologic, settlement, energy, and economic systems. However, this is a long description. It represents a typical degree of complication that a data scientist must mentally track when working with a robust system and trying to build a model of it.

The Big Island is about 4,000 square miles in size, and still adding-on. It was created by the amalgamation of five shield volcanos with three still being active. A shield volcano forms by the stacking of flows of basalt (a hard, black, extrusive igneous rock). The northern most one, Kohala, is the oldest and has a peak elevation of 5,480 feet. Mauna Kea is also dormant, snow-covered through most of the year at 13,803 feet. Hualālai is on the west side and last erupted in 1801. When one lands at the Kailua-Kona airport, the view out the plane's window is of the lava from this eruption and stretches up to the peak of 8,271 feet. Mauna Loa is another snow-capped 13,000+ foot peak that last erupted in 1975. Finally, Kīlauea is the southernmost volcano and has been erupting continuously since 1983. The peak at 4,091 feet is on the rim of the active caldera.[4]

Being an island that is about 2,000 miles from the mainland implies a closed system by definition. Being only 26 miles from its nearest neighbor (Maui) implies certain aspects are of an open system. However, the three active volcanos continue to expand the size of the island, to add nutrients to the soil, and to impact the atmosphere with Sulphur Dioxide gas emissions. Usually, land is a closed system because there is a finite amount. However, all along the Pacific Rim there are volcanos that continue to increase the land.

Biology is a more complicated system. The ecosystem is more open on the Big Island than the Galapagos Islands that Darwin first investigated. The oceanic ecosystem surrounding the Big Island is mostly closed for breeding species, but larvae and accidental species do drift in. However, there is a huge migratory component to that ecosystem. Well, the species are huge – Humpback whales. The whales use the ocean to the west of the island for breeding and calving.

The bird life component of the terrestrial ecosystem is even more open. In addition to the numerous migratory species, there are many endemic birds, with the honeycreepers being unique because they have evolved into distinct species on each of the major islands. How the native birds ever arrived is a true mystery. Because of human settlements, a number of species from around the world have been introduced. This is most obvious at sunrise on Kauai when the roosters awaken. On the Big Island, many of the introduced species occupy niches in the ecosystem that resulted from extreme mortality to the endemic species, others are escaped pets that thrived, and some were introduced as game birds for hunters. While disease and competition played a role, it was a change in the predator-prey relationship that had the largest effect.

Mammals did not play a role in the ecosystem until the Polynesians arrived. While dogs and hogs had some impacts, it was during the expansion to large scale farming that rats and the

---

[4] From Wikipedia.

mongoose were introduced to the island. Since there had been no predators, many of the endemic bird species adopted ground-based nests that had the benefit of lower mortality from fledgling flights. However, these nests were eradicated by the mongoose. A classic example of quantitative analysis of systems is the predator-prey model[5] that calculates what the sustained populations will be under different circumstances.

The final aspect of the terrestrial ecosystem is the vegetation. This is an open system since logs and fruit are transported by ocean currents and seeds are carried by migrating birds resulting in the initial colonization of the islands by plants. There also is colonization on fresh lava with the most dramatic growth occurring on the west side of the island. Polynesian settlers introduced many food plants, some of which spread from the farms into the wild. More recent settlers introduced many ornamental plants that resulted in changes, especially when substantial landscaping was performed, or irrigation applied for the creation of golf courses. However, the biggest changes resulted from the species man introduced for planting in mono-culture agricultural fields, like rice, sugar cane, pineapple, macadamia nuts, coffee, eucalyptus trees, and marijuana. These commercial efforts resulted in many changes, including insect infestations and chemical spraying of insecticides, which had the unintended consequence of decimating indigenous species. This change to a man-controlled aspect to the ecosystem is the true start of the Anthropocene.

There is not a corner of the planet that has not been affected by man's activities. While much work has been done to model what has changed, more robust system models are needed to guide future activities so that desired goals can be achieved. The heart of building these models will be the quantitative analysis of data that describes the system to obtain high resolution and trustworthy results.

Human settlement is a definite open system component. The history of settlement by the Polynesians is fascinating and a testament to the courage and adaptability of the human race. They brought numerous changes to the localized ecosystems, mostly by the addition of nutrients from manure, and even to the landscape by their building campaigns. The Big Island was not significantly changed. I do not know of any computer games that allow the player to simulate colonization in the south Pacific. Simulations of this kind are the experimental laboratory for systems analysis and are only as good as the data used for validation and verification of the workings of this model.

Even though the Spanish galleons of the Manila-Acapulco trade route went around the Hawaiian Islands, western interactions are first credited to Captain Cook's explorations in 1779 and the start of colonization. Settlement and large-scale agriculture did not become significant

---

[5] For more information, see https://en.wikipedia.org/wiki/Lotka–Volterra_equations.

until steam ships became reliable and included refrigeration. This brought about the era of Western settlers, the transport of Mexican cattle to the Big Island, the importation of forced labor to work the fields, and the debilitating effect of western disease on the Polynesian people. With the advent of airplane travel to the island, the isolation that led to the closed system was erased and cultural changes accelerated. Large military bases from World War II and the all-reaching influence of telecommunications truly were forces of homogenization between the island and the mainland.

The most modern aspect to the systems analysis of the Big Island involves energy. It is the life force of the modern world. The original Polynesian inhabitants only had access to a limited supply of firewood which influenced their food preparation choices (think luau). Since the island chain is remote, it became a repository for fossil fuel to restock steamships, which also became abundant for use on the island. Automobiles and air conditioning quickly became wide-spread. The electricity supply grid is now heavily supported by wind and solar farms. The island lies along the course of the trade winds allowing wind turbines to produce electricity 24-hours each day. Someday, when green generating capacity is increased on the Big Island, there might be a massive change to electric cars and boats and the cessation of the import of gasoline. However, diesel will still be needed for the big trucks that use Saddle Pass Road to go between the two sides and for construction equipment that is pitted against raw basalt. While some bio-diesel can be produced, the need to import will always exist. It seems to be a straightforward decision to change the energy system to use more renewable sources. A barrier to develop this change concerns the visual impact of a turbine farm. More analysis is needed to determine the requirements as to what is not visibly pleasing and what is minimally acceptable in the Anthropocene.

How should one quantitatively compare wind turbines to plantations of eucalyptus or macadamia trees or to a giant cruise ship dominating a harbor? This analysis will be quite different from similar analyses made for the small island of Kauai and the crowded island of Oahu. One approach is to consider the painters concept of landscape. This concept captures the sequence of landscapes 'painted' by man on a 'canvas' provided by geology. Pryor (2010) loosely defines landscapes as the "products of human efforts, combined with natural features of the terrain, which together give certain regions a distinctive character." It is a four-dimensional dynamic concept since the landscape changes over time with the coastal views of native temples being replaced by views of anti-aircraft batteries and then being replaced by views of algae farms. Decisions made regarding the future of the landscape always requires tradeoffs between issues leading to consensus between the stakeholders.

The Big Island system is a microcosm of all human settlements. The modern world demands certain essentials (housing, food, water, electricity, and connectivity). In order to realize these essentials, people need to work to earn the money. The challenge of the economic aspect of the system is how to make the work meet the major requirements of sustainability and health while honoring liberty. There always is a big question concerning how to measure economic activity. What data are available for free and what for purchase? How will increases in e-commerce help define the quantified view so that predictions can be made, and models can be optimized and prescribed so that none of the goods shipped to the Wal-Mart in Kona are disposed rather than sold? Island economies are sometimes used as an experimental laboratory due to the closed system aspects. However, modern connectedness through the internet and the airline networks blurs some of the independence of observation for the isolated locations.

This case study captures the many aspects that are simultaneously considered by a systems thinker during a system analysis, the need to build models of many kinds to guide and support decision making, and the critical importance of high fidelity data to support the models. This book explores methods and approaches, tools and techniques, and cultural attitudes and mores, required for the quantitative analysis for system applications.

Other system characteristics

Complexity is a limiting factor on the understanding of a system. Care must be taken not to confuse complexity with complicated (many moving parts) or obfuscated (cryptic language that bars understanding) portrayals. Strong indicators that complexity exists are feedback loops and nonlinear behaviors. Ludwig von Bertalanffy, in his book *General Systems Theory* (1968), provided the basic explanation of many characteristics of systems. Regarding complexity, he wrote:

> *It seems, therefore, that a general theory of systems would be a useful tool providing, on the one hand, models that can be used in, and transferred to, different fields, and safeguarding, on the other hand, from vague analogies which often have marred the progress in these fields.*

> *There is, however, another and even more important aspect of general systems theory. It can be paraphrased by a felicitous formulation due to the well-known mathematician and founder of information theory, Warren Weaver. Classical physics, Weaver said, was highly successful in developing the theory of unorganized complexity. Thus, for example, the behavior of a gas is the result of unorganized and individually untraceable movements of innumerable molecules; as a whole it is governed by the laws of thermodynamics. The theory of*

unorganized complexity is ultimately rooted in the laws of chance and probability and in the second law of thermodynamics. In contrast, the fundamental problem today is that of organized complexity. Concepts like those of organization, wholeness, directiveness, teleology, and differentiation are alien to conventional physics. However, they pop up everywhere in the biological, behavioral, and social sciences, and are, in fact, indispensable for dealing with living organisms or social groups. Thus, a basic problem posed to modern science is a general theory of organization. General systems theory is, in principle, capable of giving exact definitions for such concepts, and, in suitable cases, of putting them to quantitative analysis. (page 34)

In short, he describes classic scientific analysis. The analysis takes time, energy, and, most critically, focus. There is much overlap and repetition for all business systems as they involve humans and work, so not all subsystems must be analyzed if standardized processes are used. Beruvides and Cantu (2003) modernized Bertalanffy's sentiments regarding an organization by comparing the concepts of organism and mechanism. They present the case of a symbiotic relationship resulting in a mechano-biological organization. This view allows for complexity to be embraced in the building of digital models and is the one adopted throughout this book.

Ecosystems are a special case when man is not in control and actions are simple responses to the conditions present. For example, rain on asphalt evaporates quicker than rain on grass. Does Mother Nature measure? Nature does not track processes using numbers. Sometimes it appears as if nature is measuring, but this is only an artifact of man's attempt to analyze. The static laws govern how a process will finish as shown by this example from hydrology. It is a truth of nature that water runs downhill. We all have observed this most simple hydrologic principle. Most people understand that downhill flow applies to groundwater flow, too.

A disparity is seen when an artesian well is constructed, such as the irrigation wells around Roswell, New Mexico, and the water rises to the land surface. An artesian aquifer is a series of rocks that holds water under a pressure head between two layers of impermeable strata. An artesian aquifer is the classical example in geology of a closed system and serves as a segue to understanding pressure gradients in petroleum geology (think of the gusher from Spindle Top).

When too many wells are drilled, or the wells flow too much water, the pressure generated upstream is less than required to maintain the artesian system. There is no measuring by nature – just a cause-and-effect relationship across a pressure gradient that is measurable by man. Overall, systems are the complicated things in real life and the tools and techniques discussed in this book can be applied to the analysis of them.

How to analyze real systems

The Big Island case study identified the diversity of aspects that must be quantified and assessed as part of a systems analysis. There is no shortcut for thinking, or reasoning to use a more philosophic term. Deductive reasoning (top-down) starts with the general theory and then predicts behavior at the individual or micro level. Inductive reasoning (bottom-up) starts with observations then assembles commonalities to form theories that can be used to explain universal behavior. When applying logic to a real system, the data scientist toggles between the two modes in conjunction with forming and evaluating hypotheses as part of an empirical study. Empiricism works if you have enough time and money. If not, computer simulations can be built to serve as *de facto* experiments. Being able to tap participants with deep familiarity and extensive experience (sometimes called *subject matter experts*) are golden opportunities for success. Many times, however, this is not the case and details have to be discovered by a slow and meticulous process.

This section describes six standardized approaches for investigating systems – think of them as *modus operendi* and realize that they work best when they are combined into an ensemble suite subject to the complimentary law of systems: any two different views of a system will identify aspects that are neither entirely independent nor entirely compatible. Most of the approaches described are classic ones that focus on spatial and chronological relationships, behaviors of the participants, decomposing the system into smaller pieces, and quantitative analysis. The first one presented is more philosophical because it is based on a unifying theory of four interconnected dimensions. Finally, this section ends with a discussion of common pitfalls.

Theory of Profound Knowledge

J. Edwards Deming is the father of the Total Quality Movement of the 1980s and 1990s that lead to Ford placing stickers on their cars that said, "quality is job number one." His work experiences are fascinating and include managing production for Western Electric (an AT&T subsidiary) in the 1920s, studying agricultural statistics in England in the 1930s, supporting US logistics planning during World War II, and, most famously, helping Japan develop modern manufacturing capabilities after World War II. He developed a philosophic approach (reinforced throughout this book) based upon a unifying theory – the Theory of Profound Knowledge (TPK) described in the book, *The New Economics*. Some people find his writing to be a bit stuffy and pretentious. However, he considered himself to be quite the hipster during the roaring 1920s and never changed his style. It helps to watch an old silent movie before tackling his writings.

TPK is a paradigm for shaping how a system is to be analyzed. Once it is learned, it provides a filtered view of the world, like rose-colored glasses. The TPK consists of 4 pillars:

- Knowledge of Systems – defining the subject of interest and the processes and transactions that operate therein,
- Knowledge of Variation – fundamental measurement and analysis of the quantitative nature of the system,
- Knowledge of Knowledge – philosophical inquiry along the lines of:
    o What do I know?
    o How do I know it?
    o How sure am I?
    o Who has the right to see it? and
- Knowledge of People – for whom, by whom, funded by whom, according to whom, and despite whom.

All of these aspects must be considered individually and in relation to the other three pillars.

While conceptually simple, the implementation of the TPK becomes extremely complex when applied to real-life systems, because of the myriad details and nuances encountered. Deming did not invent his approaches from nothing. His mentor at Western Electric was Walter Shewhart, the inventor of Statistical Process Control charts in 1924, and whose 1931 book has concepts that are still too advanced for the modern manufacturing line. Walter is also credited with the Plan-Do-Check-Act cycle that is the foundation of all continuous improvement paradigms with the Six Sigma movement trending now.

Spatial

Spatial analysis is the realm of geographers. Maps, network diagrams, and detailed tables have been used for centuries to anchor our knowledge concerning *where*. Tools invented by geographers are primarily for visualization, but also broadly include census taking, polling, and database storage developments. While some consider the list of 220 countries to be a big data challenge, all agree that knowing where to find the soon-to-be ten billion humans on this planet (and our cell phones) is a true big data challenge. Software and database tools are critical for analysis and display of spatial data. Simple mapping capabilities can be found in Excel, Tableau, and JMP Pro. ArcGIS is the premier mapping software and has hooks that allow easy interface with all major database types. It is worth hiring experts with this software to generate accurate maps quickly.

Chronological

Chronological analysis is a powerful tool that can be applied in many situations. The history of a system explains what happened when, by whom, in accordance with what, where, and how. Transactions can also be viewed in detail. The detailed view captures the actual working time for each step as well as the waiting time between steps (this is called the *Value Stream Mapping* tool by continuous improvement practitioners). Timelines are a powerful tool and best built using sticky notes on a board, so they can be expanded as needed. Once completed, they can be memorialized in a document or even entered into Microsoft Project, which serves as an off-the-shelf temporal driven database that can be easily changed, extensively annotated, and exhaustively embellished with new fields. There are even add-on modules that allow hierarchical views when required. Working with the quantitative aspects of time is a challenge because it is not Base 10 like most measurement systems of length, weight, volume, and amount. Care must be taken when moving from 60 seconds to 60 minutes to 24 hours to seven days to 12 unequal months and usually 365 days. Year-over-year analysis is the straightforward aspect.

Behavioral

What motivates you? This is the fundamental behavioral question that can be applied to every stakeholder associated with a system. There are many tools to help in psychological analysis and the masters are literary experts. When you read a novel, a mystery, or a romance, you learn a phenomenal amount of details concerning a person. Sometimes this approach is required to analyze a system to determine what the quantitative data truly relates. It becomes a challenge to capture these features when building a virtual reality representation.

How do you work? This is another aspect of behavioral analysis. Are you an extrovert who thrives in group settings, or an introvert who settles comfortably in a hidden and quiet office? How do you gather and process information, and then how do you organize, store, and synthesize it? A useful paradigm is the "six thinking hats" approach from Edward deBono. You do not have to don the hats physically while working, just be willing to change your point of view when performing design and review tasks. The six hats are:

- Black – Critical Judgement,
- White – Facts and Information,
- Blue – Systems View,
- Red – Feelings and Emotions,
- Yellow – Positive Spin, and
- Green – New Ideas.

Thinking in terms of these hats is also a good skill for team building. It gives participants a language to describe the operational point of view they have adopted in order to control the intensity of storming that the team experiences while moving through Tuckman's stages. There is a time and a place for each hat with the leader being responsible for smooth transitions during the course of the project.

Decomposition

Decomposition enters the quasi-engineering realm – taking things apart to see how they work. Sometimes this leads to back-engineering a solution, other times to the invention of a replacement, and always to a deeper understanding of the system. It starts with a common terminology: a system harnesses multiple subsystems containing multiple components consisting of various parts made of various materials that have physical interfaces and respond to physical, electrical, optical, magnetic, and logical interactions that result in defined and known outcomes. This becomes very complicated quickly with the details usually captured in a technical manual. Decomposition is addressed by the realm of systems engineering and specific tools for configuration management, requirements identification, specification writing, adherence to standards, and inspections. While expensive, it is powerful in that risk can truly be defined and optimized into a minimal state.

Quantitative

My favorite superhero is the movie character Neo from *The Matrix*. Try as I might, I cannot visualize everything in my world as strings of constantly-changing numbers. That world is one giant computer program where everything is scripted. If real, how much data would be required to make a virtual representation of our entire world? Eventually, this data will exist for all human systems and optimization routines will replace decision making. In the interim, data scientists will continue to gather and interpret data from many systems to aid the decision-making process. Systems can be defined down to the finest degree by measurements and counts aspects, but there is an art for how this is done. The majority of this book consists of a series of deep dives into the many aspects of quantitative analysis needed to support decision making. If you are looking for something lighter, I suggest Annalyn Ng and Kenneth Soo's 2017 book *Numsence! Data Science for the Layman (no math added)*. Table 1-2 is a recipe for getting started.

As quantitative data are collected, challenges arise because of the VUCA elements that complicate the analysis. In addition to a thick skin and an indomitable drive by the leader, management support is needed to keep the effort on the rails. Most participants in the quantitative analysis of a system freely share their information. When participants are extremely resistant to release the numbers, this should be interpreted as a red flag indicating that something is wrong, especially when the unit of measure is money. It also warrants consulting with accountants and possibly bringing in the green-eye-shade wearing auditors or even Agent Smith from *The Matrix*. Their tools are designed to unravel a broken quantitative system.

Table 1-2. Steps for getting started with data analysis.

Six Steps for Data Collection
1. Clarify goals.
2. Develop operational definitions and procedures.
3. Validate the measurement system.
4. Begin data gathering. Timing set according to a rational plan and not adrenaline.
5. Define VUCA (Volatility, Uncertainty, Complexity, and Ambiguity).
6. Sustain efforts and monitor for effectiveness, which usually ensures compliance because you "get what you measure." Be vigilant against efforts to "fiddle the figures" or "juke the stats" or "cook the books."

Another restorative approach taps into physics and applies the law of conservation of matter and energy. Or, to say it simply, "do things add up?" In systems speak, audits represent a feedback loop with the results leading to actions to better achieve the goals. Correction of a problem always results in an increase in confidence regarding the system, because that aspect that was fixed is known not to be wrong.

Pitfalls

Regardless of which system analysis approach is taken, there are several classic pitfalls that are encountered. General George Casey said "VUCA environments thus become invitations for inaction – people are befuddled by the turmoil and don't act. And to succeed, you must act." One of the terms associated with not acting is *analysis paralysis*. This is addressed by the parable of Buridan's Ass and the dilemma the beast faces when released halfway between the hay and the water. Due to the inability to make a choice, the beast starves to death.

Occam's Razor is named for a 14th century English philosopher, William of Occam, and states that when given two tools that can be used for a situation, always select the simpler one. While software vendors praise the power of their wares, always question whether the job can be done using Excel in a transparent manner. This concept leads to the Law of Requisite Parsimony, a design adage that guides choices to the simplest possible, but not too simple, given the inherent complexity. In India, tinkerers make solutions that are called *jugaads* – the creation of something new using meager resources.

A Machiavellian fallacy exists in the digital world because information is **not** power. Information is force. Power is force over distance, so information must be disseminated, consulted, and used for the power to be realized. This is evident from Wikipedia and its encyclopedic holdings. It is nowhere as powerful as Facebook, where that information is continuously being applied by the two billion users. If nobody is using the stored information, should it even be collected?

How to analyze IT systems

While virtual reality mimics real life, it is never as complex or as nuanced, and is transient. Sometimes what existed yesterday is part of the ether today. Information Technology (IT) builds virtual or digital systems and takes a team, like the teams supporting a real NASCAR entry: designer, tire technician, mechanic, transmission mechanic, decal technician, and driver. The IT team has system architects, vendor support that pushes updates, programmers, database administrators, cyber-security specialists, users, and data scientists. The most important team member is the Chief Information Officer (CIO) as he or she makes the critical decisions.

Some of the systems are fast-paced data warehouse operations, like the checkout line at 8,000 Wal-Marts worldwide that process over 1,000,000 transactions per hour. Others are small scientific models that have five users over a 20-year history. Most of the analysis falls under the decomposition approach described above, along with the verification and validation activities described in Chapter 4. IT is not glamourous when it comes to making all the "1's and 0's" stay in line. IT systems are just another engineered system, like your car or your cell phone.

In the future, there will only be one database as shown in Figure 1-3. Every aspect regarding a system will be accessible to those who have proper credentials when they log-in. The vendor SAS already supports this approach with the JMP Pro software being a primo package that will allow data scientists to investigate all questions.

This matrix will reside in "the Cloud." There are many IT systems in use that have not reached this maturity, but they are all being consolidated. This book does not address the details of the legacy systems or the migration, but rather focuses on the capabilities applicable to the data scientist.

Figure 1-3. "The Matrix" for <u>all</u> signal and transactional data.

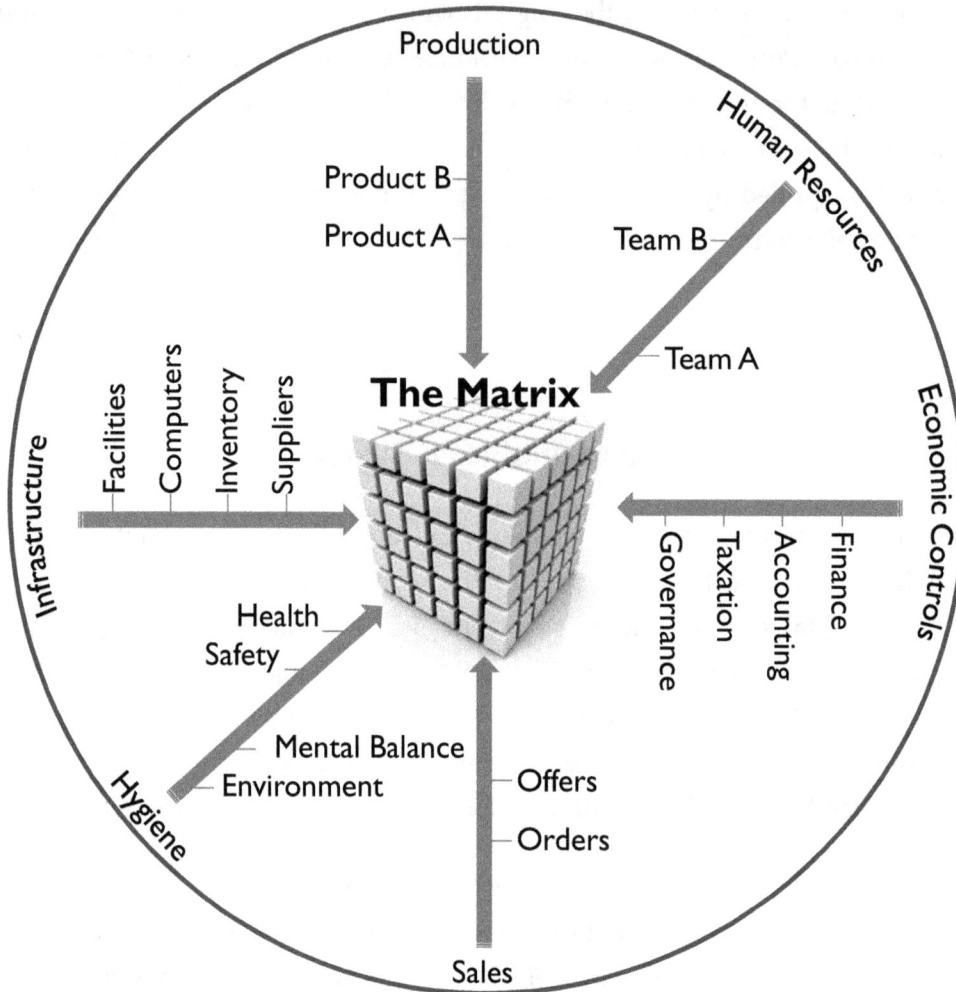

This conversion about having everything in your life in "the Cloud" has already started for the youngest generation. My daughter showed me the application on her phone called "baby tracker" that she is using with her one-week-old son. She clicks an icon every time she feeds him and changes a diaper. The benefit is that her sleep-deprived brain does not need to track this data and she can share it with the pediatrician at the next appointment. Poor little guy...he has already been assimilated.

Another example is the Public Library in Fairfax County, Virginia. The county owns all the books and the books move around the 23 branches constantly. When a book is returned, it is simply scanned, then shelved. The location is visible in "the Matrix." When you want a book, you log-on to your account and select the book and the branch where you want it delivered.

An email is sent when the book is ready for pickup. The Oakton branch even has a drive-up window for the patron's convenience.

Figure 1-4 shows the distinction between the digital world of the data scientist and the analog world of the engineer as a Venn diagram. While both use big data, the focus and goals are quite different. The engineer must truly have a system view to balance the tradeoffs that come when changing time, cost, quality, and safety, in order to avoid unintended consequences.

Figure 1-4. Focus of engineering role in big data processing systems.

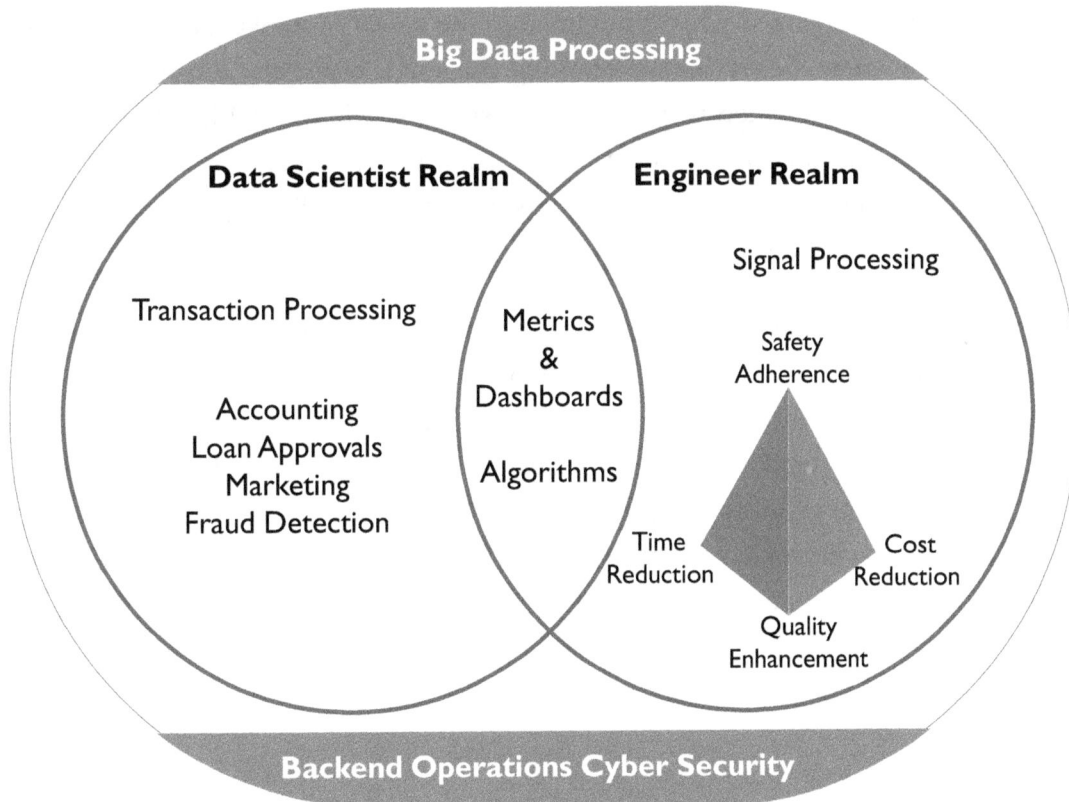

The detailed technical world of IT focuses on six areas that are unique to virtual systems that have their own key themes:

- Schemas – robustness and speed of query,
- Legacy – how to migrate old data into new schema, fill holes, and accommodate change,
- Hardware and connectivity – what is available and what is coming next,
- Inputs – what is being generated where, and then, where should it go for storage and use,
- Interfaces – how will the inputs be processed and analyzed, and
- Outputs – what are the stakeholder expectations?

In addition, there is an overarching layer of computer security that interacts with all six themes. These areas make the realm of IT systems as complex and challenging as the Big Island of Hawaii for the largest instances of the digital realm such as the US financial system. Without them, big data would be too complicated and quantitative analysis tools too slow and labor-intensive to provide utility. This book does not address how all of these areas are achieved but assumes that they are done correctly. If not, then that dreaded acronym must be applied: GIGO (garbage in = garbage out).

I have been lucky to have worked in the three major technical eras: mainframe, distributed, and web-based. Everything changed when I migrated to the newer systems, and some of the older systems never went away. Legacy is a continuing challenge, especially when paper forms are still in use rather than automated workflows. In systems-speak, this achievement of the same end regardless of the starting point is called *equifinality*. Will there be a fourth technical era in the future, or will technical innovation reach a steady-state equilibrium, like gasoline-fueled automobiles have been for the last 100 years?

These IT activities, while critical to data science, are not data science. They must be done right for data science to be successfully applied such as in the creation of a predictive model. The people who do the different functions have very different skill sets, just like the NASCAR team. When done right, the data science solutions can be provided to IT via a feedback loop and then deployed within real-time applications. This book focuses on the many skills that are necessary to quantitatively analyze a system, including building predictive models (which are just including another IT system).

CHAPTER 2

Analytic Approaches for Big Data

Analysis is analysis. Once analytical skills are learned, they can be applied to many situations by simply having a questioning attitude and following the scientific method. Typically, more questions are generated when answers are obtained: some being profound discoveries and others being heart-wrenching show stoppers. This chapter provides a foundation that defines analytics and big data, then details a few approaches showing how analyses can be performed.

Analytics and big data defined

What is analytics?

"Four major influences act on data analysis today:

1. The formal theory of statistics,
2. Revolutionary developments in computers and display devices,
3. The challenge, in many fields, of more and ever larger bodies of data, and
4. The accelerating emphasis on quantification in an ever wider variety of disciplines."[6]

These words apply as much today as they did when they were written in 1965! The current muse of analytics, and its use with big data, is Thomas Davenport.[7] Davenport defines

---

[6] Tukey, J.W. and Wilk, M.B., (1965). "Data Analysis and Statistics: Techniques and Approaches" in E.R. Tufte, (1970) *The Quantitative Analysis of Social Problems*, Addison-Wesley Publishing Co., pp. 370-390.

[7] *Analytics at Work: Smarter Decisions, Better Results* (2007, 2nd Edition 2017), *Competing on Analytics: The New Science of Winning* (2010), *Judgment Calls: Twelve Stories of Big Decisions and the Teams That Got Them Right* (2012), *Keeping up with the Quants: Your Guide to Understanding + Using Analytics* (2013), *Big Data at Work: Dispelling the Myths, Uncovering the Opportunities* (2014), *Only Humans Need Apply* (2016). All are published by the Harvard Business Review Press.

analytics as "sophisticated quantitative and statistical analysis and predictive modeling supported by data-savvy senior leaders and powerful information technology. Every data analytics project is different, not just because of the data, but also because of the client's culture, decision-making process, and maturity of the data collection and storage technologies."

Whose data is it?

There are many ethical concerns raised by the innovative technologies of big data. Concepts being investigated involve privacy, civil rights and liberties, intellectual property, and proprietary holdings. Andreas Wiegand's 2017 book titled, *Data for the People: How to Make Our Post-Privacy Economy Work for You*, introduces many concepts that are hotly debated. There is also the debate between secrecy and transparency that involves differing points of view for the individual, the government, and the corporate world. Complications arise due to obfuscation, complexity, opacity, and ignorance. The first big modern challenge resulted from the invention of portable cameras and subsequent rise of the paparazzi. According to Supreme Court Justice Brandeis who wrote in 1890:[8]

> *That the individual shall have full protection in person and property is a principle as old as the common law: but it has been found necessary from time to define anew the exact nature and extent of such protection. Political, social, and economic changes entail the recognition of new rights, and the common law, in its eternal youth, grows to meet the new demands of society. … The design of the law must be to protect those persons with whose affairs the community has no legitimate concern, from being dragged into an undesirable and undesired publicity and to protect all persons, whatsoever; their position or station, from having matters which they may properly prefer to keep private, made public against their will. It is the unwarranted invasion of individual privacy which is reprehended, and to be, so far as possible, prevented.*

Discussions today about logging of locations of cell phones, tagging of people in Facebook pictures, and the plethora of security cameras to track our every move need to be considered in light of common law. When a person "checks-in" on line, they declared their location at a distinct time in a public forum and, thus, surrendered any right to privacy in that application. Then there is the ability of your Smart TV to track what you watch and provide intelligence so that targeted advertisements can be sent to your other connected devices.[9] This is all allowable

---

[8] Warren and Brandeis, 1890, "The Right to Privacy", *Harvard Law Review*, Vol. IV, No. 5.

[9] Sapna Maheshwari, "How Smart TVs in Millions of US Homes Track More than What's on Tonight", New York Times, July 5, 2018.

because users click the box that they read the terms and conditions. Their privacy was lost because of ignorance. So, what is better: an encyclopedic collection of every fact or a lake of tagged transactions that can be used as precedents to support decision making, like approval to go buy a new car. Time will tell!

No, seriously, whose data is it?

That is a hard question to answer. While the data might be owned by a company or a government, there is a human in the loop that takes ownership. In most large organizations, there are roles such as database administrator, gatekeeper, or curator, that are filled by a subject matter expert who takes personal ownership of that data. This ownership involves periodic inspections of the quality of the data and control of read/write access to the database. Other organizations, Amazon for one, only exist because of the control they have of the data.

The amount of information related to one item for sale is mind-numbing and the types of data overwhelming, such as description and specifications, images, sales statistics, customer reviews, and videos. Amazon makes sales because of information the buyer can glean from this trove of data. Basically, the owner of the data is the person who pays the electric bill for the server the data resides on and they decide if access is free or if it is fee-based. Again, time will tell what the final answer is. And then there is the concept of metadata. These are the insights, constructs, and amalgamations that the analyst, statistician, or data scientist, generates in the process of deriving an answer to the original question. There is a profound sense of ownership of the resulting information and knowledge that was generated, even though someone else's data were used.

Why use analytics?

There are multiple conceptual models that tie to the analytics paradigm. The purpose of performing the analysis is to achieve a goal, but sometimes it takes substantial work to define the goal itself. Figure 2-1 is a scoping tool that helps focus efforts based on the intended end state.

Figure 2-1. Categorization of analytical approaches based on level of constraints.

		Means	
		Loose	Tight
Ends	Loose	Discovery	Science
	Tight	Quest	Planning

Without selecting a target from the start, an analysis project can wander aimlessly. The first split depends upon the end and the means[10] that will be followed. Discovery approaches can be further subdivided based on the time and money for analysis. It also is guided by the depth of *a priori* knowledge the user currently has. The concepts I present are not to be confused with the formal discovery phase performed by lawyers before a trial. The five subdivisions are:

- Reconnaissance – first pass review to determine scope and nature of the data,
- Investigation – detailed review focused on the utility and validity of the data,
- Analysis – structured review to identify relationships to aid in model development with exploratory analysis represented by a single pass and observatory analysis represented by collection over time with guided refinement of developed models,
- Research – process where hypotheses are developed and tested by experiments,
- Decomposition – elemental review where all components are uncoupled, and failure limits determined using simulation tools.

Other conceptual models exist and are controlled by deductive versus inductive approaches, macro versus micro views, or if a prototype analysis is being performed. Prototyping, which is typically a labor-intensive effort, will either prove or disprove that a meaningful answer can be obtained. Davenport, in *Analytics at Work*, states:

> *Fact-based decisions employ objective data and analysis as the primary guides to decision making. The goal of these guides is to get at the most objective answer through a rational and fair-minded process, one that is not colored by conventional wisdom or personal biases. Whenever feasible, fact-based decision makers rely on the scientific method – with hypotheses and testing – rigorous quantitative analysis. They eschew deliberations that are primarily based on intuition, gut feeling, hearsay, or faith, although each of these may be helpful in framing or assessing a fact-based decision. (page 176)*

The overall reason as to why to use analytics is simply pragmatic since it will generate the answer you need.

How is big data defined?

The world of computer processing is at the start of a revolutionary period – the computational era. Retailers and service providers have adopted big data as a key element in ensuring profitability, and also have added a new job title – data scientist – for the people who perform this work. Much of the technology developed for business use can be applied to solve

---

[10] Julian Birkenshaw, 2010. *Reinventing Management: Smarter Choices for Getting Work Done.* Jossey-Bass.

government problems. In many ways the definition of big data is an intangible concept like beauty or quality.

One definition of big data embraces four dimensions: volume, velocity, variety, and veracity:

- **Volume**. For US Federal Government work, the lower definition of big data volume starts when values related to all 3,143 counties in 50 US states must be addressed. The upper end has not been defined, but I approximate it to the average three phone calls per day made by the seven billion people on the planet for ten years (7,980,000,000,000 or eight trillion records). As data moves into "the Matrix", the upper limit becomes a moot point. Technology does not pose a limit for big data.

- **Velocity**. Velocity refers to how fast data are generated, such as the number of barcodes scanned at Wal-Marts worldwide every day. Real-time analysis is much harder in dynamic settings with "snapshots" of a limited timeframe commonly analyzed by a discovery approach. Velocity is not a factor when annual batches of data are processed in a static manner. As computerization becomes more prevalent, event streaming technologies are being developed so that analytics can be generated in near-real-time.

- **Variety**. Variety is captured when a data set has high dimensionality. That is, a large number of fields are populated differently. Variety is a challenge for most government data sets. One challenge results when many separate data sets are merged together with each one being unique and possibly changing in format between years. Another challenge is a lack of standardization such as Customs and Border Protection receiving airline reservations (Passenger Name Records – PNR's) directly from the airlines in 130+ different versions. The process to extract, transform, and load (ETL) is done mostly manually now, but ingest becomes easier when the data-owning agency itself upgrades. It is anticipated that the variety of data sets will continue to change over the next few years and will require incrementally less manual processing/programming until stability is met. In anticipation of a stable end point being reached, all supporting technology will be refreshed by the end of the decade. (Amazon does not allow variety to exist for either its holdings catalog or the user profiles as it tightly controls both structure and content for all data.) All information must fit the required format and records are not created unless all required fields are populated.

- **Veracity**. The final dimension of the big data definition, veracity, is commonly cited as being the one most critical for government work. Accuracy, and subsequently truth, are required for all agencies by all stakeholders. Most companies have similar needs for truthful data to convince stakeholders, for regulatory-required reporting, and in the event of a lawsuit. The nuances of how each data set is processed and incorporated into a decision becomes the answers that either satisfy or enrage the stakeholders and, thus, solutions should be simple, transparent, and fair.

Periodic refreshes of technology (both software and hardware) are necessary, but the refresh itself is not always sufficient. Technology refreshes simply focus on tools that provide support rather than on a grand solution. There are other key elements that must be addressed and together they shape the proper culture and mindset that will transform an organization into a data-driven operation.

The first key element is defined business processes so that consistency can be assured between locations in one year, and between years. The second key element is deploying the cadre of essential personnel who understand the nuances of the data and share the same point-of-view with respect to all stakeholders. The third key element is having a team of personnel who can support processing in surges as well as bringing subject matter expertise for the data sets themselves and for the fundamental big data skills. The fourth key element is a willingness within the organization to learn, grow, and continuously improve to reflect new data sets, new technologies, and new requirements from policy makers within the constraints of how fast and how much change can be absorbed by the stakeholders. The fifth key element involves the interactions with the political sphere itself and anticipating the impacts from large-scale trends and shifts that result after each election and budget cycle. In total, big data are defined by each organization in the terms that reflect the hardest and most critical activities that they perform.

What are the key functions involved with analytics?

Data science is not a new field. Sir Isaac Newton invented Calculus as a predictive tool for plotting the moon's course in its orbit based on the big data he had collected by observation. Modern computer systems combine vast amounts of raw data with today's tools allowing data scientists to drill in from their desktop, or even their phone.

Six major areas that constitute the body of knowledge for data scientist's daily jobs are described below. It is the rare individual who has mastered them all, but all data scientists must be competent in each. The six areas are: 1) Database Management, 2) Extraction/Transformation/Loading, 3) Data Mining and Descriptive Statistics, 4) Visualization, 5) Inferential Statistics, and 6) Modeling. Together, these six areas form a specialized type of tradecraft that supports the traditional business analytics model.

An inventory of individual competencies is shown in Table 2-1. This is not to be confused with an individual's strengths which will be discussed in Chapter 8. This graphic was used for a team building purpose with a diverse group of analysts and data scientists. It identified who had the most competency in what area. We had long discussions of who was the true data scientist, but never made a conclusion. It also served a conflict resolution role because person EEE had an overbearing attitude and focused on labor-intensive trivial issues.

Table 2-1. Team evaluation matrix.

Person	Ranking in Stars					
	Database Management	Extraction/ Transformation/ Loading	Data Mining and Descriptive Statistics	Visualization	Inferential Statistics	Modeling (Predictive & Perscriptive)
AAA	★	★★	★★★★★	★★★★	★★★★★	★★★★★
BBB	★★★	★★★	★★★	★★★	★★★	★★★★
CCC	★★	★★	★★★★★	★★★★★	★★	★★
DDD	★★★★	★★★★	★★★	★★★	★★★	★★
EEE	★★★	★★	★★★★	★★★	★★	★
FFF	★★★	★★★★★	★★★★	★★★	★★	★★
GGG	★★★★★	★★★★★	★★★★	★★★★★	★★★★	★★★
HHH	★★★★	★★★★	★★★	★★★★	★★★	★★★
III	★★★	★★	★★★★★	★★★★★	★★★★★	★★★★★
average	★★★	★★★	★★★★	★★★★	★★★	★★★

Database management

This area is the traditional realm of the IT world and database administrators. They speak a different language involving Third Normal Form (a database structural term) and tuples (elements from tabulated lists). Challenges arise from hardware, software, and security constraints. Typically, the data scientist harvests the data, rather than working with the raw data itself so that authenticity is assured in the database. For discovery projects, data may be exported for use in Excel in chunks of up to one million rows and handled as a sample. For recurring and routine work, the database itself can be automated to generate the required outputs that the data scientist reviews, approves, and presents. This is an area where subject matter experts need to be brought to the team. I have generated many prototypes in Excel and Access®, but when the holdings got large, I went and found help.

Extraction/Transformation/Loading (ETL)

Typically, data will be obtained from diverse sources, different databases, and different timeframes into one working file for processing and analysis. This involves the classic IT operations of extraction, transformation, and loading. Depending on the data source, a substantial amount of time can be spent on cleansing and conditioning the data. The user needs to be mindful of the processes that created the data itself and any systemic issues, patterns, and trends that might be observed during ETL activities.

Data mining and descriptive statistics

Once data are of requisite quality, it is ready to be transformed into information that requestors can start using for decision making. The tools and techniques are shared by all science disciplines and taught in basic statistics classes. The specific details of how to work with different kinds of data are presented in Chapter 5. For completeness, this subfield includes measurement system design and reporting (periodic, automated routine, and ad hoc). This area, along with ETL and Visualization, typically consume 85% of a data project's efforts. Data are easily summarized and tabulated using software tools. Table 2-2 was generated in Excel and presents the typical statistics used for descriptive purposes. Always. For every project. For every field. Failure to assess the basics first always results in rework and lost time. If using Minitab, there is an app that goes way beyond this that includes a graphical view using a bell curve for comparative purposes. A more detailed discussion is provided in Chapter 6.

Table 2-2. Table of most common descriptive statistical measures (aka univariate analysis).

Count	70
Minimum	-1
Average	12.4
Median	14
Maximum	38
Range	39
Standard Deviation	9.4
Coefficent of Variation	76%

A spin-off of this stage is the use of automation to issue alerts, to generate metrics, and to update dashboards. In the late 1980s, I was running a data-intensive operation where we sent samples to a lab who sent back paper reports. My team of scientists had to type the results into the database. It was tedious, but they became quite aware as to what the expected vales were. We started modernizing and were able to receive 5 ¼ inch disks, then 3 ½ inch disks, which were uploaded to the database. I issued my programmers a challenge: run a script at midnight that evaluated newly updated data and report to me if the values were a first detection, an all-time high, beyond the mean plus or minus three standard deviations, or greater than a value in a look-up table. The script went into production in 1996 after seven years of effort. I checked recently, and it is still in use. Some might call this Artificial Intelligence, but I simply considered it as putting technology to work.

Visualization

Displaying the data in graphs, tress, networks, and maps is quite easy in all software packages. However, doing this effectively is an art since many design concepts must be addressed such as color palette, font size, scale, and level of detail. Media also presents design challenges because the printed page differs from granting web access to an HTML view where the reader can customize and interact with the data. Numerous examples are provided below. The goal of visualization is to tell a story with pictures[11] – like a comic book or a manga novel. Sometimes the goal is to be persuasive. Other times it is to sell an idea with the principles of an infomercial

---

[11] See Cole Nussbaumer Knaflic (2015) *Storytelling with Data: A Data Visualization Guide for Business Professionals* or Martin Ewing (2017) *Once Upon an Algorithm: How Stories Explain Computing.*

being a good starting point. They are: emotional well-being, self-image, and relationships. There are some classic rubrics that can be used:

- FAB – Features-Advantages-Benefits to present innovative ideas,
- DESC – Describe-Express-Suggest-Consequences to deliver bad news, and
- DAR – Describe-Approve-Reward to give praise up the management chain.

Another approach is to use the Record Album Architecture:

- Opener,
- Chart Buster,
- Love Song,
- Blues,
- Ballad, and
- Upbeat Finish.

It is surprising how well this approach resonates in the business world. The chart buster piece is the real nugget of actionable knowledge that requires a decision. Quicker pieces are best, except for the ballad. Top managers are quite busy and sometimes need a bit of time to reflect before making a decision. That is when having a long ballad is useful. The ballad is where I decompose data from different parts of the system or go into details of how the data were obtained. If the manager does not want to spend that time, she will cut you off with "Let's drill down later" or "Get with so-and-so to go over the details." It is a good sign if they are writing notes while you endlessly drone. Most times, a decision is made based on the data in the chart buster with implementation steps identified before the meeting ends.

But back to visualization. There are many sources that provide fitting examples of cutting edge techniques. The most recent, entitled "A Practitioner's Guide to Best Practices in Data Visualization" is by Camm and others, and provides useful rules applicable to all data-rich situations. Every September, I carefully watch the NFL pre-game shows to see the graphics (as well as the current men's clothing styles) and frequently update my work products to stay current. Newspapers are also a useful source of new ideas such as *The Economist*, the *New York Times*, and the *Dallas Morning News* (again in September for football coverage). The large paper allows for some quite dense portrayals when using a complete center-folded page. It makes sense that big data requires big paper. I started college as a geography major. I have always had access to large plotters and have utilized them throughout my career no matter what assignment I was given. There is nothing like showing 50 graphs at the same time on one sheet for making a point. That is when I became a fan of Goode's projection of the world. Also, if the report is going to be delivered as a PDF, paper size can be mixed and matched and the user has the capability to zoom or to display on a larger screen.

An over-the-top approach to visualization can be found at the Imperial College in London. They have developed a Data Observatory[12] that consists of 64 monitors on a circular wall that is eight feet high and covering 313° of view. The total screen resolution is over 130 million pixels. It allows for an immersive and comparative display that enables decision makers to derive new implications and actions from interrogating data sets in an innovative, unique environment.

Inferential statistics

"Should I believe these results?" That is the question managers ask during every data presentation. Through the use of statistical techniques, significant results (the signal) can be separated from the noise contained in the data. Many textbooks demonstrate how these tests are performed, but selection and explanation are the largest stumbling blocks. Software does the majority of work either as standalone packages, as capabilities embedded in a product suite, or as add-in modules for other products (such as a SAS link within Excel). In addition to IT alternatives, costing alternatives are rapidly changing and include deployments on user-owned servers, software installed on desktops, cloud solutions with processing done remotely (SaaS), and pay-per-use models for tools that are only rarely accessed by advanced users. The outcome of performing inferential statistics is that an answer can be given that includes an associated quantification of the degree of significance. You have proof that the answer is correct!

Predictive and prescriptive modeling

Once good data are obtained and understood, it can be put to the ultimate test. Can it predict the future? Once inherent patterns, trends, and relationships are unraveled, numerous tools, including regression analysis, optimization routines, Monte Carlo[13] simulation packages, agent-based simulations, recommendation engines (similar to those used on e-commerce sites like Amazon), algorithms (like those for predicting weather), and self-learning classification trees, can be applied in order to achieve predictive knowledge. Once demonstrated, scenarios can be tested to see if changes for the better can be implemented. Sometimes, multiple models are applied to a situation like hurricane tracking models, with the composite output called an *ensemble model*. This is the realm of *big data analytics* in the current IT sales literature with SAS being the market leader in this realm. However, an adequate foundation must be set by the first five subfields to ensure success.

---

[12] https://bit.ly/2OYvIMC.

[13] Monte Carlo is the term used for a simulation that is repeatedly run using random numbers for the input in order to generate the range of output for all combinations of inputs. The output is a probabilistic representation. It is the tool of choice for understanding complex systems.

The chart in Figure 2-2 shows the results of a Monte Carlo simulation run in the program Crystal Ball® (now part of the Oracle analytical suite) that allowed inputs to vary by ±10% and shows a dramatic shift from the deterministic (base case) solution of a nonlinear model. The model was a two-term, coupled probability tree. The first term had an 80% success rate. The second term had a 75% success rate. The example is for a success on the first term and a failure on the second term. Base case was calculated as 0.8 x 0.25 = 0.20. However, when the ±10% was applied to the success rate the second term failure percentage varied widely. (Given 10,000 simulation trails, the Monte Carlo error itself is ~1%.)

Figure 2-2. Bell-curve generated from a Monte Carlo Simulation in Crystal Ball.

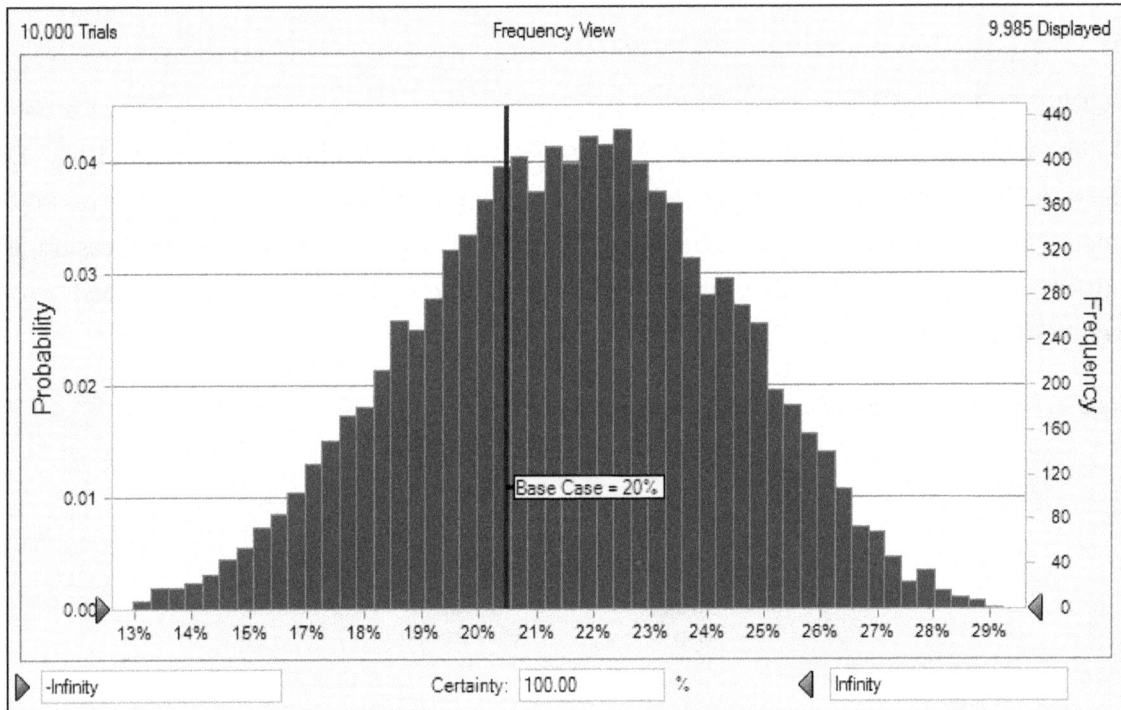

The output figure shows what the highest and lowest values would be for extreme conditions. Most importantly it shows that the majority of results are considerably higher than the base case itself. From a design prescriptive, a value of 24% would be a much more conservative factor than the base case of 20%. This ability to experiment using simulations is one of the most powerful benefits of applying quantitative tools to the analysis of a system.

Data-centric approaches

Is data mining an individual activity that is a reflection of the skills and abilities of the individual data scientist? Is the data scientist like a prospector or dowsinger who explores endlessly and occasionally gets lucky? I think not. I think the activities can be captured by a procedure and implemented in a repeatable manner by focused and disciplined data scientists. I am not the only one to think along these lines. A number of standardized methodologies have been developed, like the popular CRISP methodology.

CRISP

The CRISP-DM paradigm[14] shown in Figure 2-3 is becoming a recognized standard for use on analytics projects. This methodology involves six steps: Business Understanding, Data Understanding, Data Preparation, Modeling, Evaluation, and Deployment. I call the review for Business Understanding a *First Look* analysis. The data scientist is left with an impression as to the utility of the data and the ultimate chance of success. I have recommended project cancelations after taking this "First Look."

Figure 2-3. General approach of the CRISP-DM approach.

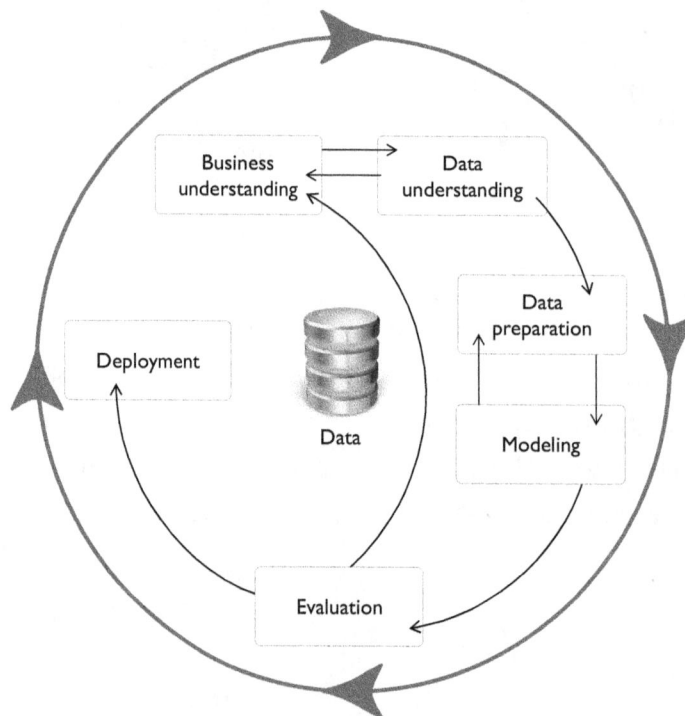

[14] For more information see F. Provost and T. Fawcett (2013) *Data Science for Business: What you need to know about data mining and data-analytic thinking.* O'Reilly, pp. 25-34.

Figure 2-4 shows each step can be further divided into actionable tasks. This rigorous structure has stood the test of time to be the quickest way to achieve a high-fidelity model. Shortcuts lead to errors which lead to rework. There are other paradigms that can be used, but some are gimmicky, some are slogans, and some are training tools. CRISP is a time proven universal approach that always can be applied.

Figure 2-4. Detailed steps for achieving the CRISP-DM approach.

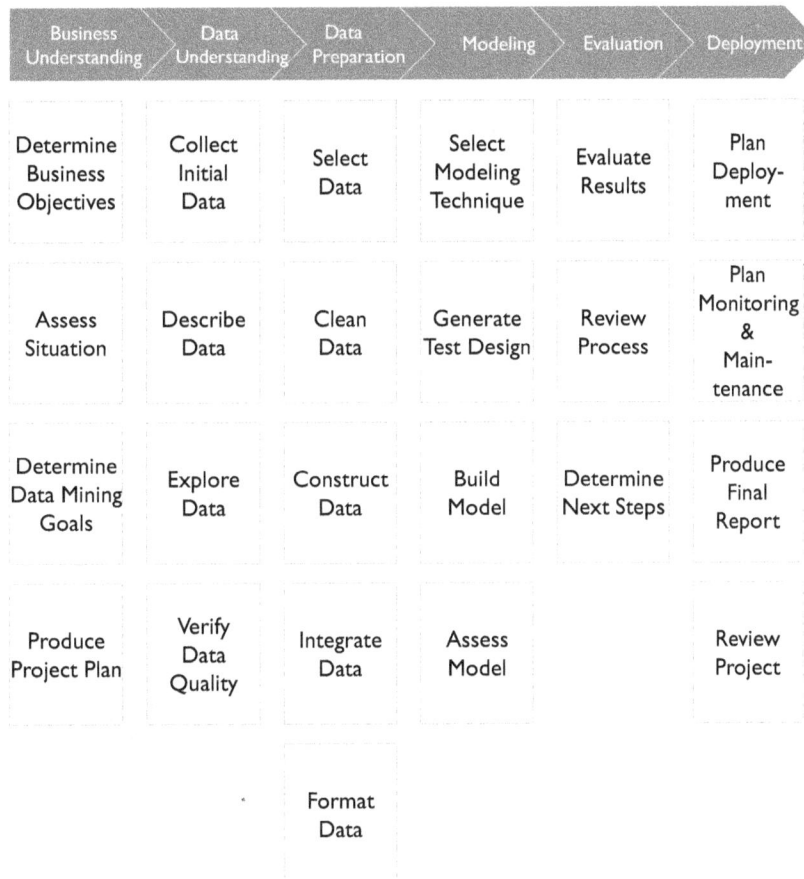

Business Understanding	Data Understanding	Data Preparation	Modeling	Evaluation	Deployment
Determine Business Objectives	Collect Initial Data	Select Data	Select Modeling Technique	Evaluate Results	Plan Deployment
Assess Situation	Describe Data	Clean Data	Generate Test Design	Review Process	Plan Monitoring & Maintenance
Determine Data Mining Goals	Explore Data	Construct Data	Build Model	Determine Next Steps	Produce Final Report
Produce Project Plan	Verify Data Quality	Integrate Data	Assess Model		Review Project
		Format Data			

There is more data than I can handle. How do I do anything with it?

The flippant answer would be: the same way you eat an elephant…one bite at a time. The real answer is you must have a plan. The CRISP-DM approach provides a tactical approach, but a strategic plan adds structure and forces the creation of documentation. The main element of a strategic plan for data analysis is a plan of analysis (POA). This plan not only describes the broad sequencing of activities, but also lists activities that will not be performed with an explanation of why not. It is a good place to review the modes shown in Figure 2-1 and state which is being pursued and to determine if a sample will suffice or if the whole population must be analyzed. This backdrop will enable future flexibility if warranted.

The activities must be grouped under key intelligence questions (KIQs) that will provide answers of interest to the key stakeholders. The answers for the KIQs may not be the easiest to find in the data, but, if the easy finds are not aligned with needs, they represent expensive noise that must eventually be filtered. Sometimes new questions arise that require the same data set to be investigated. These "special requests" need to be documented as they can cascade into a whole new line of questioning. The whole POA must be framed in terms of the nexus of the data to the concept being reviewed. Again, unaligned answers simply become noise. Finally, the answers must be disseminated for the contained power to be released.

In a fast-paced environment, multiple products of many kinds will be produced. The POA needs to document when integrative documents will be prepared such as assessments, annual summaries, and other planned production pieces. These documents then become corporate knowledge so that specific studies can be re-run, appended, modified, or even avoided, in the future. In a word – *planning*. Planning is the key to successful data management. Analytics projects are not suitable for Agile methodologies as Agile applies to writing code when requirements are undefined, or capabilities are not well understood.

10 T's for Data Mining

Data mining is "the exploration and analysis, by automatic or semiautomatic means, of large quantities of data in order to discover meaningful patterns and rules" (Berry & Linoff, 1997). These tools and techniques are applied in the same manner to sciences, marketing, production, finance, and human resource management. That is, anywhere a data-driven decision is required data mining is an option. However, the same warnings apply for all instances of use. Many kinds of data are confidential and become sensitive when personally identifiable information is concerned. Did you know that by disclosing gender, birth date, and zip code, that 87% of the people in the US could be pinpointed by name?[15] There are also legal requirements that must be honored such as annual tax fillings, banking fundamental data submitted per Sarbanes-Oxley, and drug study results presented to the Food and Drug Administration for approval. Another commonality is that processing the data takes considerable time. During detailed reviews, the data scientist spends prolonged periods of time interfacing with computers during the "Meditation of 10,000 Mouse Clicks" to clean and transform the data. Failure to take breaks and stretch can lead to repetitive strain injuries and high A1C levels.

---

[15] Stephen Baker, 2008. *The Numerati*, Houghton Mifflin Harcourt. p.13.

There are three temporal aspects to data mining: historical data analysis, real-time feedback and control, and results from predictive modeling. The same tools and approaches work, but the mindset of what it means, how it can be used, and how it is presented, differs substantially. The outcome of any data mining endeavor is the discovery of new knowledge – true science. Performing data mining is a discovery process that falls into one of five broad types:

- Reconnaissance,
- Investigation,
- Analysis (Exploratory and Observatory),
- Research, and
- Decomposition.

Reconnaissance simply involves a cursory viewing of data such as taking a windshield tour when we visit a new city. Investigation is a more involved step and involves interacting some with the data. Imagine a prospector panning for gold. There are two levels of analysis that can be performed on data holdings. Exploratory analysis involves a single pass drilldown of the holdings to identify the nature and extent of the data distributions for each field and determine the strength of interactions between the data fields. Observatory analysis is the realm of scientists and involves the collection of data over time for the identification of patterns and trends and to identify if complex interactions are displayed. Research is a more formal process that involves the development of a hypothesis and the performance of structured experiments so that the significance of relationships can be stated. Decomposition is the process used in model building where all components and factors are uncoupled and allowed to vary independently. The model outputs define the predictability of the expected responses as well as determine the failure limits for the system in question. The selection of which process to follow is a function of time, funds, consequences, and the robustness of the data set itself.

Therefore, the data scientist needs to take approaches that make him work smarter, not harder and longer. The 10 T's is one such approach. These are listed in Table 2-3 and are a grouping of concepts that guides the data scientist in all data mining activities. A longer discussion is provided in Appendix A.

Note that the scope of data mining is different than IT production activities or detailed quantitative analysis. It is a universal skill, such as technical writing and public speaking. There are rules and styles that each data scientist prefers, but this is a modern world skill applicable to the digital representation of a system.

The 10 T's are neither a tactical nor a strategic approach. They are values used to create a mindset of the data scientist. The common denominator across them is a questioning mind. The structured presentation in Appendix A also reflects this mindset because analysis must be thorough and structured. Yeoman's work is never glamourous, but it always leads to a solid foundation.

Do you remember the TV show *Lost in Space*? In the middle of a crisis, Robbie the Robot would roll up and point out an incipient danger because only he was looking. As a data scientist, you still need to be mindful to what is going on around you – like when the CEO who loves data is replaced with a data-phobe. I was caught by this situation once and ended up consigning two months of work to the ether as that study did not make the new list of priority projects.

Table 2-3. The 10 T's for data mining.

The T's	Objective	Purpose
Time (Attention)	Long, focused sessions.	To strategically evaluate and plan how the analysis will be performed. Mining data is not like changing a tire. A long reflection period is required to determine how to proceed.
Time (Duration)	Make multiple passes.	To prepare to spend sufficient time to re-work approaches after false starts are made. It takes long focused sessions. MRI research has shown it takes the brain 15 minutes to refocus once interrupted by a phone call or checking email. So, the old computer lab with a ban on phones still makes sense.
Transparency	Be open – no hidden agendas.	To assure all stakeholders of the scope and purpose of the data mining activities.
Tenacity	Data must be integrated from multiple, fragmented sources.	To alert the data scientist that not all the processes were built to handle a rigorous review and that the interim results will guide the method and duration. If the answer was already known, you wouldn't be doing this. The easy stuff is already done!
Testing	Need for cynical and skeptical views while striving for the truth in a non-judgmental manner.	To look at the results both cynically and skeptically. Sometimes, a mess is found because of incompetence, loss of control, willful manipulation, or the desire to deceive. These instances are not fun. People typically lose their jobs. Data scientists need to be a fair and truthful and assertive enough to report findings up the management chain.
Type	Determine the true nature of the data holdings.	To categorize and classify the data contained in each field. This is a true IT function so that data types and widths can be assigned for transformation and loading functions.
Tools	Evaluate and select what is needed from the available options.	To assess holdings and determine what is need for ETL and the First Pass analysis (Excel, SAS/JMP, R, Specialty Packages).

The T's	Objective	Purpose
Treatment	Practical, graphical, and analytical reviews.	To critically and systematically digest everything that is obtained. This processing is done in conjunction with each pass and is the true realm of the data scientist. Remember practice makes perfect! It also gives serendipity a chance to find the giant golden nugget.
Transport	Presentation of the results.	To explain to stakeholders what was obtained and how it can be applied.
To Learn More	Keep up-to-date with evolving technology.	To evaluate the state-of-the-art to maintain your currency.

Case Study: Mining of data.gov holdings for US international air travel

The US government is a major player in the realm of big data. Most holdings are open and transparent with data.gov being the best starting point when searching for a specific interest. This is the first of the detailed examples in this book. The Excel file that was used for this example can be found at https://technicspub.com/qasa to allow you to experiment and work with a big data set. This data set is from the US Department of Transportation, specifically the Research and Innovative Technology Administration (RITA) for Form 41: Schedule T-100(f) – Foreign Air Carrier Traffic Data by Nonstop Segment and On-Flight Market. This form is required for every commercial flight to or from the US with postings delayed six months. While not suitable for real-time analysis, it is the data of record. This data set is an authenticated source and should always be revisited when a new analysis is performed (versus working with data stored locally) so that the most recent data are used. It is mandatory that the date that the data were downloaded is recorded. The Venn diagram that was shown in Figure 1-1 was generated from this data set. While not all the fields are as nuanced as the example, a questioning attitude needs to be applied to each mouse click.

To obtain the T-100 data, go to https://bit.ly/2vNQgQ7 and click on "T-100 International Segment (All Carriers)".

Table 2-4 is a snapshot of the data that is available in the spreadsheet with the column names being identical to the fields on the selection screen. The data in Table 2-4 were downloaded on 20141017 (October 10, 2014) and represents the three-year period from January, 2011, to April, 2014. Passenger count data are aggregated by month by carrier by departure city and by arrival city. A total of 250,920 rows were downloaded. Figure 2-5 is a summary graphic that compares the total number of flights to the US to the total number of passengers. The dot size represents the population of the departure country with China being the largest and the dot for The Bahamas not being distinguishable. The website version of this example has a graph of all countries that can be examined at 400% magnification, or printed as a poster size image, or re-graphed to exclude the top 20 countries that account for 77% of all passengers. The average lines serve as an aid for comparison when dealing with an unknown condition. If targets exist, those can also be displayed.

There are several major questions that can be asked of these data, such as:

- Which US airports had the most international travel?
- Which foreign airports had the highest volume of travelers to and from the US?
- Which carrier had the most passengers? Note that other data can be downloaded for number of flights by day and size of plane.
- What are the year-over-year trends of these three data fields as shown on a pivot chart?
- Are there differences between US arrivals and US departures?
- What patterns are readily apparent?

Table 2-4. Snapshot of extracted data.

PASSENGER	UNIQUE_CARRIER_N	ORIGIN_CITY_NAME	ORIGIN_COUNTRY_NAM	DEST_CITY_NAME	DEST_COUNTRY_NAME	MONT	Year
49	Falcon Air Express	Guatemala City, Guatemala	Guatemala	Phoenix, AZ	United States	9	2011
81	Falcon Air Express	Guatemala City, Guatemala	Guatemala	Miami, FL	United States	9	2011
77	Falcon Air Express	Havana, Cuba	Cuba	Miami, FL	United States	9	2011
391	Falcon Air Express	Houston, TX	United States	Guatemala City, Guatemala	Guatemala	9	2011
269	Falcon Air Express	Houston, TX	United States	San Salvador, El Salvador	El Salvador	9	2011
85	Falcon Air Express	Houston, TX	United States	San Pedro Sula, Honduras	Honduras	9	2011
17	Falcon Air Express	Mexico City, Mexico	Mexico	Phoenix, AZ	United States	9	2011
436	Falcon Air Express	Mexico City, Mexico	Mexico	Tucson, AZ	United States	9	2011
14	Falcon Air Express	Miami, FL	United States	Barbados/Bridgetown, Barbados	Barbados	9	2011
98	Falcon Air Express	Miami, FL	United States	Camaguey, Cuba	Cuba	9	2011
127	Falcon Air Express	Miami, FL	United States	Guatemala City, Guatemala	Guatemala	9	2011
78	Falcon Air Express	Miami, FL	United States	Havana, Cuba	Cuba	9	2011
313	Falcon Air Express	Miami, FL	United States	Port-au-Prince, Haiti	Haiti	9	2011
95	Falcon Air Express	Miami, FL	United States	San Pedro Sula, Honduras	Honduras	9	2011
270	Falcon Air Express	Port-au-Prince, Haiti	Haiti	Miami, FL	United States	9	2011
33	Falcon Air Express	San Salvador, El Salvador	El Salvador	Phoenix, AZ	United States	9	2011
34	Falcon Air Express	San Salvador, El Salvador	El Salvador	Miami, FL	United States	9	2011
49	Falcon Air Express	San Pedro Sula, Honduras	Honduras	Phoenix, AZ	United States	9	2011
77	Falcon Air Express	San Pedro Sula, Honduras	Honduras	Miami, FL	United States	9	2011
135	Falcon Air Express	San Antonio, TX	United States	San Pedro Sula, Honduras	Honduras	9	2011
17	Falcon Air Express	Santo Domingo, Dominican Republic	Dominican Republic	Miami, FL	United States	9	2011
3596	Falcon Air Express	Tucson, AZ	United States	Mexico City, Mexico	Mexico	9	2011
560	Caribbean Sun Airlines	Cienfuegos, Cuba	Cuba	Miami, FL	United States	9	2011
0	Caribbean Sun Airlines	Willemstad, Curacao	Curacao	Miami, FL	United States	9	2011
2618	Caribbean Sun Airlines	Havana, Cuba	Cuba	Miami, FL	United States	9	2011
466	Caribbean Sun Airlines	Holguin, Cuba	Cuba	Miami, FL	United States	9	2011
24	Caribbean Sun Airlines	Maracaibo, Venezuela	Venezuela	Miami, FL	United States	9	2011
485	Caribbean Sun Airlines	Miami, FL	United States	Cienfuegos, Cuba	Cuba	9	2011
0	Caribbean Sun Airlines	Miami, FL	United States	La Desirade, Guadeloupe	Guadeloupe	9	2011
2150	Caribbean Sun Airlines	Miami, FL	United States	Havana, Cuba	Cuba	9	2011
365	Caribbean Sun Airlines	Miami, FL	United States	Holguin, Cuba	Cuba	9	2011
44	Caribbean Sun Airlines	Miami, FL	United States	Maracaibo, Venezuela	Venezuela	9	2011
230	Caribbean Sun Airlines	Miami, FL	United States	Santo Domingo, Venezuela	Venezuela	9	2011
559	Caribbean Sun Airlines	Miami, FL	United States	Tobago, Trinidad and Tobago	Trinidad and Tobago	9	2011
143	Caribbean Sun Airlines	Santo Domingo, Venezuela	Venezuela	Miami, FL	United States	9	2011
708	Caribbean Sun Airlines	Tobago, Trinidad and Tobago	Trinidad and Tobago	Miami, FL	United States	9	2011
4	Kenmore Air Harbor	Seattle, WA	United States	Tofino, Canada	Canada	9	2011

The last question is a hard one. It involves staring at the data, writing observations, generating hypotheses, and investigating the validity of the hypotheses. It is very important to write your thoughts when you go through your review because those fleeting thoughts are easily lost in the data and never recover from a delayed glimmer. Many times, these glimmers arrive at 3:00AM. I get out of bed, scribble them down, and then go back to sleep. If I don't write them down, I rarely fall back asleep because I dwell on them. When working in Excel, I open a word document on a second screen and toggle back and forth with notes being jotted for every thought I have. Minitab is the most convenient package. It keeps a history of every command entered with the corresponding results as well as containing a notepad panel. This notepad can be saved as a .txt document and opened in Word. The file gets quite large when pop-up screen graphs are aggregated on the notepad.

When looking for patterns, I follow the advice of Agatha Christies' superstar detective, Hercule Poirot – apply order and method and let the little gray cells work. Chapter 6 provides discussions of the fundamental tools for performing quantitative analysis. It is during these activities that the first glimpses of patterns become visible. As mentioned previously, jotting a note of these observations is essential so that they can be explored more during a latter session. If you give serendipity a chance, fact patterns will emerge that allow you to derive the truth as demonstrated by the Three Princess of Serendip from the old fable.

Figure 2-5. Comparisons of US-bound passenger to flights to total population by country 2011-14 using Excel.

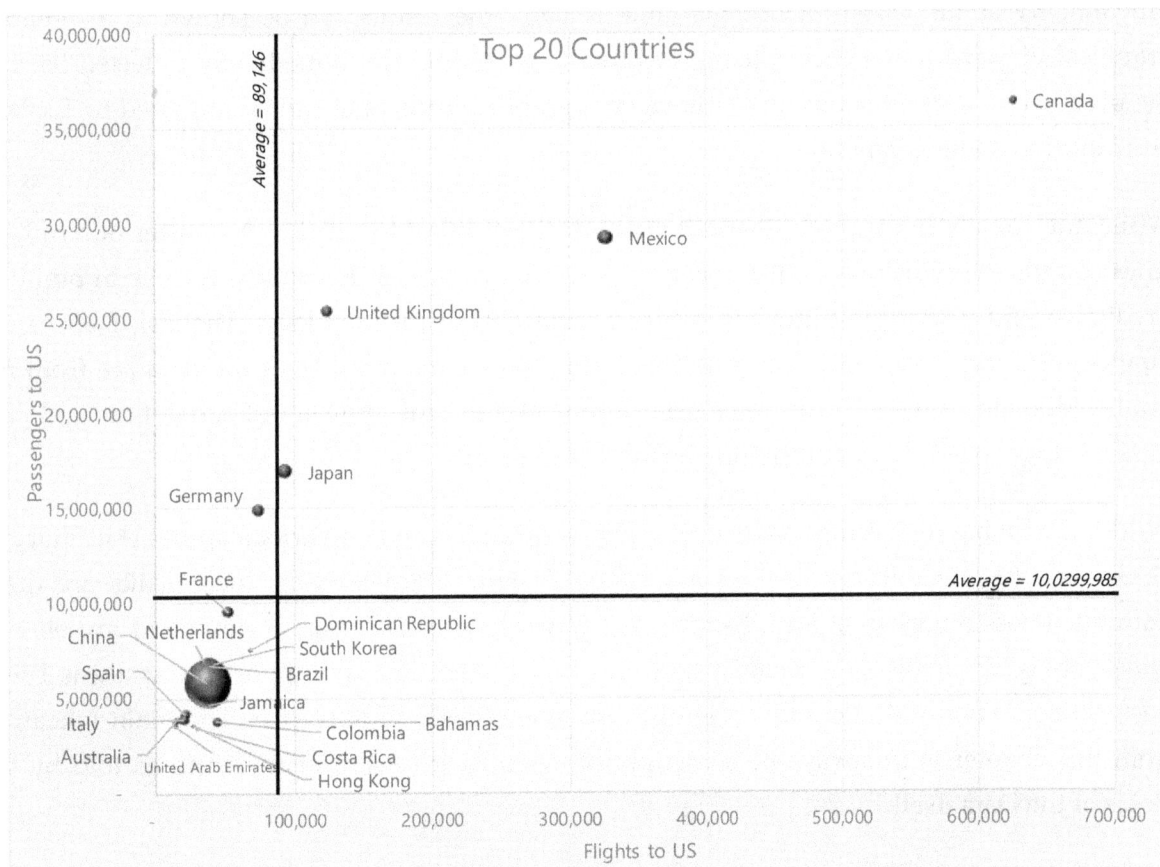

There are two tools in Excel that are invaluable for analyzing these data: auto-filters and pivot tables. These are the work horses for data mining. Auto-filters represent what geeks call *online analytical processing*. By activating the filter mode, the user can select (or deselect) categories or ranges of values to view. These filters can be applied in succession in support of drill down analysis. If one highlights certain values using a fill color, those colors can be filtered. (However, since you will want to be able to filter on this flag, it is much better to add another column to the spreadsheet, label it as a flag for the specific feature being highlighted, and then entering a simple x.) Sorting can also be done by using the auto-filter capability with the creation of an index in the first column so that the original order can be regained. If the resultant table becomes too messy, the filter function can be uninstalled, and the original view regained.

Pivot tables are a magical feature rumored to have been invented at Hogwarts! It is a dynamic tally sheet that is infinitely changeable and can display counts, percentages, sums, maximums, and many other representations of the data. Geeks talk about hypercubes and cross-tab tables when describing this functionality. To use one, the data must be normalized – that is, each row needs to be a complete record that has the appropriate value for each field. It is like when I worked with punch cards in the 1970s. The right hand of the screen is a GUI interface that allows one to drag and drop fields (data, rows, columns, or filters). The filters have the same functionality as auto-filters. Once the table is built, the results can be shown as a dynamic graphical view with one click. Changes to the pivot graphs are immediately reflected back in the pivot table. If pivot tables do not have enough power, the data can be migrated to Tableau and additional views generated.

While not directly highlighted, this example addresses all of the 10 T's. A quarter of a million rows sounds overwhelming but is easily processed by Excel. It is suitable for use in building predictive and prescriptive models following the CRISP methodology. That is, if all of the nuances and steps were documented. There are no ownership issues as the data are from the public domain. Overall, this example demonstrates the power of analytics and the methodology to follow to start getting answers and results.

Working with big data can be a challenge. This chapter focused on various approaches that can be applied to any data set and its associated system. Once learned, these skills are quite portable across a variety of jobs. I know, as I have done it. For years, I described myself to a wide variety of clients as a scientist first and that I used data to support the analysis I was performing. When the title "data scientist" became stylish, I adopted it. The main takeaway from this chapter is to always be a disciplined scientist in search of the truth as told by the Voice of the Data itself.

CHAPTER 3

Foundations of Data Management

Data are the critical component for quantitative analysis. There are many ways they can be obtained – some honorable and some not. There are many aspects of ownership as discussed previously. In order for data to be useable, there are certain foundation stones that must be touched correctly: organizational readiness, mature IT environment, adequate cyber-security protection, and software selection. While these factors will not guarantee success, failure can be assumed if they are not adequately addressed.

This chapter presents two themes. The first theme is a deeper dive into the IT aspects of systems presented in a blog-style discussion. The second theme introduces a powerful tool, the Million Row Data Audit, that deploys the strength of the data scientist to provide the IT team with a rapid prototype view to support the design process.

Virtual systems fundamentals

Once the real-life situation is mastered and data collected, there are a few foundational elements that must be incorporated into an IT system in order that accurate quantitative analysis can be performed. Many of these elements are unique to the data science function itself and of no interest to the process owners or the end users.

How does an organization implement data analytics?

As with all efforts at organizational change, the key to success lies with consistency of purpose and enactment of a shared vision. Small initial successes need to be celebrated and communicated. Initial failures are just as important. A formal Lessons Learned report can be developed and the behavior of management will define the environment for future innovation initiatives. First, an assessment needs to be made of capabilities and strengths. Then a roadmap can be designed to fill areas of weakness. Typically, change is accomplished in steps such as those shown in the Capability Maturity Model shown in Figure 3-1.

Figure 3-1. Stages of maturity in an organization's deployment of analytics.

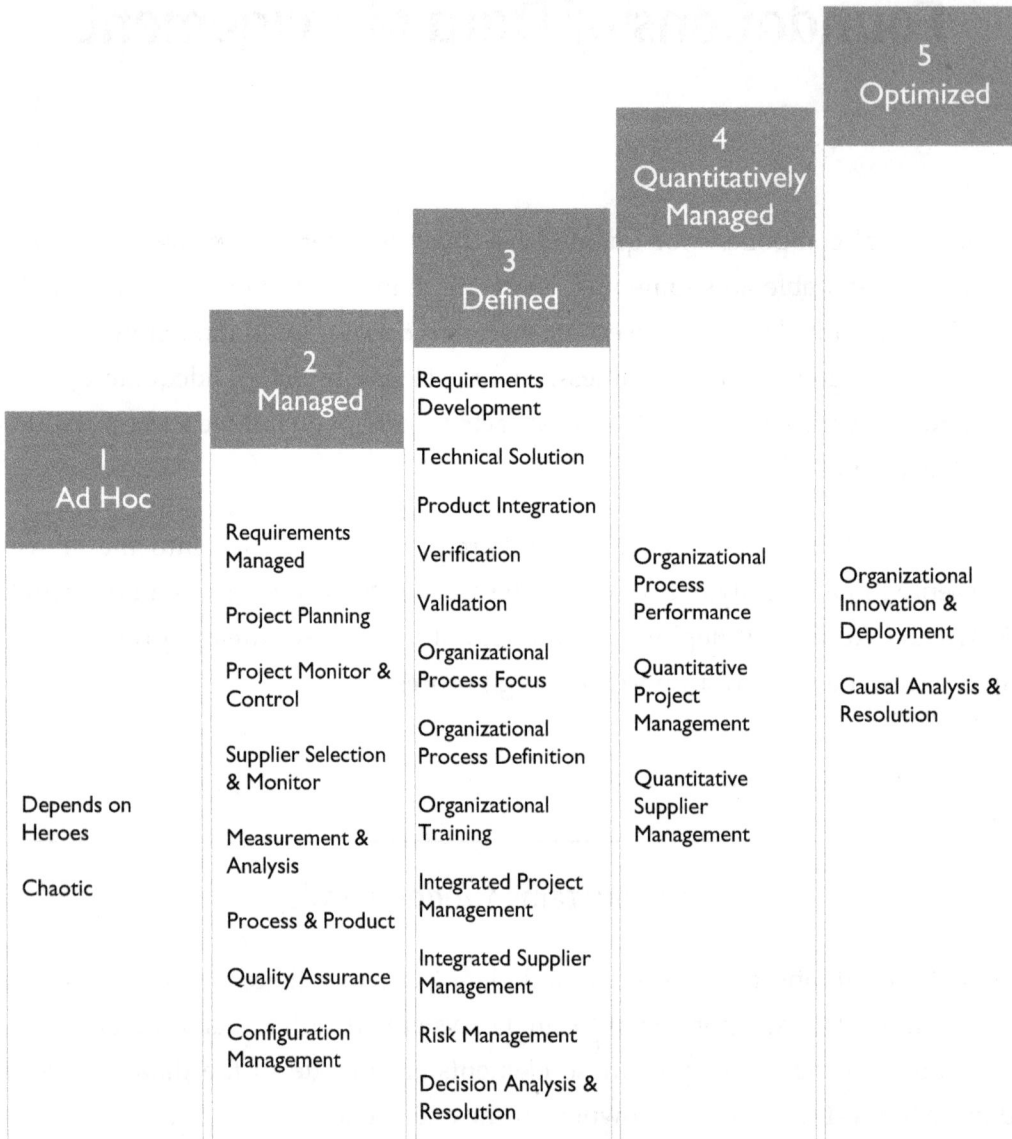

1 Ad Hoc	2 Managed	3 Defined	4 Quantitatively Managed	5 Optimized
Depends on Heroes Chaotic	Requirements Managed Project Planning Project Monitor & Control Supplier Selection & Monitor Measurement & Analysis Process & Product Quality Assurance Configuration Management	Requirements Development Technical Solution Product Integration Verification Validation Organizational Process Focus Organizational Process Definition Organizational Training Integrated Project Management Integrated Supplier Management Risk Management Decision Analysis & Resolution	Organizational Process Performance Quantitative Project Management Quantitative Supplier Management	Organizational Innovation & Deployment Causal Analysis & Resolution

Typically, an organization makes a leap every 18-24 month. CAPEX spending on hardware and software will also govern the speed of adoption. There are many templates that can be used for the design basis, many of which have assessment surveys available to aid in making a customized design for implementation. INFORMS has a well-respected assessment tool[16] that asks questions about organizational structure, analytics capabilities, and data and

---

[16] https://analyticsmaturity.informs.org/.

infrastructure maturity providing scores for people, leadership impact, measures, and processes.

While called *data science*, there is an element of art-of-the-practice involved in the implementation that is most easily achieved by staffing a core center of excellence that balances programmers and visual artists, modelers and observant natural scientists, whiz-kid mathematicians and battle-scarred engineers, and risk-adverse legal staff and results-focused management. A mantra Einstein is credited with saying that "not everything that can be counted, counts," should fuel a healthy dose of cynicism and skepticism when excellence is pursued.

Do I really need to think like a Chief Information Officer?

In September, 2017, Equifax was hacked and reported lost records for over 170 million accounts. That event begs the question of what the CIO had been doing. When analyzing large amounts of data, the data scientist needs to periodically think like the CIO. As a data scientist, you will be working constantly on computers and your success will tie to how well they run. Failure to run a tight operation will lead to excessive time spent troubleshooting problems, re-doing work because of lost files, and not having access to the newest capabilities found in the latest software versions. The most common areas of CIO responsibilities to evaluate are:

- **Backups**. When were they last done? When were the backups actually exercised in a mock recovery event? How robust is the off-site storage capability?
- **Cyber-security**. What tools are in place? Are they up-to-date? When were the defenses last tested?
- **Technology Refresh**. What versions of all software packages are in use? Are there uninstalled updates available? What testing is required to ensure continuity if updates are installed?
- **Hardware Replacement**. How old are the boxes? How hostile has the working environment been? Would migrating to new equipment cause a significant impact?
- **Disruptive Innovations**. What is coming down the pike that would benefit this endeavor? If our current vendors went under, is there something better I would want to deploy?

When looking at a digital representation of a system, one must be aware of the key functions applied to data: convert, store, protect, process, transmit, and securely receive. These functions rely on hardware, software, and communications channels. All this technology has unique concerns in the following areas: operations, interactions, training of users, suppliers, investments, and management. Another area of concern is knowledge management, especially on the software development side. There are tools (like the Atlassian Suite) that allow requirements, code, test results, studies, and all communications to be stored in a central location that is retained forever.

Table 3-1 presents a checklist of key aspects of the CIO responsibilities that you can apply to your home computer. Just think about how aggravating it can be when you make a mistake with your system and then try to project that feeling to the management team at Equifax that lost 170 million accounts.

Table 3-1. CIO Realm Checklist.

Key Aspect	Rating (5 is perfect, 1 is at risk)
Backup Frequency	
Backup Redundancy	
Recovery Testing	
Virus Protection Currency	
Networking	
Equipment Age	
Memory Capacity	
Storage Capacity Flash Drive Hard Drive Cloud	
Power Supply	
Power Conditioning	
Connection Speed	
Bandwidth	
Operating System Currency	
Standard Office Applications	
Specialized Software Applications	
Knowledge Management Suite	
Printing Costs	
Disposal by Recycling	

Then there are the laws and regulations that apply to the data themselves, like professor's test banks, Personally Identifying Information, HIPPA requirements to health care data, Sarbannes-Oxley rules to financial data, classification of government data, proprietary holdings of corporations, legal documents with attorney-client privilege, and plain-old embarrassing stuff.

All these standards are uniquely situational, but the approach to protection is rather common and involves many tools of the cyber-security realm, which also fall under the span of control of the CIO.

What is the best way to protect data?

This brings us back to thinking like a CIO. It is not so much about how much data, but more about how many users. As I am the only author of this book, and the only user of this laptop, I have complete control. Except that I store it in "the Cloud" with real-time backups to my hard drives. (Yes, three copies at all times!) As more users are added, they tend to have different roles and needs and may not need access to every field of every record. (This is like the examples for this book residing in "the Cloud" that are read only files.) Once you open them, you can save them as you wish. But what happens if there are thousands of users? Do you control access at the server, at the folder, at the file, or at the data level? All are possible, and the right solution depends on how much money is available, consequences of a breach, need to know, and maturity of the users. It is always best to lock-down authenticated sources so that every query returns the same data. Once data are in "the Cloud", like in a Hadoop table, each field in each record can be tagged and an authorization-based routine implemented that limits access to individuals with the proper credentials.

How important is software selection?

Software is only a tool, not an answer nor a solution. This question is answered in an analogous manner that we all use when we buy a new car or phone or other technological product. The starting point involves identifying the requirements of what is expected of the technology. Then the lower limit of acceptability must be established so that a level that is "good enough" can be identified. From there, products are reviewed, trade-off analyses are performed, and decisions are made. Regarding big data and analytics, the user may not have much input to the software selection process and needs to be prepared to learn the available packages that are authorized for use in "the Matrix." Sometimes the Chief Information Officer may make a global decision based on security or architecture or contracts and only the selected software can be used. Workarounds exist, such as the use of a dedicated standalone machine that is separate from the network and which requires very large efforts for administrative and security compliance. For complicated applications, it might help to have a technical advisor, or even a system engineer, who can support the selection of vendors and products for a particular application.

The workhorse package is Microsoft Excel. It is always available to use on any system and almost everyone has used it. The mantra "when in doubt, just right click" usually works. If not, asking Google may provide the answer. Visual Basic (VBA) programming can make it very automated. Tables (or worksheets to use the proper name) typically serve as databases that can handle 1,048,576 rows and 16,384 columns. Most databases will export tables as .xlsx or .csv or plain old .txt files that typically transfer automatically. Sometimes it might take a lesson in the "The Meditation of 10,000 Mouse Clicks" to make the new table correct, especially if reworking a PDF file. Auto-filters and pivot tables are pure magic and are the equivalent tools to the pick-and-shovel of mining (or, for big data, the tools are a digger and a dump). The technical terms are *Online Analytical Processing (OLAP)* and *hypercubes* to describe these functionalities. As a user, you cannot become "too good" at Excel. Even when you move to more advanced software, training many times takes a compare and contrast approach to bridge the gap between Excel and the new package. The latest addition for Excel 2016[17] is a Business Analytics module, Power BI Desktop, a cloud-based business analytics service that brings advanced capabilities to the most commonly used tool of the data scientist.

A powerful, widely-used, and open-source package is called *R*. It allows the user to have unlimited control over every feature. However, this also means the user must control every feature. As it is open-source, the many users add to the capabilities, but the user has the responsibility for testing and acceptance of these capabilities. Commercial packages are pre-tested, warranted, and the vendor can be sued over extreme defect conditions. R also runs from a command line – like a hand-held calculator. This package is great if you are a mathematician performing research. This package is also a great tool from which to learn computer programming of mathematical functions or if you do not have money to buy software from a for-profit company. It is not the best tool to use in a fast-paced production environment because commands are cryptic and easily forgotten if not routinely used, inputs constantly change so that scripts have to be edited, there are only a limited number of programmers, and peer review sometimes involves redoing the entire analysis. Since it is an open-source software package, the user must be prepared to serve as the expert witness regarding all settings on all features. Python is another package that is a widely used open-source application.

Alternatively, commercial-off-the-shelf (COTS) packages provide the expert support, especially if default settings are used, as well as push updates to fix identified problems. SAS is a major player, especially when tables contain a billion rows. JMP Pro is an instance of SAS that is a desktop memory-resident sandbox that allows one to write scripts against snapshots of data that can be later migrated to SAS in the production environment for real-time alerting. JMP Pro has all of the capabilities that a data scientist will need in their career. SPSS, Matlab, and

---

[17] See https://bit.ly/1KoMf6k.

Minitab are also major vendors of packages that focus on advanced mathematical aspects. Numerous minor vendors also exist with many specialized products. Minitab is the software package most commonly used in Six Sigma continuous improvement deployments.

What are the major challenges I will face?

Conducting a big data study is like the clown at the circus who keeps nine plates spinning on sticks simultaneously. The nine plates that the data scientist needs to watch are shown in Figure 3-2. One of the ways to keep everything moving is to generate prototypes and conduct First Look analyses. When a defect is found, the root cause can be determined, and a solution implemented. This is a very iterative process that takes time but is still faster than a traditional approach of formal requirements analysis. Prototypes also provide results that can be shown to management and stakeholders. The results from this preliminary work will show where there are holes and to manage the customer's expectations of what information can be derived from the available data. The goal of the CRISP-DM model shown in Figure 2-3 is to prevent any of the plates being dropped.

Figure 3-2. Multi-dimensionality associated with all data analytics projects.

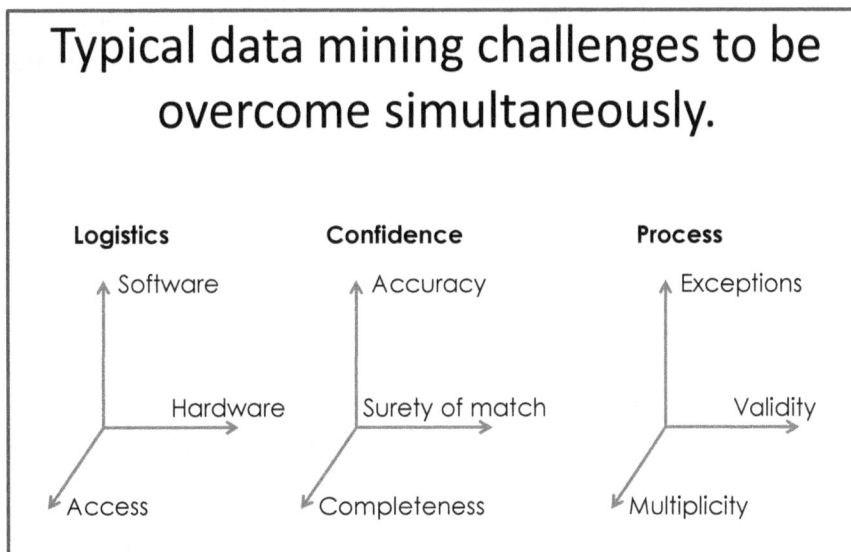

The logistics dimension challenges involve time and money. While some software is open-source and can be downloaded for free, it usually requires buying another license or adding a seat to an existing license. The frequency of use must always be explored before buying hardware, with big ticket items, such as a data observatory, being best established in a lab setting for multiple users. Access challenges are due to security rights and privileges. There is always a disciplined process that requires signatures, training, and the issuance of passwords.

You might be in a hurry, but the administrators must make sure all the wickets are successfully navigated.

The confidence dimension challenges relate to the database structure, the tools being used for ETL steps, and the rigor shown during the input stage. How do you know you have all of the records? It helps if you can access multiple systems and you achieve surety when the counts match for both fields and records. The main problem with being unsure of a match is when formats change. Dates are a vexing problem. When you encounter 3-4-2013 this is interpreted as being March 4th in the US, but some parts of Europe interpret it as April 3rd. Accuracy challenges are typically experienced when a field is truncated during a migration and sometimes when a mapping proves to have been made in error. While hardware failure is frequently questioned, it rarely is experienced.

The process dimension challenges refer to having a deep knowledge of the actual work being performed that is being recorded in the database. IT people sometimes refer to this as *provenance*. Multiplicity is a challenge in older databases that do not use normal forms of data storage and end up capturing entire records every time an additional field is populated. The bane of all data mining activities is when exceptions to a process are captured, rather than a separate process being followed. While this work around did not cause problems during the pen-and-paper days as humans were in the loop and could adjust, exceptions wreak havoc in the digital age. Automated processing requires discrete inputs and are unbending. Validity enters the soft ground of approved transactions. Just because a record was created does not always mean it was acceptable and fully processed. The aid to ensuring validity is to always question assumptions about a process.

Who should be involved?

The application of analytical tools is as much art as it is science. There also are maturity and awareness aspects that are inversely correlated. An experienced data scientist may know how to avoid the pitfalls but is not an expert in the cutting-edge technology. While a junior data scientist knows how to run the latest software, the lack of experience typically leads to many false starts and partial answers. Teamwork is an obvious solution. The main benefit of assembling a team is to minimize the time required to establish a capability in place that will provide output of known precision to be used in the data-driven decision-making process. The team must consist of members that cover six roles: customer, project manager, subject matter experts, data scientist, programmer, and vendor. Each project is unique in that sometimes two people may fill all the roles and at other times multiple people, including multiple specialized subject matter experts, may be assigned to each role.

This discussion has focused on the man-machine interface. It describes that the data scientist can only be as good as the technology being used. The good practices described need to be made permeant to sustain effectiveness. Two of the habits Stephen Covey presented in his book, *The Seven Habits of Highly Effective People*, are good guides. The first one, "Begin with the End in Mind," requires a holistic view of the situation prior to planning. The last one, "Sharpen the Saw," when applied to quantitative analysis, stresses the need for being mindful of changes to data, approaches, and capabilities.

Data audits – A rapid prototyping approach for small operations

In Chapter 2, an answer of planning was provided to the comment "there is more data than I can handle." Here is an option of using a sample when confronted with big data sources that delves into the realm of statistics and the use of samples. The size of the sample and the rigor of the analysis are governed by the requirements and specifications previously identified. Regarding big data, a sample of a million records is easily processed in Microsoft Excel and gives good results to the third decimal place. So how big are a million rows? If you take one column of 10-point font data in Excel that is a million rows long and generate a PDF to print 47 rows per sheet, it will take 21,227 sheets (albeit with only 28 rows on the last sheet). This requires 43 reams of paper or just over four boxes. The printer cartridge will need to be replaced at least twice. A million rows of data can be found in the case study to be discussed shortly. The file is over 40 megabytes in size and takes nine seconds to save. I do not recommend printing it, but you should scroll through it at least once just for the experience.

The Million Row Data Audit (MRDA) is a great discovery tool and also serves as a powerful first-pass prototype builder supporting projects using spiral development,[18] verifies accuracy and integrity, and many times leads to the discovery of Personally Identifying Information and other sensitive data that should be protected. Again, it is a technique that helps keep the nine plates spinning as shown in Figure 3-2. This information allows for detailed planning on how to use the data and how to proceed in implementing a big data solution. When the MRDA concept is applied to a random and independent binomial category like gender, results are obtained that present a measure of probability (such as 61.1% male ±0.12% with a 99% level of confidence). A manager can safely state that this population is about 61% male with little worry of being disproven. The precision may exceed that required by the question, but the surety is a benefit that justifies exploring big data. The MRDA approach will generate a "code book" (or

---

[18] For more information, see http://en.wikipedia.org/wiki/Spiral_model.

schema) that will allow programmers to quickly write scripts because most of the values for each variable will have been identified as well as provide lessons learned from the data cleaning step. The main benefit is that a "no go" conclusion can be quickly reached if the data are not suitable for providing the required answer. The graphs and tables developed are prototypes that can be wired into routine reports from "the Cloud" holdings using packages like Graphana.

The main benefit of the analysis is the identification of the structure and framework, not the content, so that a degree of confidence can be assigned that then becomes a bounding limit on the interpretation of the content. Once the MRDA is complete, an assessment of the as-is schema can be made, and a decision reached on how the should-be schema needs to be structured. There are many benefits that can be realized: creating a preliminary data dictionary, assessing the completeness of each field or the Rolled Throughput Yield for sequential actions, having a basis to estimate the cost for a follow-on project, and generating a cost-to-benefit ratio in support of decision making.

When a reluctance to share data is encountered, it may be indicative of bigger problems such as poor quality, inadequate capabilities, and incompetence. Therefore, it is better if a senior data scientist does this work rather than a junior one, because the probability of conflict is high, and the ramifications are serious (like someone getting fired). The senior person must have the necessary life skills to handle these sensitive situations.

The selection of the million rows needs a slight modification for chronologically-ordered data that involves taking subsets or stratified samples. The statistical term for making subsets is *stratification*, which can cause consternation to some managers. The sample interval is determined by dividing a million by the number of periods of interest to determine the sample size. For a 10-year time span, a sample of 100,000 records will be needed from each year. For monthly samples across one year, division by 12 results in a sample size of 86,667.

As you use samples, you need to be mindful of the significance of results when the sample size is reduced. If you desire to be 99% confident of less than a 1% chance of error, a sample size of 16,512 is needed. This limit is met with five years of monthly data (60 months) where n = 16,666/month or one year of weekly data (52 weeks) where n = 19,230/week. Options exist if longer timeframes are needed at these frequencies:

- Reduce confidence levels,
- Process more than one million rows,
- Randomly subsample the periods (every other month for a ten-year period or every other week for a two-year period), or
- Perform a time-comparative study.

The time-comparative million row data audit is used to: (1) determine if content is changing as a function of time, (2) assess if underlying processes have changed over time, and (3) survey a large data set in a more randomized manner. There are multiple approaches that can be followed depending on the key questions being asked of the data. A two-sample approach could be applied where the first million rows and the last million rows are processed separately and then statistically compared using a Two-Sample Student t-test for normally distributed data or a Mann-Whitney test for data that is not normally distributed (see Chapter 6). Another approach is to process a million rows with half from the start of the database, and half from the end of the database. You could also randomly sample a million records across the database and then perform a time series analysis of the results using the ANOVA test and other advanced statistical tools discussed in Chapter 6.

Appendix B is a detailed procedure that describes how to perform a MRDA and how to present the results. The following example is a demonstration of a data set suitable for such an analysis with the Excel file available at https://technicspub.com/qasa.

Case Study: Million Row Data Audit (MRDA) – Building HVAC example

You do not want to know how much data wrangling went into making this data set. There are a million rows! The file itself is over 40 megabytes in size and takes nine seconds to save. It was definitely a Zen practice session of "The Meditation of a 10,000 Mouse Clicks." This effort was to assure that patterns and idiosyncrasies are included. It is totally fabricated data, so there are no ownership issues other than my title to it. Here is the scenario.

Imagine a ten-story office building with the first story underground (basement or garden level). Imagine a new HVAC control system that records temperature data every minute. The building is oriented north-to-south and there are five offices on either side. The east side has a temperature gain in the mornings, while the west side has a gain in the afternoons. There are no differences in the gains between cloudy and sunny days, other than the outside temperatures themselves. There are three outside thermometers: one on the roof and one on both the east and west sides. Each floor has 11 thermometers: one in each office and one in the hallway that is not affected by solar gain. This results in a total of 113 thermometers as shown in Figure 3-3. Temperatures are logged each minute from each thermometer as:

113 thermometers * 60 minutes * 24 hours * 365 days = 59,392,800 data points

That is a lot of data! It is more than can be processed on one Excel sheet. However, does every data point have meaning? There is not much change minute-to-minute and January is quite similar to both December and February. This suggests that data can be subset to generate a representative sample. After a little bit of noodling, a solution was found that using a reading from every 20th minute for four months would result in 1,000,728 data points, which is below Excel's maximum of 1,048,576 rows. The sample represents 1.7% of all available data. Data were selected from January, April, July, and October to represent the four seasons.

Figure 3-3. Wireframe schematic for distribution of thermometers in building.

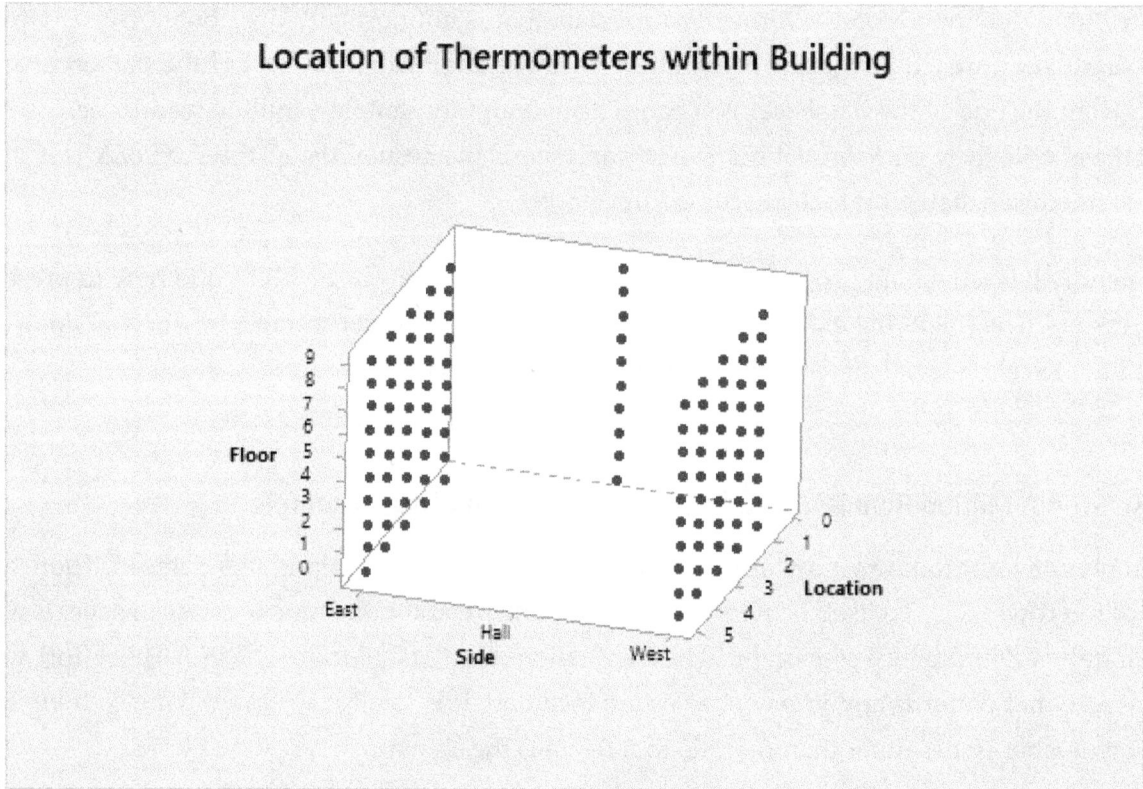

So, what does the data tells us? Figure 3-4 is a simple pivot graph made from all million rows. It shows the average temperature for the outside and for each floor for the four months of interest. Notice that the outside temperatures show a large degree of fluctuation, while the inside temperatures are more moderated. There appears to be an insulation problem the higher one goes in the building with January showing a steep decline and July showing a steep rise. These changes are more discernable in the tabulated data shown in Table 3-2. The tabular view shows that the basement temperatures are extremely moderated.

The pivot table allows for rapid drill down of results for one floor as shown in Table 3-3. Note the slight difference in average temperatures between the east and west sides. Note the 1° spike in temperature in July for the 5th office on the west side relative to the other west side offices.

Figure 3-4. Pivot chart view of average temperature by floor and month.

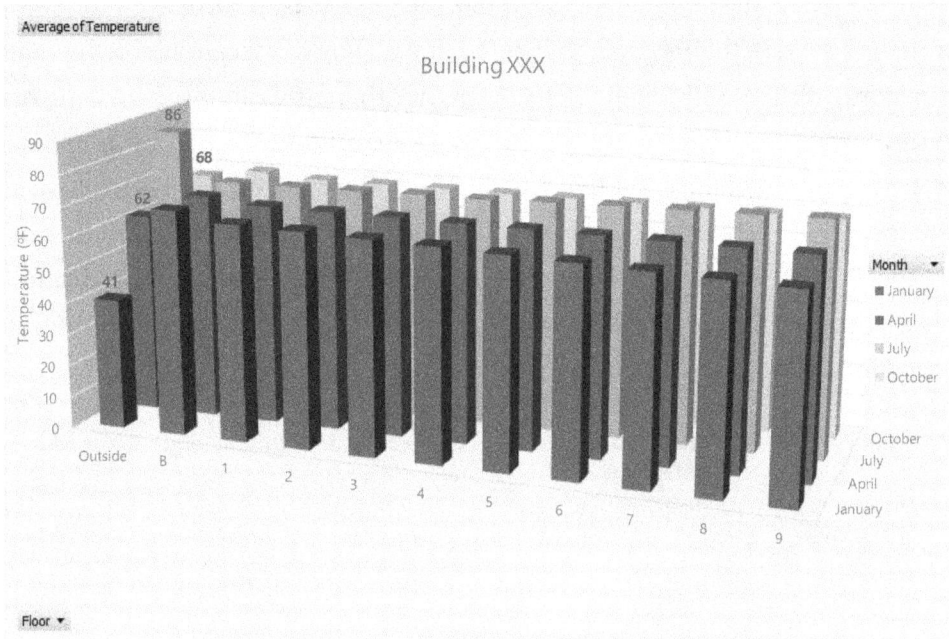

Table 3-2. Tabulated view pertaining to Figure 3-3.

Average of Temperature	Column			
Row Labels	January	April	July	October
Outside	41	62	86	68
B	70	70	70	70
1	67	69	71	69
2	67	68	71	69
3	66	68	71	69
4	66	68	71	69
5	65	68	72	69
6	65	68	72	69
7	64	68	72	69
8	63	67	73	69
9	63	67	73	69

Table 3-3. Maximum temperatures for each room on the 7th floor for each month.

Floor	7			
Max of Temperature	Column			
Row Labels	January	April	July	October
⊟ Hall				
(blank)	72.0	72.5	85.3	73.4
⊟ E				
1	72.0	72.5	85.3	73.4
2	72.0	72.5	85.3	73.4
3	72.0	72.5	85.3	73.4
4	72.0	72.5	85.3	73.4
5	72.0	72.5	85.3	73.4
⊟ W				
1	72.0	72.9	85.7	73.8
2	72.0	72.9	85.7	73.8
3	72.0	72.9	85.7	73.8
4	72.0	72.9	85.7	73.8
5	72.0	72.9	86.7	73.8

Table 3-4 is a drill down for just office 7W5 in July by day and by hour. The value presented is the average from the value registered at 20-minute intervals. Upon examination, the daily average for July 7 is 2.5° higher than the other days. Further examination reveals that the temperatures from 3 PM to 6 PM were 12 to 14° higher than any other time. While hard to read that difference in the sea of numbers, the change is readily apparent in a graphical view. In Excel, there is a feature called *Conditional Formatting* that allows numbers above and below user-defined thresholds to be highlighted in distinct colors. This automation can help flag exceptions, especially when you are tired. This is just one example of a nuanced situation. Have fun exploring for the others! Please contact me if you would like the complete list of idiosyncrasies that can be found in this data set.

Table 3-4. Hourly average temperatures for office 7W5 for the month of July.

Month	July
Floor	7
Side	W
Unit	5

Average of Temperature — Hour

Day	12 AM	1 AM	2 AM	3 AM	4 AM	5 AM	6 AM	7 AM	8 AM	9 AM	10 AM	11 AM	12 PM	1 PM	2 PM	3 PM	4 PM	5 PM	6 PM	7 PM	8 PM	9 PM	10 PM	11 PM	Grand Average
1	71.1	71.3	71.0	70.8	70.4	70.4	70.4	70.6	71.0	71.6	72.1	72.7	72.8	73.0	73.7	73.9	73.7	73.6	73.3	72.8	72.1	71.6	71.4	71.1	71.9
2	70.8	70.7	70.9	71.7	70.8	70.4	70.5	71.1	71.6	71.9	72.5	72.9	73.3	73.9	74.4	74.3	74.0	74.0	73.7	73.1	72.6	72.3	72.0	71.6	72.3
3	71.4	71.4	71.5	70.9	70.5	71.0	71.0	70.6	71.2	71.9	72.3	72.9	73.2	73.4	73.9	73.8	74.0	74.2	74.0	73.4	72.4	71.8	71.5	71.4	72.2
4	71.3	71.0	70.8	70.7	70.9	70.7	70.4	71.2	71.9	72.4	72.7	73.1	73.5	74.0	74.7	74.6	74.7	74.7	74.5	73.6	72.8	72.3	72.1	71.9	72.5
5	71.5	71.5	71.8	71.9	71.7	71.5	71.3	71.7	72.4	72.8	73.0	73.4	73.6	73.9	74.5	74.4	73.3	72.4	72.3	72.3	72.1	71.8	71.7	71.8	72.4
6	72.0	71.9	71.7	71.4	71.0	70.8	71.0	71.4	71.8	72.2	73.0	73.7	74.0	74.3	75.1	74.9	73.8	73.4	72.6	72.3	72.6	72.2	71.5	71.3	72.5
7	71.2	71.1	71.2	71.4	71.2	71.0	71.0	71.4	71.9	72.4	73.1	73.6	74.2	74.9	75.5	86.6	86.6	86.5	86.1	74.3	73.5	73.0	72.6	72.1	74.8
8	71.7	71.4	70.8	70.8	70.9	70.9	71.1	71.4	72.0	72.1	72.4	73.0	73.2	73.2	73.6	73.8	73.9	73.2	72.8	72.0	71.2	70.9	70.8	70.6	72.0
9	70.4	71.7	71.0	70.7	69.7	68.7	69.3	70.7	71.4	72.0	72.4	72.9	73.1	73.5	74.1	74.1	74.0	73.9	73.6	73.1	72.4	72.1	71.7	71.5	72.0
10	71.4	71.4	71.1	70.9	70.7	70.5	70.7	71.0	71.6	72.3	72.9	73.4	73.6	73.8	74.3	74.3	73.9	73.8	73.6	73.3	72.7	72.3	72.0	71.9	72.4
11	71.9	71.8	71.6	71.4	71.2	71.1	71.1	71.4	71.9	72.4	73.0	73.6	74.1	74.6	75.2	75.6	75.7	75.3	75.0	74.2	73.4	72.9	72.6	72.3	73.1
12	72.1	72.0	71.9	71.7	71.5	71.4	71.3	71.4	72.0	72.4	73.0	73.5	73.9	74.2	74.9	75.3	75.4	75.4	75.3	74.5	73.6	73.1	72.7	72.3	73.1
13	72.1	71.9	71.8	71.6	71.4	71.4	71.3	71.5	72.2	72.8	73.2	73.8	74.2	74.7	75.4	75.5	75.0	74.2	74.0	73.9	73.4	72.8	72.5	71.7	73.0
14	71.7	71.6	71.5	71.5	71.2	70.8	70.8	71.4	72.0	72.5	73.1	73.6	74.1	74.6	74.7	74.3	73.4	73.1	73.0	72.8	71.7	71.1	70.7	70.6	72.3
15	71.1	70.7	69.7	68.7	68.0	68.0	68.0	70.8	71.0	71.5	72.1	72.4	72.8	73.4	74.1	74.2	74.2	74.2	73.9	73.5	72.7	72.3	68.6	69.4	71.5
16	69.8	71.0	69.3	68.0	68.0	68.0	69.0	71.1	70.9	71.4	72.0	72.4	72.7	73.0	73.7	74.0	74.1	74.0	73.6	73.0	72.1	71.5	71.4	71.1	71.5
17	71.0	70.9	70.9	71.1	70.9	70.6	70.5	70.8	71.4	72.1	72.5	72.9	73.3	73.5	74.2	74.3	74.3	74.2	73.9	73.1	72.1	71.6	71.4	71.1	72.2
18	70.8	70.5	70.9	70.0	68.7	68.0	69.5	70.8	71.5	72.1	72.5	72.9	73.2	73.4	74.0	74.0	73.9	73.9	73.6	73.0	72.0	71.5	71.2	71.1	71.8
19	70.9	71.0	70.7	69.7	69.3	70.0	70.5	70.9	71.5	72.1	72.4	72.6	72.9	73.1	73.5	73.8	73.8	73.6	73.4	72.9	72.0	71.7	71.6	71.4	71.9
20	71.0	70.8	70.7	70.7	70.5	70.5	70.7	71.1	71.5	72.0	72.5	72.8	73.1	73.4	74.0	74.1	74.0	73.9	73.6	73.2	72.4	71.9	71.7	71.4	72.1
21	71.2	71.1	71.2	71.4	71.2	71.1	71.0	71.3	71.8	72.1	72.6	73.0	73.3	73.5	74.1	74.2	74.2	74.2	73.9	73.4	72.8	72.3	72.0	71.7	72.4
22	71.7	71.4	71.1	70.7	70.6	71.0	71.0	71.1	71.9	72.4	72.8	73.2	73.5	73.6	74.1	74.2	74.3	74.2	73.8	73.4	72.5	72.1	71.9	71.5	72.4
23	71.5	71.3	71.1	70.9	71.2	70.7	70.4	71.2	72.0	72.4	72.8	73.3	73.4	73.5	74.1	74.2	74.3	74.2	73.9	73.4	72.8	72.1	71.5	71.4	72.4
24	71.5	71.4	71.1	70.7	70.5	70.5	70.7	71.5	72.1	72.7	73.0	73.4	73.6	73.8	74.2	74.2	74.3	74.2	74.0	73.4	72.7	71.9	71.6	71.4	72.4
25	71.3	71.1	71.0	70.9	70.7	70.3	70.2	71.1	71.9	72.5	73.1	73.5	73.7	73.8	74.2	74.5	72.6	72.4	72.4	72.3	71.7	71.4	71.2	71.1	72.0
26	71.1	71.0	70.7	70.4	71.7	70.7	70.5	70.8	71.5	71.9	72.7	73.1	73.4	73.5	73.9	73.4	73.5	73.6	73.3	72.7	71.7	71.4	71.2	70.9	72.0
27	70.6	71.1	69.7	69.0	69.0	68.5	69.9	70.8	71.7	72.4	72.8	73.1	73.3	73.5	74.1	73.9	73.4	72.8	72.6	72.2	71.4	71.1	71.0	71.0	71.6
28	70.8	70.3	70.0	68.2	67.6	67.5	68.5	70.7	71.5	72.0	72.4	72.7	72.9	73.2	73.8	73.9	74.0	73.8	73.6	73.3	72.1	71.9	71.7	71.5	71.6
29	71.3	70.9	70.7	71.1	70.7	68.7	69.3	70.9	71.5	72.0	72.4	72.7	72.7	73.1	73.6	73.7	73.7	73.6	73.5	73.1	72.6	71.9	71.3	71.3	71.9
30	71.0	71.0	70.0	70.3	70.3	69.7	71.0	71.3	70.5	71.1	71.9	72.1	72.4	72.7	73.5	73.7	73.7	73.6	73.4	72.8	71.6	71.2	71.0	70.9	71.7
31	70.9	70.7	70.9	71.1	71.1	71.0	70.9	71.1	71.4	72.0	72.3	72.6	72.9	73.3	73.8	74.0	74.1	74.0	73.7	73.4	72.8	72.2	71.9	71.8	72.2
Grand Average	71.2	71.2	70.9	70.7	70.4	70.2	70.4	71.1	71.6	72.1	72.6	73.1	73.3	73.7	74.2	74.6	74.5	74.2	74.0	73.1	72.4	71.9	71.5	71.4	72.3

CHAPTER 4

Data Quality – How Do You Know Your Numbers Are Good?

Only four types of organizations need to worry about data quality:
Those that care about their customers,
Those that care about profit and loss,
Those that care about their employees, and
Those that care about their futures.

Thomas C. Redman (2006)

Bad Data Costs the US $3 Trillion Per Year.

"I call the added steps the "hidden data factory." Companies, government agencies, and other organizations are rife with hidden data factories. Salespeople waste time dealing with erred prospect data; service delivery people waste time correcting flawed customer orders received from sales. Data scientists spend an inordinate amount of time cleaning data; IT expends enormous effort lining up systems that "don't talk." Senior executives hedge their plans because they don't trust the numbers from finance." T.C. Redman, Harvard Business Review. September 22, 2016.

As Redman repeatedly points out, there are systemic problems with data quality. Many of these are legacy problems. Legacy problems plague many US Government systems. Some databases are like that radio station that only plays the best from the 80s, 90s, and today. This is because some of the data still resides in old databases or even on paper forms. Amazon never reports data quality problems because they built their system to only have top quality data by controlling user profiles and product listings to strict standards. These records are populated by the owners, much like when you pick your toppings at Subway, and they are responsible for ensuring the quality of the data. They are the ones who will know if it is wrong.

Are my data good? How do I know that? How can I continually prove it? These questions bring us back to the discussion in Chapter 1 about Deming's Knowledge of Knowledge. Many times, quality comes down to what you pay for it – like the difference between a BMW ("the ultimate driving machine") and an entry-level Chevy. The difference can best be described by some terms from the Japanese school of quality management. *Atarimae hinshitsu* refers to everyday quality, the "compliance with requirements" type of quality. The owner expects the car to start, to brake straight, and for all the electrical subsystems to function, at least until the final payment is made.

Miryokuteki hinshitsu is the type of quality that leads to the proverbial "A-ha" moment when the customer truly experiences delight that "things have gone right" with their purchasing decision. These moments are typically experienced during the use of the product and represent that the designers were truly expert in what they did. The first time I experienced this with respect to data was on my 401k website when my brokerage firm added a new green button that was labeled "Download to Excel." I ended up programming all night in Visual Basic so that, with one click of a button, I could transpose from a daily download to the visualization screens that I had created. I have nothing against the excellent tools my broker makes available, but I like the customization that highlights the strategic and tactical approaches I have implemented. The miryokuteki type of quality allows the product to be taken to the next level of use.

This chapter tackles defining data quality, as well as exploring classic scientific aspects to aid in the collection of good data. Four tools are also presented that support quality activities: verification and validation, measuring performance, checklists, and daily management protocols.

Data quality dimensions

First, quality is defined as:

1. a: peculiar and essential character :nature, b: an inherent feature :property, c: capacity, role.
2. a: degree of excellence: grade, b: superiority in kind.
3. social status: rank.
4. a distinguishing attribute: characteristic.[3]

This wide range of concepts has caused many misperceptions in the eyes of the beholders. Quality is sometimes compared to beauty with success achieved in the eye of the beholder only. Other times quality is user-defined when there are detailed requirements and measurable specifications against which conformance can be judged, zero defects are achieved, or even when goods and services are continuously improved (which implies defects are tracked and root causes addressed).[19]

Wikipedia cites data quality as:

The condition of a set of values of qualitative or quantitative variables. There are many definitions of data quality but data is generally considered high quality if it is "fit for [its] intended uses in operations, decision making and planning."[20] Alternatively, data is deemed of high quality if it correctly represents the real-world construct to which it refers. Furthermore, apart from these definitions, as data volume increases, the question of internal data consistency becomes significant, regardless of fitness for use for any particular external purpose. People's views on data quality can often be in disagreement, even when discussing the same set of data used for the same purpose.

Wikipedia also states, that according to the Government of British Columbia, Canada, data quality can be defined as: "the state of completeness, validity, consistency, timeliness, and accuracy that makes data appropriate for a specific use." These five dimensions are shown in Table 4-1 and are a useful framework for building a working system that describes the utility of the digital holdings related to a physical system. I once performed an extensive literature review on this topic and determined that all other schemas could be mapped back into these five dimensions.

Table 4-1. Dimensions of data quality.

Dimension	Intention
Completeness	Capture every field from every transaction including units of measure.
Validity	Review inputs against the required formats or listed values.
Consistency	Test formats from all data feeds ensuring nothing has changed when compared to the implemented schema for storage.
Timeliness	Ensure that data are available for analysis in accordance with the pre-defined specifications.
Accuracy	Deliver exactly what was loaded when data are extracted.

Note the caveat: "for a specific use." Attempting to use the same data for a different use may be risky in that a false degree of quality is assumed. Measurements for these dimensions are best when decided on a case-by-case basis so that budgetary and customer expectations are met. A

---

[19] Robert Pirsig provides an interesting exploration in his book, *Zen and the Art of Motorcycle Maintenance.*

[20] Redman, Thomas C. (30 December 2013). *Data Driven: Profiting from Your Most Important Business Asset.* Harvard Business Press.

powerful approach is to gather the questions that will be asked of the data regarding the quality and then assign them to one of the five dimensions. In some cases, subdimensions will need to be assigned to aid in tracking. For each question, measurement approaches can be defined, action levels established, and standard responses generated. Once implemented, this approach can be refactored into any standardized methodology for which the customer is willing to pay.

A more formal assessment tool has been created by Steve Hoberman – The Data Model Scorecard.[21] This assessment tool works by asking a series of questions (over 150 total questions in the methodology) and providing ratings over the following ten categories:

- Correctness
- Completeness
- Scheme
- Structure
- Abstraction
- Standards
- Readability
- Definitions
- Consistency
- Data

Armed with these results, you can assess the maturity of your data model, develop a plan to fix deficiencies, and schedule reviews for the opportunities for improvement that have been identified. This is a universal approach applicable to all data holdings.

The bottom line for assuring good quality data is check and recheck. The war cry needs to be "Remember the Orbiter," NASA's ill-fated Mars probe that crashed because conversions from the metric system were never made.[22]

---

[21] https://stevehoberman.com/data-model-scorecard/.

[22] Paul Krugman, *The Excel Depression*, New York Times, April 18, 2013.

The science part of data science

The progress of science has always been a journey laden with measurements and data. Challenges exist as to where to store these data, how to share it, and how to monetize it. The essential element is an inquiring mind – a desire to not only know what, but also how and why. The scientific method starts with a question, like, "I wonder." Then a hypothesis is generated, and a test plan designed. Sometimes there is a delay if the measurement tools do not exist, or are too coarse, to produce a testable answer. The test results are evaluated, and the initial hypothesis is either supported or disproved. With enough conceptual knowledge and data, models can be built for predictive purposes.

The poster boy is Sir Isaac Newton. He tracked the location of the moon in the sky for years and noticed a pattern existed. Rather than staying up all night, he developed a predictive model that allowed him to know where the moon would be at any time in the future. However, to build this model, he needed a new tool and developed Calculus. Once the model was built, it was applied to many similar situations. In systems-speak, this extension and duplication of a concept is called *isomorphism*. All of the branches of science have developed specific tools that can be applied elsewhere with the common factors applicable to all systems discussed below.

Data collection

The fundamental item for performing a quantitative analysis of a system is its data. Values should be gathered from every aspect. It is far better to have too many data points. The closer the design is to the atomic level of the system the better. However, the database can become too wide and cumbersome. Remember that Excel allows 16,384 columns of data in every worksheet (or fields per record if using database speak). I am still searching for a data set that will challenge this limit. This collection activity is not a wild west adventure or a photo shoot for Shark Week where unknown crises must be continuously tamed. It is a classic management function consisting of robust processes implemented in a disciplined manner, as limited by a budget. As with all management endeavors, there are five key functions: Planning, Organizing, Staffing, Directing, and Controlling. The actions required for each function are listed in Table 4-2.

Table 4-2. Actions to drive key management functions for high quality operations.

Function	Action
Plan	20 Questions (shown in Table 4-3)
	Concept of Operations View
	Data Governance Document
	Deliverable Timeline
	Gantt Chart with Critical Path
Organize	Procedures
	Forms
	Workflows
	Standardized Methods
	Equipment
	Access
	Requirements Matrix
	Configuration Management Strategy
	Records Management Plan
	Reporting Guidelines
Staff	Competency Inventory
	Responsibility Assignment Matrix
	Training Needs
Direct	Poke Yoke Design (mistake proofing)
	Management by Walking Around
	Checklists
	Daily Management
Control	Duplicates
	Blanks
	Known Standards
	Known Defects
	Repeats
	Measurement System Analysis Study
	Performance Measures
	Dashboards and Reports
	Reviews
	Audits

These activities all come naturally to scientists. The first experiences they had were performing experiments for the science fair at school. Their mindset is captured with the following joke: did you know that alligators can grow up to 15 feet, but usually stop at 4? This is their favorite kind of humor since it mixes units of length measurement and the name of a body part. Scientists are always vigilant against confusion of the units of measurement in their analyses as it is not a joking matter. Hence the mirth captured by this joke. In the work world, their projects are much more complicated in the depth, breadth, and scope of investigation, but rarely are they more complex than what they first learned. Remember, just because fundamental scientific principles are followed, does not mean success is guaranteed. However, if fundamental principles are not followed, failure will be assured.

Table 4-3. Inquiring questions to support the planning function.

Question	Answer
1. Who cares?	
2. What are the benefits?	
3. What are the consequences?	
4. How much will this cost?	
5. How much money is available?	
6. Who can do this?	
7. What is the expected duration?	
8. Where will the work occur?	
9. Where will the data reside?	
10. What is the frequency of the process?	
11. When must results be reported?	
12. Has anyone ever done this before?	
13. Can the data and documentation be found for this previous work?	
14. What are the requirement targets for precision and accuracy?	
15. What are the realities for precision and accuracy?	
16. How will the results be reported?	
17. How will this project look if it hits the headlines?	
18. What is the expected duration?	
19. Why?	
20. WIIT-FM! What is in this for me?	

Data reporting

Reporting data is governed by a golden rule – whoever has the gold, sets the rule! There are costs associated with extracting the data, tabulating and charting the data, preparing presentations (including reviewing to make sure the words are spelled correctly), and delivering the presentations. For some situations, the report might be a real-time dashboard in a control room, most times it is a weekly or monthly or quarterly management review, and occasionally it is a polished annual report. First and foremost, the report must present the facts in a defensible manner. Interpretations are sometimes presented, but there is no place for speculation. However, oddities are always questioned by inquiring minds, which leads to the generation of hypotheses and the need for further testing. The scientific method functions because of the documentation chain and is implied to be active in any endeavor termed "data science." While explorations and proof of concept projects may deliver actionable results, they do not deliver results that are repeatable or sustainable.

With respect to data quality, there is a traditional body of knowledge that transcends all branches of science. Descriptive statistics are generated by lot and batch and compared against each other and the design specifications. When discrepancies are observed, corrective actions are deployed that can entail re-calculation, re-running, re-processing, or even re-sampling. These sub-processes entail their own raft of documentation. Then there are specific quality measures associated with the measurement system itself, like calibration runs, accuracy, precision, percent recovery, relative percent difference, and task specific compra-ratios. These are generated, reviewed, dispositioned, and documented. When measurements are critical to success, some organizations have an autonomous Metrology department that allows accuracy tracing back to NIST standards. For new measurements, there is a formal approach for measurement systems analysis that involves gauge R&R (repeatability and reproducibility) studies. One of the most important equations looks at the variance of the system as:

$$\sigma^2_{system} = \sigma^2_{process} + \sigma^2_{measurements}$$

where σ^2 represents the variance.

A good rule of thumb is that the variance due to the measurement system should be less than 10% of the total variance. When more, it is hard to separate the signal from the noise. When less, costs typically take a quantum leap for every order of magnitude gained. For databases and supporting models, there are numerous assessment tools that are presented later in this chapter.

Appendices are the foundation for any report. They are a place to capture massive amounts of systemic documentation and analysis. There is an art in reducing raw signals and measurement data into concise summary information that becomes actionable knowledge in support of a decision. The key is knowing what the stakeholders want and desire, and what they think they paid for. Chapter 8 discusses the elicitation process that is useful for focusing the reporting goals. While tempting, it is never acceptable to use the "bury the bastard in bullshit" approach because personnel turnover rates are always high, and you never want to make a bad first impression with the replacement.

Data recording

The first question to ask when recording data is, "what is the chance we will get sued and must defend this work in a court of law?" There is nothing like giving a deposition in front of a bench full of attorneys and experts as a means to validate a data collection system. At the heart of the presentation you give are records in support of the data that includes log books, paper copies, electronic records, database authentication checks, external sources, read/write access authorities, custody control protocols, physical control for sample storage and transport, and photographs and video documentation, to name a few. The keys for managing data recording activities are documentation, reviewing of generated documents, documenting all discrepancies and subsequent corrective actions, and, finally, never granting an exception for any step in any procedure. Staff may complain it is boring and redundant to perform all of this work, but it only takes a few minutes in a court battle to experience the fear and terror associated with having inadequate documentation to understand its true value. This is an activity worth the Zen-like journey to perfection.

Electronic database management is a new player in the world of science. The earliest systems were designed for commercial use, such as the SABRE airline reservation system in the early 1960s. These systems were applied to science projects very quickly when the benefits of having one, authenticated and controlled data repository being intuitively obvious because the numbers will always be locatable and unchanging. One of the most intense data recording applications involves drug trial submissions to the US Food and Drug Administration. The current state-of-the-art systems have data stored redundantly in "the Cloud" and access controlled using virtual machines that can neither input nor export anything – a closed system. This extreme control ensures ownership, while preventing hacking and disruption. It does not come cheaply, quickly, or smoothly.

Optimization

As time progresses and the data holdings increase, there are numerous refinements that can be made so that the amount of information obtained is maximized for the funds being expended. The Capability Maturity Matrix shown in Figure 3-1 can also be applied to the entire data quality function. A traditional assessment method is to conduct a cost-to-benefit analysis. A benefit of this analysis is that some factors are identified that are truly priceless. Relationships are identified between expensive measurements and cheap measurements so that ratios can be used as indicators, and the adoption of frequent cheap measurements and long-cycle expensive measurement for confirmation. Those factors that are determined to be priceless (sometimes called *Key Process Input Variables*) represent the leverage points where meaningful changes can be made. Time also identifies what measurements can be halted when the body of knowledge only contains null values and when supported by detailed process knowledge. Again, documentation is required in support of all optimization activities.

The following differential equation is another way to demonstrate optimization of a measurement system. It defines the complex relationships between many terms associated with a data collection system as:

$$\frac{\delta_{measurements}}{\delta_{locations}} d_\$ + n \frac{\delta_{requirements}}{\delta_{opinions}} d_{time} = Q_p - \xi$$

where n = the number of audits, Q = quality, Q_p is the perception of quality, δ is mathematical symbol for partial derivatives, and ξ represents chaos that is external to the system portrayed by the data (the uncontrollable elements). This is a conceptual equation, because gathering relevant data would be extremely tricky. Also, the impact of external chaos is hard to scale, because quality itself is hard to scale.

Can you have too much quality? That is an interesting question and leads to the concept of robustness. As they say in the movies, "overkill is underrated." Robustness can also be applied to this equation because all enhancements to Q_p will have a beneficial impact in the long term. The equation also leads to stakeholder outrage when an operation has been in existence for a substantial period, with a large amount of funds expended on it, and audits reveal an extreme problem with quality. The *a priori* expectation was that quality should not be an issue. When a continuous improvement to quality mindset exists, efforts will always be expended in evaluating and upgrading quality to prevent surprises during an audit.

So, the optimal solution to this equation will prove that the resulting implementation will be good, fast, and cheap! And, yes, I used this calculation for that purpose. This was at a previous position that was being fundamentally realigned because of external events. The customer was requiring financial analysis in support of a zero-based-budget process and I was being asked every other day what if we cut you 10%, or what if we bumped you up 25%. I grew tired of the drill and just inserted the equation on the form. I was shortly summoned to the manager's office to explain it. Once he read it, he started laughing out loud. The lead for the budget team did not understand it. I received a wink and was told that my input was perfect, and I was exempt from the rest of the budget drills. This is another example of data science being used to generate a model (conceptual in this case) that traces its roots back to the example set by Newton circa 1700 when he invented calculus to predict the orbit of the moon. In the end, my budget was left unchanged.

How do I know my results are good?

This question can be the most challenging of any data-intensive project. It applies equally from data collection through the design of databases all the way to the building of models. Without credibility, the most detailed analysis will not lead to action by the decision makers. In order to prove that results are good, the entire data processing system must undergo verification. This includes that the user's needs are being met in the design specifications as well as being fully implemented, and that the correct answer is being provided (validation).

This section describes six traditional approaches applied to data-intensive products, so they can be verified and validated prior to operational use. They are:

- Cascading Test Plans,
- Pilot Study Demonstrations,
- Peer Review Inspections,
- Red Team Analysis,
- Change Management, and
- Sensitivity Analysis.

The reason for verification and validation is to ultimately support decision making. The reason for assessing the quality of data is to establish a measure of confidence in the certainty of information generated. These assessments also illuminate the areas of uncertainty. Armed with this knowledge, the decision maker can proceed with confidence. Risk is realized when a less than rigorous analysis is performed.

Given the diversity of typical analyses, and the breadth of tool types anticipated to be developed, none of these approaches are prescriptively recommended. Given that development of IT tools traditionally follows a spiral development approach with sequential prototypes, it is possible that a different approach could be taken for each prototype. The outcome would be a high degree of confidence as to the robustness of the tool. The commonality of all six approaches is that user involvement is integral in each approach to concurrently verify that the user needs and compliance with specifications are met and validate that the correct answer is being delivered.

Regardless of the approach taken, there are seven critical elements that must be addressed during the verification and validation process. These elements, and the reason they are important are:

- **Reliability**. All results must be from a known and stable process and stored in a controlled manner. Acceptable limits of reliability are user-defined as part of the development of the quality assurance envelope related to the context of the application with a typical default specification of a 95% confidence of less than a 5% chance of error.
- **Validity**. All input and output data must satisfy the general criteria of accuracy, representativeness, and credulity. Extrapolation of a result beyond its design context can only be performed when the limits of these criteria are known.
- **Bias**. All processes must be free of the influence of the data scientist.
- **Replicability**. The output from a product must be the same regardless of operator, platform, or software version.
- **Representativeness**. Data sets used for verification and validation must resemble the data stream for which the product was intended to support. Ideally, a subset of the target data should be used for verification and validation.
- **Sensitivity**. The data scientist must maintain a healthy skepticism of the output from all models and software tools until the sensitivity of the model itself is assessed in terms of responsiveness to changing parameters.
- **Limitations**. All data products have known limits. These limits must be documented, and a demonstration made that shows the consequences when the limits are exceeded.

In the simplest form, verification is achieved when desired functionality is demonstrated using select values: 0, 1, minimum input, maximum input, normal value (mean or median). It is akin to taking a new vehicle on a test drive and putting it through its paces. When working in Excel, verification is quite easy. One selects the Formulas tab, then clicks on the Show Formulas icon. This displays every formula in every cell. Yes, it takes a while to read through all of them in a big spreadsheet, but it is an easy way to make sure the coding was down properly. When I receive a spreadsheet created by someone else, I always spend a few minutes checking it out to see how it is built. There are a number of articles that complain that Excel results in bad models. These articles were written by amateurs who do not understand the need to verify every model no matter what programming language was used. So, up your game to the professional level and always verify what you are doing.

A benefit of a verification and validation study is that several graphic prototypes can be generated and shared with the customer in order to elicit preferences about color, legend, and style. It also ties to the CRISP-DM model discussed in Chapter 2 because verification is an evaluation of the model with the subsequent feedback refining the business understanding of the system. By sharing these preliminary results, which demonstrate functionality and presentation formats, stakeholders will be much better prepared to interpret results from the initial model runs.

In the simplest form, validation is achieved by running a known situation from a previous version and comparing the results. If the results do not meet expectations, rework is required.

Cascading test plan approach

The use of rigorous test plans is most compatible with material solution products designed using a capability-based acquisition approach. The test plan is started in parallel with the identification of the end use of the product. The test items identified at this stage are of a validation nature and will be typically tested last. When the design process becomes more detailed, the test plan becomes more specific with a focus on verification of specifications.

The design progression from end user needs, to system requirements, to subsystem requirements, to component requirements, to part requirements, while applicable to a manufactured product, become fuzzier with software products where commercial-off-the-shelf (COTS) tools are incorporated. Therefore, a graded approach must be followed to ensure the appropriate level of rigor is applied to fresh designs like VBA macros or R subroutines, versus COTS products like Excel. An integral part of developing the test plan is addressing interfaces and interactions between all elements of the product. For database functions, testing is accomplished by separately assessing the data (source), the aggregator (tarball), and the map (module) for each release. Ideally, a test plan element will prove the function of a higher-level design by exploring the operational limits of the next lower level and simultaneously prove the functionality of the interfaces and the interactions. An added benefit of the test plan approach is a cross-verification of operational requirements and overall design.

As the product is realized, the test plan is implemented during the assembly process. When problems are identified, resolution, repair, and re-documentation in the product support manuals must be accomplished before further testing can be performed. If the problem cannot be resolved, the test is canceled, and design activities are restarted. When satisfactory results are obtained for each test element, conclusions can be drawn that the product meets specification and user needs.

Pilot study demonstration approach

A demonstration of the utility of a product is a strong means of verifying that user needs have been meet and simultaneously proving that specifications have been met. A representative data set must be used, and caveats need to be in place expressing that the results are preliminary until the final review has been performed. This approach lends itself to rapid prototyping development cycles.

Peer review inspection approach

Verification and validation can be accomplished by an independent review of peers using the ensemble of testing strategies displayed in Table 4-4. The tests fall into four categories: unit (functionality), integration (operability), system (verification), and acceptance (deployment to users for validation). Developers need to know which tests will be performed so that appropriate documentation can be prepared. As the name implies, this approach uses independent reviewers with conflict between individuals a frequent consequence. Seasoned managers are needed to moderate this approach with blind reviewers and third-party presenters being common tactics to minimize conflict.

Table 4-4. Typical testing approaches used on IT projects.

Unit Testing	Integration Testing	System Testing	Acceptance Testing
Black Box	User Interface	Requirements	Alpha (Users)
White Box	Use Case	Usability	Beta (Real Data)
	Interaction	Documentation	
	System Interface	Performance	
		Security	

Red team analysis approach

A red team analysis is similar to the peer review inspection, except an outside group of recognized experts are used for the review. This typically involves the use of "dummy" data sets to see if the model produces an expected outcome. Four typical dummy data inputs are: all 1's, all 0's that are replaced with 1's when division is encountered, a design to generate a full suite of results of different ranks, and a test of min-mode-mean-max for each input parameter. The goal of the design of the dummy data sets is to address all perturbations and assess responsiveness across the range of values being evaluated. For the verification test cases, the term *node* is used rather than location or name to signify contrived data are being used. The main goal of the red team, is to try to "break" the model with the outcome being a measure of robustness and the limitations to which it can be applied.

Change management approach

An important step in any modeling effort is a formal approach to verification and validation regarding changes to the model. Did we do it right? Did we get the right answer? In the realm of systems engineering, the verification step is called *systems acceptance testing* with formal testing performed after every change to the programming in the software. The heart of acceptance plans simply involves loading test data for which the answer is already known and making sure the expected answer appears. Once the user is satisfied that the model is performing as intended, a second test can be performed by loading the previous data set in the revised model and inspecting that the results mimic what was previously changed and accommodations the expectations expected for the change (validation). The process described is shown in Figure 4-1 and consists of a sequence of actions to ensure that the functioning model is still viable after a change has been made to the programming. While this looks simple, much work is performed, and reperformed, when the model fails the system acceptance tests. There is usually an element of stress when deadline pressure mounts with each iteration and conflict pressure bubbles up when reviewers are independent from the programmers.

Figure 4-1. Required process steps to ensure quality is preserved during changes.

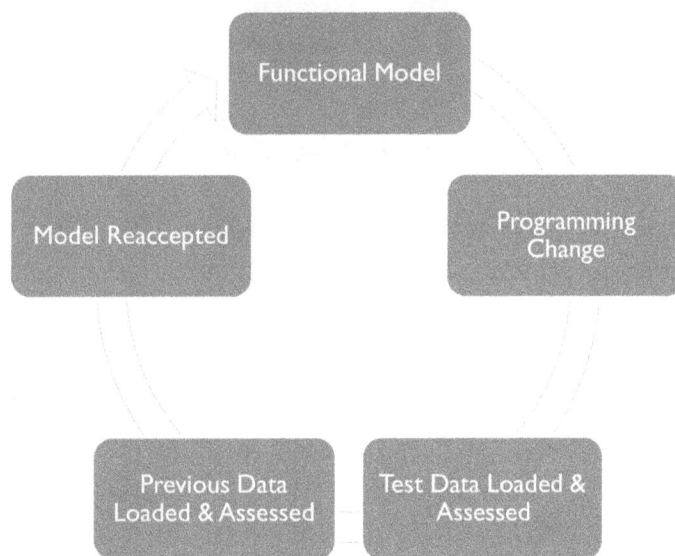

Sensitivity analysis

Sensitivity analysis is the study of how the uncertainty in the output of a mathematical model can be apportioned to various sources of uncertainty in its inputs. Figure 4-2 shows a classic oscilloscope.[23] It is a great way to explore how sensitive an output is to a change in input. The screen begins with a sine wave. The knobs allow frequency, magnitude, pitch, position, and other factors to be changed one at a time.

This approach needs to be taken for any models built so that the builder can verify that the output responds to changes in inputs like on the scope. This activity is called playing around with the model. When done right, it is an integral part of the testing plan. For complicated

---

[23] An online representation can be found at https://academo.org/demos/virtual-oscilloscope/.

situations, elaborate Design of Experiment approaches can be implemented so that many inputs can be assessed simultaneously while performing a limited number of runs. Presenting the results of a sensitivity analysis greatly increases the credibility of the model in the eyes of the stakeholder, especially when an interactive version is made available. From an empirical scientist point of view, you must determine what makes the needle move on the meter used for measurement.

Figure 4-2. Classical way to assess responsiveness – One Factor At a Time (OFAT).

The following is an example from a college class. The final grade for this example class will be determined using three assessment scores: participation (30%), mid-term (20%), and final (50%). Given these weights, a low score on the mid-term will not be very impactful, but a low score on the final will cause a major impact on the final grade. The impacts are calculated using a model:

$$Final\ Grade = 0.3 * Participation + 0.2 * Midterm + 0.5 * Final$$

Most of us have used this model to determine what grade is needed on a final once the first two are known. For a sensitivity analysis, each factor in the model is varied incrementally in order to determine the extent of change on the final grade. This was done for the above model with the scores varying from -10 points to +10 points shown in Figures 4-3 and 4-4. (Either format is acceptable and is driven by organizational preferences.) In plain English, the results can be summarized as a 10-point drop in participation will cause a 3-point drop in the final grade; mid-term, 2-point drop; and final, 5-point drop. So, one can conclude that blowing the mid-term is not the end of the world, since participation, and more importantly the final, have bigger influences on the final grade.

There is also an element of risk from the analysis itself that stems from CIO concerns. The first concern is, are files being backed up? Second, is whether Excel is the correct tool to be using. Are the users careful not to change data, or should a database be established as the authenticated source and queries run against this one instance of the holdings? Third, is configuration management being maintained for the queries, the models, the results, and the reporting elements? Fourth, are succession plans in place in case the lead data scientist wins the Mega-Millions jackpot and decides to take early retirement? These questions always need to be asked so that functionality is always ready, and managers can develop confidence that results can be obtained when situations dictate.

Figure 4-3. Spider diagram view from a sensitivity analysis of the influence of components on the whole.

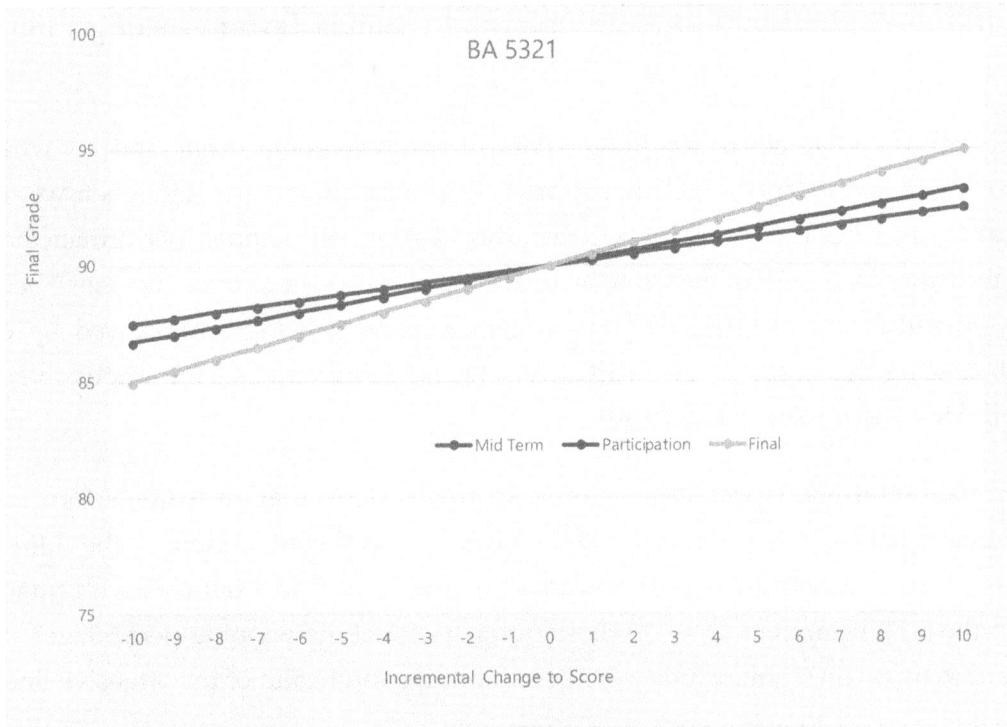

Figure 4-4. Funnel chart (tornado chart) of sensitivity analysis shown in Figure 4-3.

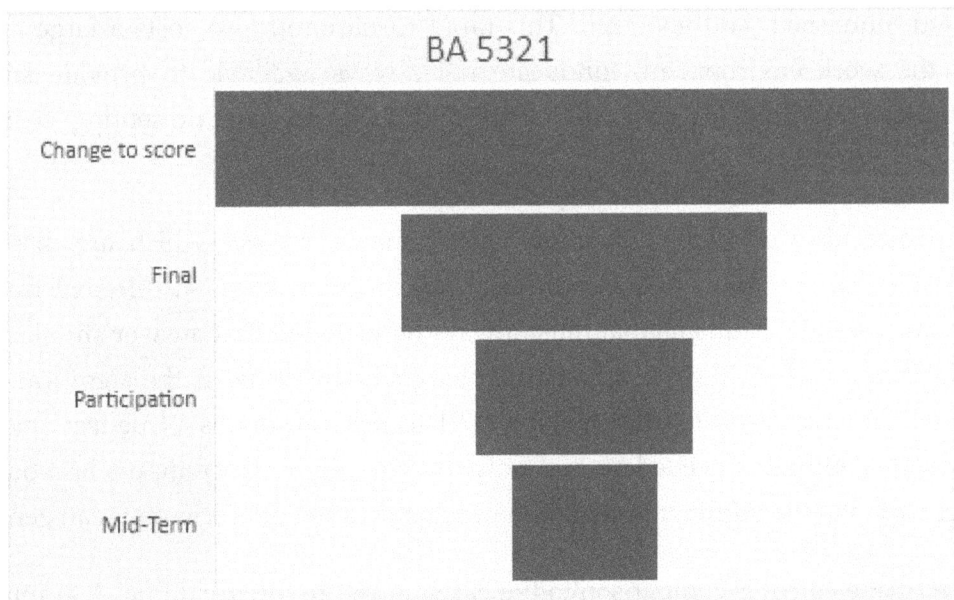

Measuring performance

If men could learn from history what lessons it might teach us! But passion and party blind our eyes, and the light which experience gives is a lantern on the stern, which shines only on the waves behind us!

Samuel Taylor Coleridge, 1831

Data are required that allow for judgements of who is doing what and to what level compliance with specifications is demonstrated. This starts at an early age when we received report cards in school. We all experience this during an annual performance review. Performance measurement is the area of quality that is bad-mouthed the most because it explicitly identifies our failures. Frustration comes in from being whip-sawed by external factors, balancing the needs of the work team, trying to accomplish the impossible, or just simply blinking and missing a key target.

Sports provide some good examples of simple scoring systems that measure performance. The count of points determines a win or a loss in a head-to-head competition, or the difference in time determines a finishing place. Sports also provide some bad examples with quarterback ratings in the NFL being one. How do those numbers affect the winning percentage? What do they mean? Why do they not include a factor for abilities and health of the offensive line?

The sports section of my morning newspaper is a treasure trove of performance metrics. Each sport has its own format that does not change year to year. It is amazing how fast these metrics are generated, published, and analyzed. This rapid turnaround also costs a large amount of money. In the work environment, funds are not always available to provide such quick turnarounds, but machine learning and artificial intelligence are presenting cost-effective solutions.

Many companies have developed their own performance metrics which are displayed on dashboards that are presented at many different organization levels at different frequencies. There are some essential elements that must always be included that answer simple questions such as why and so what? Why is this metric being shown? What is the significance of this metric and why are we spending time looking at? This last question is a true test of leadership because judicious editing is needed so that only the key performance inputs and outputs are displayed rather than the infinite number of views that creative data scientists can generate.

Another essential element is comparison to the baseline and to the trend line. A graphical view is a powerful way to show performance for the reporting period and how it compares to the most recent sequence and the historical norms. Figure 4-5 is one such example. It shows an external factor that causes downtime at many manufacturing facilities, or outside venues with large number of visitors – weather warnings.

The metric reports the percentage of time each month that the proximity alert was in effect. This downtime directly restricts the amount of throughput that can be achieved, or that requires expenditure of overtime pay to regain schedule. It is a boxplot (sometimes called *a box-and-whiskers plot*) where the center line is the median (half the values are above or below), the black dot is the mathematical average (that shows if the data are skewed), the ends of the box represent the 25th and 75th percentiles (the conservative way to show variability when the sample size is small), the bar ends correspond to about three standard deviations after adjusting for sample size, and the stars represent values that are outliers. It gives a pattern or bulls-eye of what is expected. The colored dots represent the data from the most recent year that capture the month-to-month trend and how the value compares to the baseline or historical conditions. The extreme value shown for August would imply that production quotas were missed as almost one-half of the month was disrupted. In actual practice, presenting the data against the baseline supported an easy negotiation with the customer because of *force majeure* and a subsequent re-baselining of the production schedule.

Figure 4-5. Percentage of month under weather warnings. (Generated by Minitab as part of an ANOVA.)

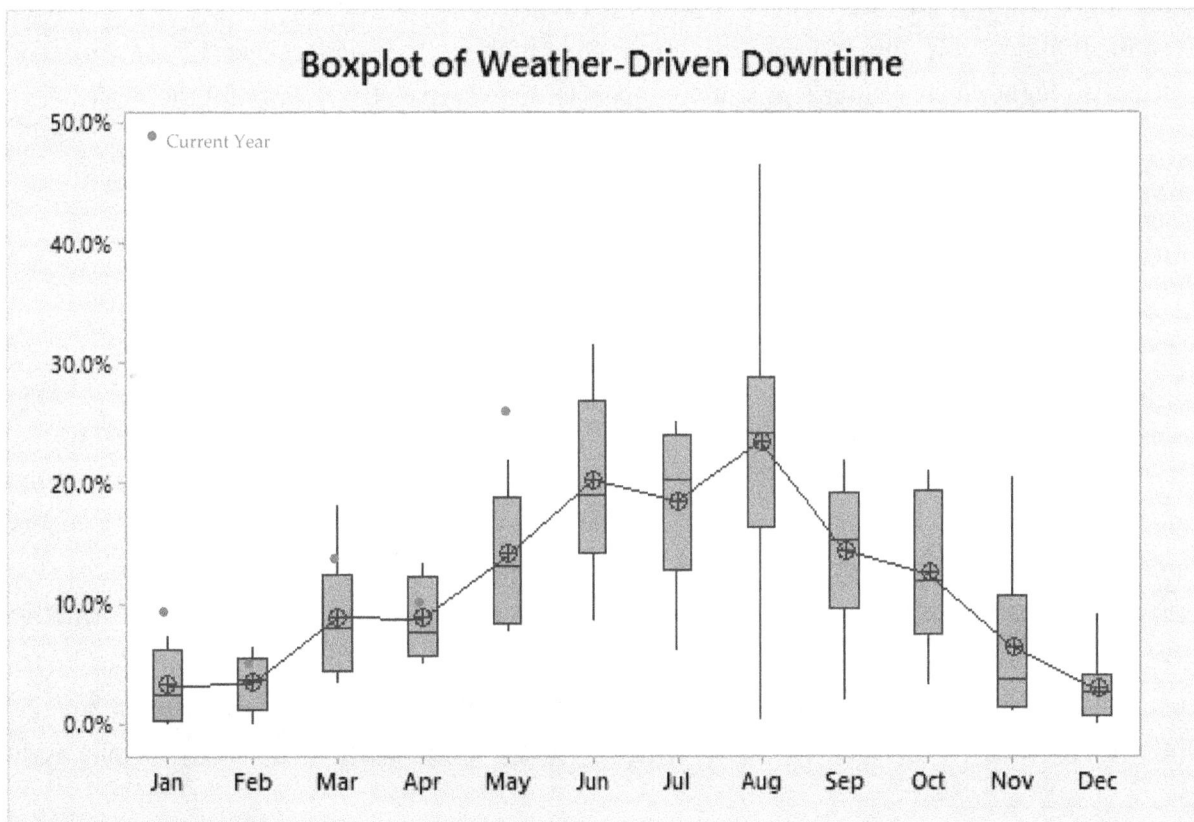

Some suites of metrics go beyond just a comparison to a baseline because they also track achieving a goal. This is repeated every quarter for publicly-traded stocks when performance is reported and compared to pre-published targets. Other examples include approaches like balanced scorecards, quad charts (a four-panel view that includes trend chart, root cause graphic, status report, and action items), and elaborate dashboards. A challenge exists for measuring performance in the digital world, when maturity is also changing as a function of time as described by Pfluger and McGowan (1990). They present a top down approach and take a hierarchical approach of goal/question/metric with the goals adjusted after they are achieved. Time is not a factor in this approach because most gains are by breakthroughs rather than by just sequential growth. It also is an approach highly compatible with JIRA[24] (or similar) issue tracking packages used to support scrum approaches to software development.

What does it mean when someone says to "check the dashboard"?

One of the hallmarks of a data-driven company is the adoption of a visual workplace that provides key process input and output data. I have seen these metrics on web pages inside the firewall, on graphs hung on the wall in the company cafeteria, and in PowerPoint slides at monthly performance meetings with each manager responsible for dissemination. This visual performance tool is simply called a *dashboard*, like the one in your car. The degree of rigor in the generation, display, and distribution is a key criterion used by auditors to assess whether an organization has truly adopted data-driven decision making, or if only lip service is being paid to satisfy a higher level of management. Here are some examples from different industries:

- Health care: https://bit.ly/2wrKA0I
- IT Operations: https://bit.ly/2vSIHa1
- Manufacturing: https://bit.ly/2OqcjDl

Take my car, for example. It has the traditional dashboard, a heads-up display, a center console computer screen, and some graphics built into the dashboard itself. It has one panel I find very useful. When I have the GPS in use, and toggle over to miles remaining given fuel consumption, I see a bar graph of the expected distance I can go given the current consumption rate and the volume of fuel in the tank. The car also displays a white dot of the distance remaining to my destination. This powerful graphic allows me to decide when I should stop for fuel with just a quick glance. Three forms of data were combined to generate an informative graphic that augments my knowledge. That is a worthy design goal for useful dashboard creation. So, what makes a good dashboard? One that is understood and used by the managers and the workers. It is a function of the culture of the organization, the resources available to

---

[24] https://www.atlassian.com/software/jira.

creating and maintaining it, and the expectations of the stakeholders. While software is necessary, it alone is not sufficient to guarantee success. Visualization tools do not represent the measurement of performance nor the interpretation in historical or futuristic contexts. Much thought and energy must be spent by management to deploy an effective suite of performance metrics as efficiently as possible.

Checklists

The human brain is amazing. How much can we remember? What is the most complex problem that we can solve? Was I supposed to pick up eggs on this supermarket run? Well, the last one is a challenge – the execution of routine tasks that include variables. If there only was a tool to help us consistently and correctly apply the knowledge we have. But there is – a checklist.

Atul Gawande has an excellent book called *The Checklist Manifesto* that describes the reason to use checklists with examples from surgeons and pilots – people who we hope do things right every time. There are certain situations that benefit from having a checklist:

- when steps are critical to execution,
- order of operations matter,
- interruptions are frequent and of variable length,
- one process can be applied to different applications,
- a repetitive operation,
- infrequent performance of a task,
- tasks that have long waiting periods, and
- a complicated situation being performed the first time and that might have to be repeated or proceduralized.

This last bullet specifically relates to data mining and analytical projects. A question is asked, data are found, extracted, transformed, loaded, determined to be wrong, and then all the steps are reworked with better inputs. The first pass does not need to be fancy. Scribbles on a notepad suffice. I typically call this annotation step "leaving a trail of bread crumbs" so that I can repeat my steps if need be. Table 4-5 is an example of a checklist that is focused on data mining activities. In addition to just checking the box, a checklist can be taken to higher levels by recording either the date and time the action was taken, the number of instances observed, or both. By making the requirement more concrete, the chances of success are magnified.

Table 4-5. Checklist for data mining activities.

Best Practice	Date	Count
Did I follow file naming guidelines?		
Did I honor repository conventions?		
How much time did I spend on transformation? Did I identify, and share, any opportunities for improvement?		
Have I personally seen the data repository and backup systems?		
Did I map the processes that generate these data from start to finish? Did I identify, and share, any opportunities for improvement?		
Did I ask the right questions of the process owners?		
How many assumptions did I make?		
Was I rushed?		
Was I personally distracted?		
Who was my peer reviewer?		
Am I comfortable betting my job on this analysis? Do I believe the results?		

Daily management

The proverbial statement rhetorically asks, "how do you eat an elephant?" The answer is simply, "one bite at a time." So, how do you analyze a year's worth of data? How about one day at a time with weekly, and monthly, and quarterly rollups all compared to baseline values? This is an especially good approach if one is trying to identify patterns and trends. The human brain is a perfect computational machine to make these discoveries, if the data are loaded slowly, kept within context, and in-line with disciplined critical thinking where hypotheses are generated and assessed.

Sometimes I miss the old days. In the 1980s, I used to have to type numbers from a paper-coy of laboratory results into my Lotus 1-2-3 spreadsheet. I was more in touch with those data sets and it was much easier to find associations and make discoveries because of my familiarity. I do like the speed and accuracy of fully automated delivery systems, but I am always struggling to engage with that data.

Most people practice a form of daily management by reviewing their checking account and credit card transactions every night to determine if there is any unexpected activity and if transactions occurred as expected. These frequent reviews allow problems to be addressed quickly in an unrushed manner. When one waits until month's end, which was done in pre-internet days, then significant blocks of time are spent performing reconciliation activities. Daily management is a technique that allows a data scientist to fully immerse within the data holdings.

Tom Redman[25] is a fan of this approach by what he calls the Friday Afternoon Measure. It involves taking a sample of the last 100 work products from the current week and doing a review of those items for conformance to requirements. It is also a very good tool to apply to "work in process". For example, lines of code. By collecting data, managers have a *modus operandi* for doing progress reviews that are fair, quick, and easily processed. It does not mean the lines of code are any good, just that they exist.

The 21st Century is all about "the Matrix". That is, the digital holdings by companies and businesses that define you, or a company. Daily management is the means to assure quality and to force review. Table 4-6 describes the steps needed to enact a daily management culture. It works best when every employee has a "channel" or an "account" assigned to them for review. Comparison to a baseline is very important because it answers the fundamental question "does this look right?"

For new data that does not have a baseline, a different mode must be followed until at least 30 data points, or even better 100, are obtained. A mean can be calculated each day and limits set using a mathematical rule. Chebyshev's Inequality [26] states that $(1-1/K^2)$% of the data, regardless of the distribution type, will fall within K standard deviations of the mean.[27] For a conservative baseline, two standard deviations are a useful heuristic with $(1-1/2^2)$ or 75% of the values being expected.

By using a simple rule, focus can be given to collecting data and not debating where arbitrary levels should be set. The goal is to obtain the Voice of the Data and not an assessment of meeting specification limits. In addition, data scientists can become buried in alarms that require investigation if the initial measurements are chosen without a baseline study when dealing with big data. If the data are normally distributed, one can say with certainty that 99.7% of the data lies within three standard deviations of the mean. Conversely, this means three out of every 1,000 data points will be beyond those limits. If a data scientist is assigned a million records to review, then one would expect at least 3,000 exceedances of the three standard deviation range. Other than counting these defects, not much analysis can be

---

[25] T.C. Redman, *Assess Whether You Have a Data Quality Problem*. Harvard Business Review. July 28, 2016.

[26] https://bit.ly/2npmOLG.

[27] The practicalities of this conservative view for small samples is demonstrated by the boxplot show in Figure 4-5.

performed quickly. Either the timeframe needs to be reduced, the process focused, or a sampling approach deployed to allow meaningful work to be performed in a reasonable time.

The key to making daily management work is to partition the data into manageable chunks! Issues in quality are quickly identified, with some root causes being data quality itself. These defects must be addressed quickly so that performance measures can be stabilized, and rational decisions trusted. Most importantly, the quality of the data must be fully understood before the quantitative tools and techniques described in the three chapters in Part II can be applied.

Table 4-6. Steps to implement a daily management culture.

Stage	Action	Explanation
Preparation	System Decomposition	Major task to determine fundamental measure for all components of subsystems. Master hierarchy for all data that could be shown on dashboards.
	Generation of List of Ideal Measures	Some measures serve multiple purposes and can be combined. Wishlist for what is missing.
	Identification of Measurement of Correct Measures	Full automation is the goal so that all data resides in The Matrix. However, there are always hidden factories, desktop databases, and off-the-books records that contain essential data that must be assimilated after the creators are rewarded.
	Assignment	Determine who is the responsible Point of Contact, ensure the reporting chain realizes the importance of the information, and assign backups and successors. It is also wise to implement similar assignments in the knowledge software suite (like Atlassian) and then require all files used during implementation to be stored therein.
	Transactional Record Review	No box checking! A fact needed by "the Matrix" must be recorded, like time, condition, name, …
Implementation	Gap Analysis	What is not being monitored? Set periodic review cycle. Exercise backup capabilities on a set schedule.
	Daily Review Checklist	Sections = Immediate Action Triggers, Performance Pitfalls, Action Tracking, BOLO lists
	Baseline Determination	Value from the annual review document. If a new measurement, a different mode relying on Chebyshev's Inequality is required.
	Daily Review Checklist	Job #1. Must be done before email is opened at the start of a day. If an emergency situation, a phone call would have been placed. When an employee returns from a vacation or an extended absence, this review will activate memory centers in the brain and make the transition easier. Create a checklist for the actions required when an alert is issued.
	Storage	Should the data be stored by the user ("work in process") or in a shared location so others can access it? Is there a knowledge management suite (like Atlassian) available and should it be used?

Stage	Action	Explanation
	Automation	Should the files be pinned to the task bar or the desktop or even to automatically open when the computer is booted? Should remote access be allowed?
Reporting	Weekly Analyst Summary	Fundamental piece of work. Rolls up daily reviews and triggered calls to action.
	Monthly Report	Presented in a meeting format with the applicable managers. More frequent if needed.
	Quarterly Briefing Package	Large session with food. Place to indoctrinate newbies (of all levels of management). Venue for cross-functional sharing of information with food prompting open and casual conversations where relationships can be made.
	Annual Summary and Baseline Reassessment	Activity to include the statistics expert, such as the Six Sigma Master Black Belt. Randomly schedule n/12 each month so that resources are not over-burdened during any one period and that touch points are built-in throughout the year to avoid surprises.
	Managerial Review Questions	Knowledge management tool where every question asked, the approach taken to generate an answer, and the answer itself are recorded. Tools such as JIRA from Atlassian are an aid, but only if disciplined processes are implemented and no exceptions granted.
Continuous Improvement	Initial and Continual Training	Classroom versus Online versus Formal Education? In-house or outsourced? Onsite or elsewhere? Competency based? Achievement badges versus Continuing Education Units (CEUs)? Level of training (manager, subject matter expert (SME), user, general awareness).
	Validation Against Level-Up in the System	A technique to validate meaning and to determine if the variable is an independent one.
	Pruning of Variables	Variables that are non-value-added or that are dependent, even if the client is paying for their collection, should be eliminated so that focus is given to the key variables. Even if the "customer is always right," does not mean that his boss is "made of money."
	Formal Measurement System Analysis	Periodic assessment of system performance of all measurement activities. Place for involvement of statistical expert.
	Quality Circle Project	What areas are detracting from performance? Who all needs to be involved to fix the problem? How will a solution be monitored after a fix is implemented? Are there too many Band-Aid patches and a full refreshment activity is need?
	Database Technology Refreshment	Linkage point to CIO. What changes are planned for IT performance, storage, and security versus what is measurable, calculable, or needs measurement. This will result in the creation of The Matrix and the assimilation of all data. Implementation will be conflict ridden with success realized when stakeholders start asking "how did we ever live without this?"

PART II

Quantitative Tools and Techniques

You cannot improve what you do not measure.

J Edwards Deming

CHAPTER 5

Characteristics of Data from Systems

This book is not a statistics textbook. There are many good ones out there. They are typically focused on a single context – which means that the practice problems all follow a theme, use appropriate jargon, and provide an introduction into the types of data that will be encountered in that field. The basic principles are the same in all of them. If you need a generic refresher, search in Wikipedia, or the National Institute of Standards and Technology Engineering Statistics Handbook online. Both facilitate access to the primary sources themselves. This book is about building the readers skills to instinctively follow the lines of thinking typical of a data scientist, especially this chapter.

Statistics is a young branch of mathematics with most fundamental research done in the 20th Century. However, the big data aspects are just now being investigated. What are the best tools to use with the First Million Row data audit approach discussed in Chapter 3? Can you use a Student's t-test to compare samples from two different time ranges?

Try scrolling down a million-row spreadsheet, like the case study from Chapter 2. It will give you a deeper understanding of how much a million is. If you generate a million records a day, it takes less than three years to generate a billion records (two years, eight months, 27 days to be exact). As the Matthew West song, *Live Forever*, emphasizes, there are 86,400 seconds a day, and if a reading is collected each second, you will accumulate a million records in 11.57 days if you have one sensor in the system. Multiple sensors will produce a million records even quicker. Big data are becoming common in everyday life.

This chapter is not about software. Because storage and processing technologies keep changing, software packages are, by definition, still beta. (Addition and subtraction have been used forever, and the abacus became a mature technology long ago.) Anyone working with big data needs to be agile and adaptive and prepared to repeat the same analysis multiple times. Keep in mind that change is the father of twins: progress and problems. Also, no two analyses generate exactly the same results. The differences always raise the question "which one is right?" The secret to making that determination requires the data scientist to calculate the answer by hand and be aware of both how many significant digits are required for the answer and whether the software can truly handle that intensity of calculation. It is surprising how many times I have applied my science background for collecting and analyzing numbers to solve complex problems.

This chapter is not a primer on the art and science of building models. There is an urban myth quote – "all models are wrong, some are useful." I contend that they all are useful for whatever purpose required their creation. The model may not be useful for other purposes, and there is no one model, nor quantitative technique, nor statistical tool that has universal power to solve all problems. We know this from geometry class. Euclid gave a powerful demonstration over 2,000 years ago when he wrote Geometry and pointed out that you must treat circles different than squares and both different from triangles:

$$\circ \neq \square \neq \Delta$$

Some models can be reapplied if conditions are isomorphic, but all assumptions must be documented so that the residual error in prediction can be understood.

The most important aspect of the output from any model are the error bars associated with the calculated answer ($\hat{y}$). How confident can I be of this result? How accurate was the measured data I used for inputs? What will it take in terms of time and money and resources to refine the model to a greater level of precision, and is that greater level of precision needed? These are the questions scientists always ask of the data they are using to solve a problem. These skills are sufficient to allow a data scientist to answer a question in general, which is necessary to assess the validity of an answer obtained from a model.

This chapter explores the types of numerical data, some fundamental concepts of statistics, and the most common generic outputs common to the analysis of all systems.

Types of data

When starting a new project, I have frequently said "numbers are numbers" when the stakeholder starts explaining why their data set is unique and complex. In response to the harrumph response that comment receives, I then say, "every one of them is different." I then say, "send me a spreadsheet view so I can take a look" (using the tools from the MRDA process discussed in Chapter 3) and then "I will come visit you to play a game of 20 Questions." I have learned to become a *tabula rasa* (or clean slate) when starting a new project.

First, by lacking preconceived notions, I have nothing to unlearn. Second, I force myself to systematically review the type and nature of the data itself. I have described this step in the process like being a chef who is inspecting the ingredients on hand. Third, I then try to match what I see to what I hear the customer describe. Sometimes they match. Sometimes they don't.

The first consideration is whether the numerical data are continuous or discrete. Can the numbers be divided into ever finer units by measurement technologies (continuous)? Or do the numbers represent something that I can 'kick and count' (discrete)? Neither is right nor wrong – it is just that each requires different tools and techniques with different outcomes possible.

The second consideration focuses on the measurement scales applied to the data. There are four types:

- Nominal is a simple categorization for binning purposes (like 1 for wins and 0 for loses). A subclass is categorical data where bins are defined by textual names rather than quantitative values but are sometimes coded with numbers to facilitate analysis.
- Ordinal involves assigning values that indicate the relationship between values with the implication that they can be ranked, like on a 10-point scale.
- Interval is most common and is a simple measurement such as height and weight where the value itself has meaning with complications presented when multiple units of measure are included in a field.
- Ratio is a scale derived by calculating differences to facilitate comparison (think dish water detergents that advertise themselves as 2x or 4x or 8x stronger).

The third consideration revolves around the context in which the data were generated, a condition the IT techies call *provenance*. This is where systems analysis comes into play. It involves entity relationship diagrams and database structures on the IT side and processes and definitions on the transaction side. And then there are the exceptions! There are always special cases that have their own rules that make programming very complicated.

A good starting place is a Venn diagram with count and proportion data (recall Figure 1-1) so that sets can be visualized. When the number of exceptions becomes large, it might indicate that the entire process needs to be redesigned, or that multiple events of different scopes are being addressed by one process.

There is a whole branch of statistics, nonparametric analysis, that has developed tools and techniques for use with all the data perturbations. One of the most common, and powerful, is simply rank ordering data. Another is a signs test when events happen in sequence and the direction of change is noted. The analysis phase frequently is delayed while the characteristics of the data are determined. Delays are much better than having to perform rework because critical elements are only illuminated when preliminary results are revealed. Detailed examples are provided below for climate and sports with the data sets available at the companion website at https://technicspub.com/qasa.

Case Study: Continuous data from rainfall results

The rainfall data shown in Figure 5-1 is an example of continuous data since the value can be anywhere between 0 and infinity. In reality, it is listed to the nearest hundredth of an inch because that is the resolution limit of the measuring system. Technically, it is interval data since it is a daily total. (Note that daily is not always a 24-hour period because the days the clocks change for daylight savings time are either 23 or 25 hours in length.) See Appendix C[28] for a more detailed discussion with graphs that correspond to tools and techniques presented in Chapter 6. The value for each year is independent of the previous year's value.

Figure 5-1. Rainfall data as an example of continuous data.

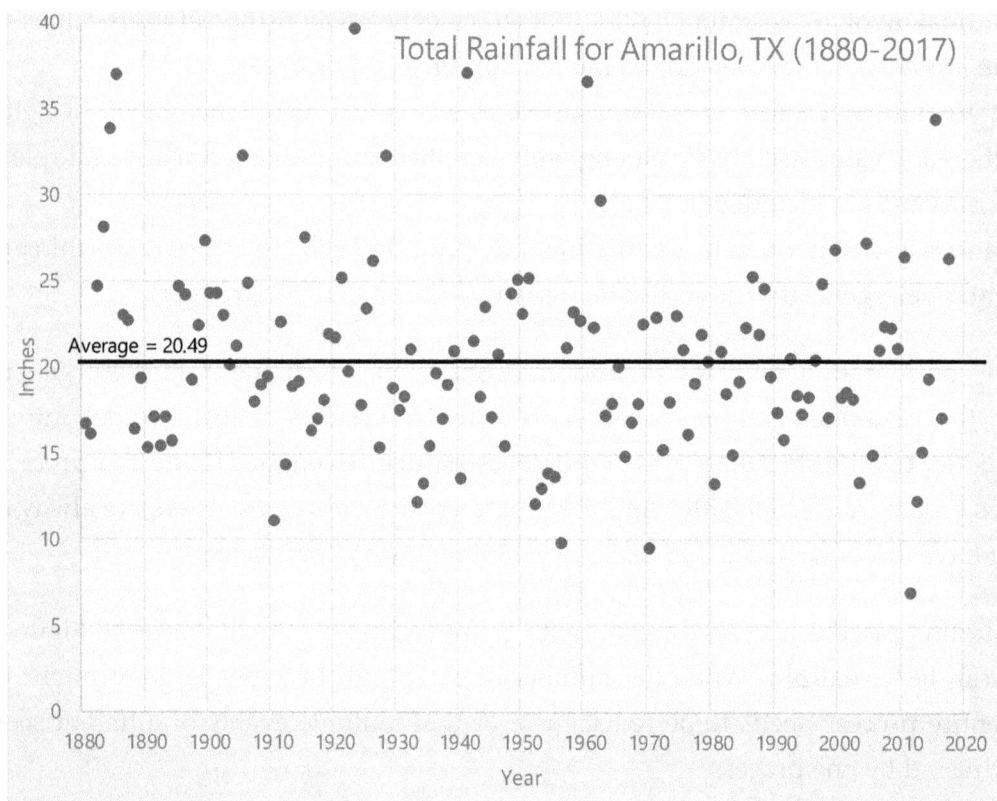

---

[28] Appendices are great places to present lots of graphs and other backup material that do not directly tell the story. My record is a report that had eight appendices with almost 500 graphs. They were never printed but rather electronically submitted. I exported them to PDF files and they supported a 50-page report that was printed. As the analyst, it was my job to look at each one of the graphs. I performed systematic reviews and presented all the observations in a table in the report. My documentation was complete enough to allow another data scientist to thoroughly review my work.

Case Study: Discrete data from soccer league results

Sports provide many examples of discrete data. Baseball has always been a hot bed of quantitative analysis from Sabermetrics that was developed starting in the 1950s to the Oakland A's "Moneyball" approach. To win the World Series, there are about 250,000 pitches that require two decisions to be made: which pitch to throw and to swing or not. The bottom line is that if players are consistent, the probability of a desired outcome can be determined with managers making decisions that hedge the odds in the team's favor.

A more universal sport is soccer (or football) that is played around the world. I am only presenting three statistics herein: number of wins, goal difference (sum of goals scored minus goals against for each game), and points (3 are registered for a win and 1 for a draw). There is an interesting option in Excel to generate spark line charts that are quite useful for portraying discrete data. They are handy since they can be included in a document as a word rather than as a standalone figure.

The following three examples are from the 2016-2017 Bundesliga season and use the results for the club Borussia Dortmund. The season consisted of 34 matches with the spark line chart consisting of 34 fields (one per match day). Dortmund registered a total of 18 wins with two long winless streaks recorded in the first half of the year. The goal difference averaged about +1 per game and a cumulative value of +32 for the season. When points are viewed , it becomes apparent that a few draws were registered during the periods when wins were not registered. These charts assume the reader is quite knowledgeable of the subject and are a convenient way to share discrete data for discussion purposes.

The most common tool for comparison of teams within a season is by tabulation, shown in Table 5-1. These tables are generated every night for printing in the newspaper (or in real-time for display on the internet). A similar approach is taken in the business world using dashboards. Most dashboard data are discrete. Many software packages allow for infinite display options and automated generation. The best ones allow a note to be entered so that narratives can be preserved for the ubiquitous unusual circumstances. Sometimes, data are more easily interpreted in graphical form. Figure 5-2 shows the relationship between wins and the cumulative goal difference for each of the teams shown in Table 5-1. Figure 5-3 shows wins and points. Both graphs show a strong positive relationship and emphasize the degree of separation between teams as well as similarities with close groupings.

Table 5-1. Bundesliga rank table for 2016-2017 season.

#	Team	Matches Played	Win	Draw	Loss	Goal Difference	Points
1	Bayern München	34	25	7	2	67	82
2	RB Leipzig	34	20	7	7	27	67
3	Borussia Dortmund	34	18	10	6	32	64
4	Hoffenheim	34	16	14	4	27	62
5	Köln	34	12	13	9	9	49
6	Hertha BSC	34	15	4	15	-4	49
7	Freiburg	34	14	6	14	-18	48
8	Werder Bremen	34	13	6	15	-3	45
9	Borussia M'gladbach	34	12	9	13	-4	45
10	Schalke 04	34	11	10	13	5	43
11	Eintracht Frankfurt	34	11	9	14	-7	42
12	Bayer Leverkusen	34	11	8	15	-2	41
13	Augsburg	34	9	11	14	-16	38
14	Hamburger SV	34	10	8	16	-28	38
15	Mainz 05	34	10	7	17	-11	37
16	Wolfsburg	34	10	7	17	-18	37
17	Ingolstadt	34	8	8	18	-21	32
18	Darmstadt 98	34	7	4	23	-35	25

Figure 5-2. Relationship between wins and goal difference for Bundesliga teams.

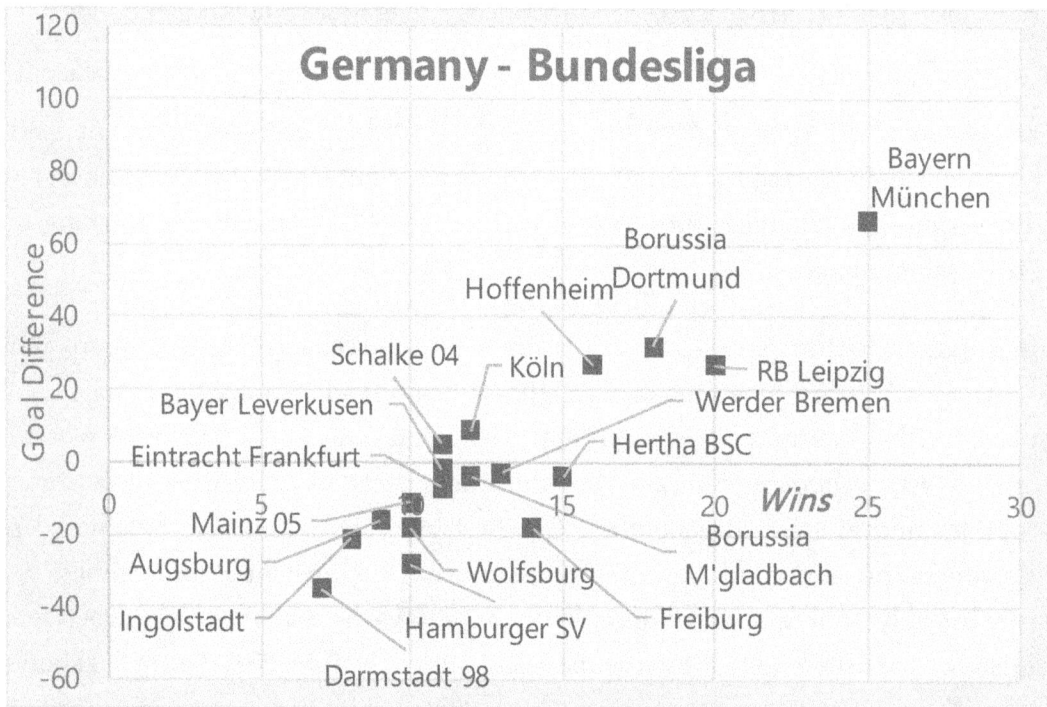

The comparison of goal difference to points was not generated because some teams win small and loose big as a strategic move by limiting the playing time by starters and maximizing the time played by junior players needing experience with Freiburg being an example. Further analysis is provided in Chapter 6 where the Bundesliga is compared to the other major leagues.

Figure 5-3. Relationship between wins and points for Bundesliga teams.

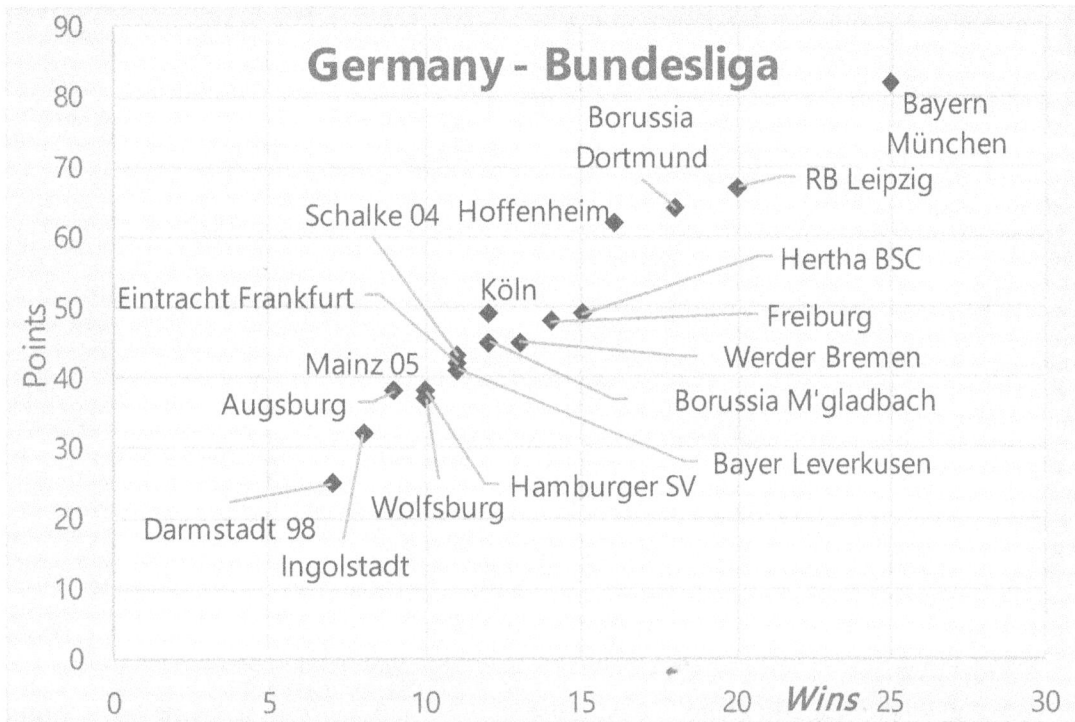

Time

Time is a special category of data. It is complicated from the start because it does not follow base ten nor is the measurement of it always orthogonal (month-length). It also has leap days and daylight savings time complications. Then there is the challenge of chunking for display and rate calculations. When I start on a new data set that involves time, I prepare myself for a lengthy battle. I never return quick answers because I always need to reflect on my actions during a sleep cycle, and then I always double check everything I have done. Excel is handy for making calculations. Many times, I migrate data to Microsoft Project to utilize the built-in calendar functions with an example presented in Figure 5-4. Note that Project is a massive database that has about 100 fields that can be customized by the user so that additional information can be related to each record. The most complicated calculations involve air travel that crosses the International Date Line. As big data becomes more encompassing, skills with time series analysis will be critical for top flight data scientists.

Figure 5-4. Typical Gantt chart[29] created using Microsoft Project.

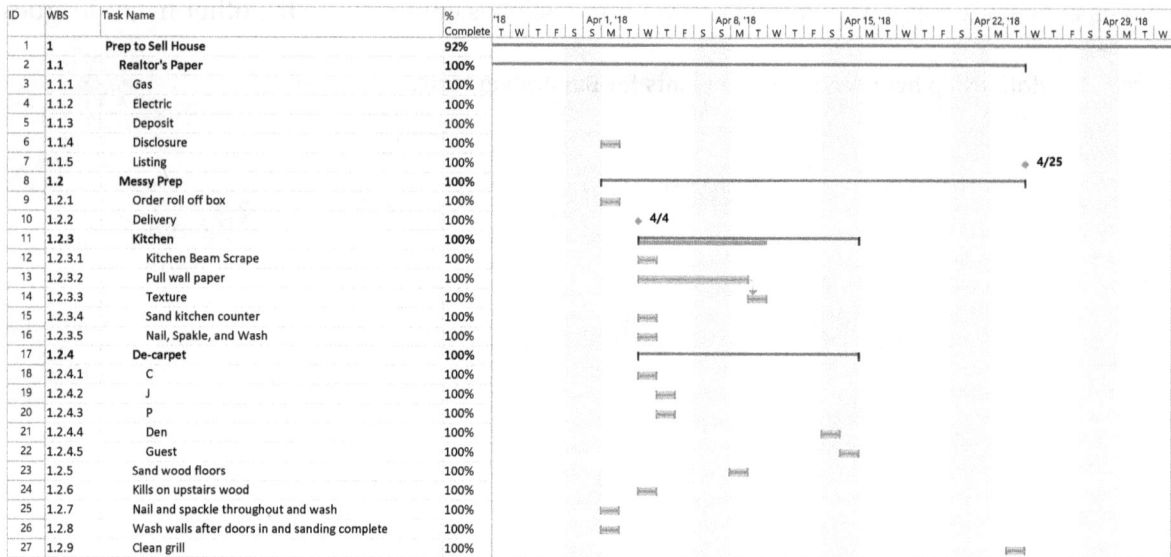

ID	WBS	Task Name	% Complete	Apr 1, '18	Apr 8, '18	Apr 15, '18	Apr 22, '18	Apr 29, '18
1	1	**Prep to Sell House**	92%					
2	1.1	**Realtor's Paper**	100%					
3	1.1.1	Gas	100%					
4	1.1.2	Electric	100%					
5	1.1.3	Deposit	100%					
6	1.1.4	Disclosure	100%					
7	1.1.5	Listing	100%				4/25	
8	1.2	**Messy Prep**	100%					
9	1.2.1	Order roll off box	100%					
10	1.2.2	Delivery	100%	4/4				
11	1.2.3	**Kitchen**	100%					
12	1.2.3.1	Kitchen Beam Scrape	100%					
13	1.2.3.2	Pull wall paper	100%					
14	1.2.3.3	Texture	100%					
15	1.2.3.4	Sand kitchen counter	100%					
16	1.2.3.5	Nail, Spakle, and Wash	100%					
17	1.2.4	**De-carpet**	100%					
18	1.2.4.1	C	100%					
19	1.2.4.2	J	100%					
20	1.2.4.3	P	100%					
21	1.2.4.4	Den	100%					
22	1.2.4.5	Guest	100%					
23	1.2.5	Sand wood floors	100%					
24	1.2.6	Kills on upstairs wood	100%					
25	1.2.7	Nail and spackle throughout and wash	100%					
26	1.2.8	Wash walls after doors in and sanding complete	100%					
27	1.2.9	Clean grill	100%					

First principles

When working with data, there are some fundamental concepts that guide all actions and activities. Again, this is one of the science aspects of data science. Most people with scientific degrees have had this information drilled into them from primary school days. I present five principles herein. Arguments can be made that some of the concepts presented in later chapters could be a first principle. I made my selection based on how the results are shared with the customer with true first principles not always being explicitly presented. The five presented are Probability, Central Limit Theorem, Law of Large Numbers, Law of Diminishing Returns, and Sampling Theory.

Probability basics

What are the odds of getting a heads when you flip a coin? One out of two or 50%. Most people are aware of this binomial choice. Most people understand that the chance of getting a specific

---

[29] Note: Diamonds represent milestones and have no duration. Colored bars represent activities with duration with the thin black line signifying duration. The bounded black line on Row eight is called a *hammock* and presents a roll-up summary for all tasks associated with that Work Breakdown Structure (WBS) element. Overall Percentage Complete is less than 100% because this view is a snapshot from data date = 20180507 of a project plan with 128 rows.

number on a die is 1 out of 6. It becomes a little bit trickier when the bookie says the odds of Manchester City beating Manchester United is five to 4. It causes one to pause. Or we hear that the chance of a specific component failing on a SpaceX rocket launch is one in a million. Or that the mean time between failures for the air conditioner compressor in our car is 3,500 hours which may or may not help you decide if you should replace it before driving the family across the Mojave Desert in July in a van with 200,000 miles on the odometer. Our brains grasp probabilities, not intuitively, but rather through learning. One of best ways to learn is by playing cards and other games. While gambling drives the point home quickly, it is not a required aspect.

Note that a different part of the brain is used for quantitative processing than for other tasks. I learned this from a neurologist who was working with my father after he suffered a stroke. In response to the doctor's questions, I informed him that my dad was an engineer and always liked to play games. The neurologist then had a hypothesis that if we could activate that part of his brain, the walking and speaking processes might be regenerated using those circuits due to the extreme plasticity of the brain itself. We started playing blackjack and poker and then moved on to gin rummy. Because there was no control case, it was not possible to prove the hypothesis, but it was easy to observe the monumental progress he made within the next week compared to the previous three.

Returning to definitions, probability is:

1: the quality or state of being probable
2: something (such as an event or circumstance) that is probable
3: a: (1): the ratio of the number of outcomes in an exhaustive set of equally likely outcomes that produce a given event to the total number of possible outcomes
  (2): the chance that a given event will occur
  b: a branch of mathematics concerned with the study of probabilities
4: a logical relation between statements such that evidence confirming one confirms the other to some degree.[3]

Probable is defined as: "supported by evidence strong enough to establish presumption but not proof."[3] What exactly does "not proof, but to establish presumption" mean for us? It means there is an element of uncertainty that must be accommodated. Probable outcomes are ones that we accept as "good enough" and not the exact solution, like 2+2=4. This implies that you do not have to measure every member of the entire population, but that you can measure a sample and determine if the majority of results meet the good enough criteria given the accuracy of the measurements.

Nothing is a sure thing, but decisions can be made to hedge against bad effects. An example is a lender to the automobile industry that has made ten million loans. That is a lot of loans and one wonders about the risk to the company. First, they use an algorithm based on credit scores and past performance. Then, they stratify the data into different ranges. They know that people with high credit scores mostly never default and thus charge a low interest rate that meets the profitability objective and a bit more so that, in aggregate, enough money will be stockpiled to offset those few cases when a default occurs. They also know that there is a cohort with low credit scores that will never make all the payments for a car, and automatically deny loans to them. They avoid risk in those situations. For the remaining applicants, a more involved algorithm is used that considers a suite of other criteria.

The lender doesn't always need to be right, but rather to set interest rates so that an adequate war chest of money (*contingency fund* if you want the technical term) can be maintained to offset defaults. This is the realm of probable knowledge informing risk when millions of decisions are made and the impact of any one error is lost in the multitude of opportunities. It works because the Law of Large numbers effectively reduces risk by arbitrage. For assessing risk in mid-size operations, like 3,500 records, it takes up to a few days to walk through every record, which must be done, because one cannot arbitrage risk away. The results represent the condition of the entire population and consequences can be directly calculated using deterministic methods. If the operation is small, or the decision is critical for the protection of human life, all of the data are insufficient for assigning a probable outcome and qualitative tools must be used to support decision making.

My software package said I have a p-value of 0.045. What does that mean?

If you read a statistics text book, you encounter an alphabet soup of specific tests. For each test you must look up a decision value for the appropriate alpha level in a table: z's, t's, F's, and χ^2's (the Greek letter chi). Once you have this value, you perform the mathematical operations to generate the test result for your data set. The critical comparison can then be made between the test result and the decision value. Then it becomes complicated to know how to explain this value and use it to show significance. A convention has been developed whereby the result, regardless of the test used, is normalized and expressed as the probability that the result is obtained from a random sample. A p-value of 0.5 means there is a 50% likelihood that the result is the same as the expectation: one of 0.05 means a 5% likelihood; .005, 0.5%. The smaller the value, the more significant the difference. This computed value is more useful than the pass/fail result associated with the simple comparison on a test-by-test basis. This convention makes the application of statistics much easier to use and explain.

Central Limit Theorem

I have gotten paid to rediscover the Central Limit Theorem (CLT) four times in my career. When you are a consultant, easy billable hours are an early Christmas present. For each of those projects, the preliminary assessment I gave to the customer was "this should be a straightforward analysis because it falls within the CLT." Each time, I heard the same response along the lines of "no, our data are special and different and not like anyone else's." I bit my tongue, saw dollar signs flashing before my eyes, and did a thorough and complete analysis. In three cases, when we briefed the report up the management chain, that manager asked at the end of the presentation – "isn't this just the CLT?" I just nodded.

So, what is the CLT? It is several different things depending on how it is applied. If you know the particulars of a population, then it can be used to determine if a sample is representative of the whole. If you have a sample, it tells you what the expected value of the population should be, with the error bars representing the degree of confidence that is an inverse proportion to the size of the sample. It is also the means to determine if two samples from the same population are different.

In a different role, it is the tendency of the sum of several random variables to result in an output that is normally distributed with the tendency becoming more pronounced as sample size (n) increases. For example, statistics is known for stating with certainty that a normal distribution with a mean of 2 and a standard deviation of $\frac{1}{3}$ (Figure 5-5), when added to a similar distribution, will yield a normal distribution with a mean of 4 and a standard deviation of $\frac{1}{2}$. Some people expect the resultant standard deviation to be $\frac{2}{3}$ because the means were added. The decrease in variation was expected because of the CLT and the tendency to normalize results. This tendency is used in many engineering designs.

There is an easy way to prove how the CLT works in this case. For both input sets, $\frac{1}{2}$ of the values are below the mean of 2 and $\frac{1}{2}$ are above. When the values from the lower half are added to the second set, there is a 50% chance the value will be below 2, resulting in a value less than 4, and a 50% chance that the value will be above 2 resulting in a value very close to 4 because the values will balance each other out. This is summarized in Table 5-2, that shows $\frac{1}{4}$ of the results will be less than 4, $\frac{1}{2}$ close to the central value of 4, and $\frac{1}{4}$ greater than 4. The difference above and below the mean net out causing the resultant mean to be reinforced. Now imagine if 4 or 8 or 16 values were being added together. The central tendency becomes even more strongly reinforced.

On the projects I mentioned above where I rediscovered the CLT, I simply had multiple fields of normally distributed data that needed to be added together. All I had to do was add them, then create a slide deck with supporting tables, figures, and graphs. The following example works with a much larger data set and demonstrates how strong the central value can be portrayed.

Figure 5-5. Statistical Representation of Uncertainty.

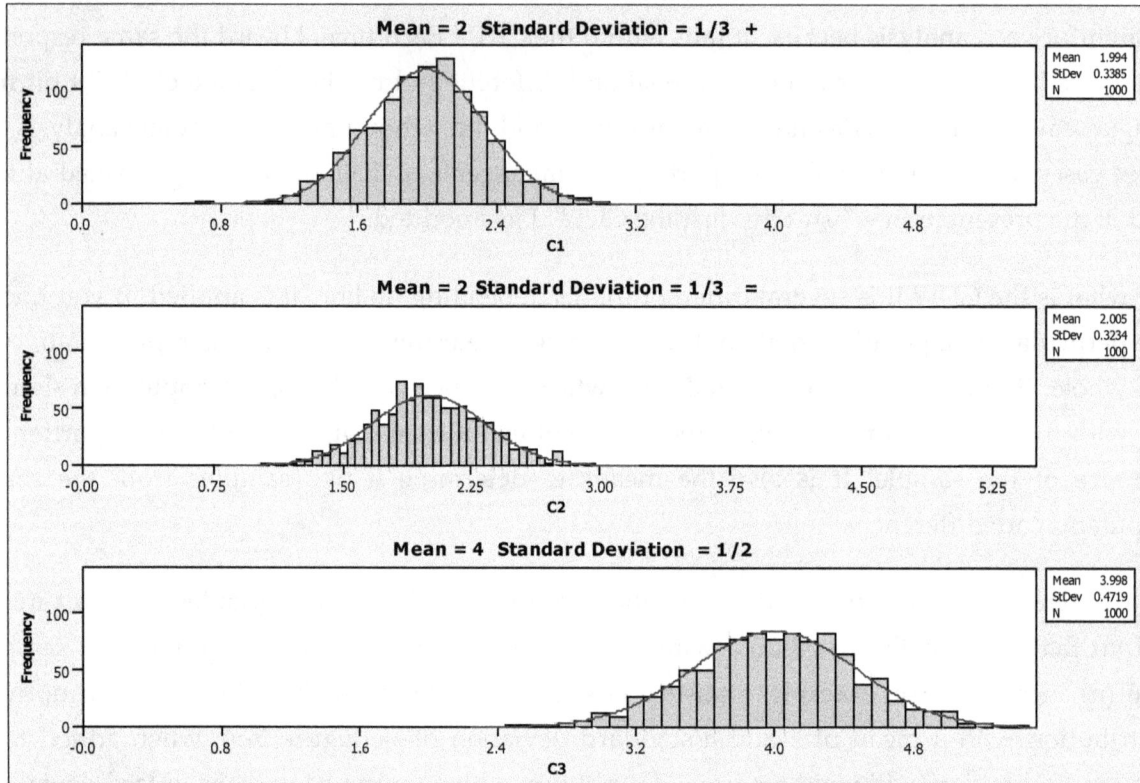

Table 5-2. Logical derivation of the CLT as shown by proportions of combined results.

			Set 2	
			Proportion Below 2	Proportion Above 2
			50%	50%
Set 1	Proportion Above 2	50%	25% approaching 4	25% above 4
	Proportion Below 2	50%	25% below 4	25% approaching 4

Case Study: Central Limit Theorem and shipping container data

This example shows the effect of sample size on the estimate of the central value of a data set. It is an application of the CLT in that an average is being calculated from a sample of the whole with the expectation being that the mean from the largest sample will most closely resemble the population average (μ). The data set itself is from trans ocean shipping and is reported as the volume of cargo containers (twenty-foot equivalent units or TEUs) for the 50 largest ports in the world for the twelve-year period of 2004 through 2015. Table 5-3 shows the population mean and standard deviation as well as estimates of those statistics using samples of various sizes. There is a large amount of variation in the means until the sample size becomes large. The large sample size limits the influence any one of the 30 values can have on the outcome (~3.3%).

Table 5-3. Effect of sample size on measure of central tendency of a data set.

	Population	n = 7	n = 10	n = 15	n = 20	n = 25	n = 30
Mean	80,426,780	60,619,501	85,211,819	92,630,438	79,330,858	74,202,001	79,107,872
Standard Deviation	76,961,565	47,863,741	83,792,028	94,378,076	81,071,810	79,068,282	73,323,317
Grand Mean				73,748,482			77,999,489
Grand Std. Deviation				78,777,932			78,399,605

Figure 5-6 is a graphical view of the estimated mean and error bars that represent ±1 standard deviations for 30 random replicates of sample size 30 that show the uncertainty of the estimates. Note that the grand mean (mean of the means) underestimated the population mean by less than 3%, while the grand standard deviation (square root of the average of the variances) overestimated the population standard deviation by less than 2%. Cutting corners by reducing sample sizes may save money, but the results do not lead to high probability outcomes.

Figure 5-6. Comparison of 30 replicates of random samples of size 30.

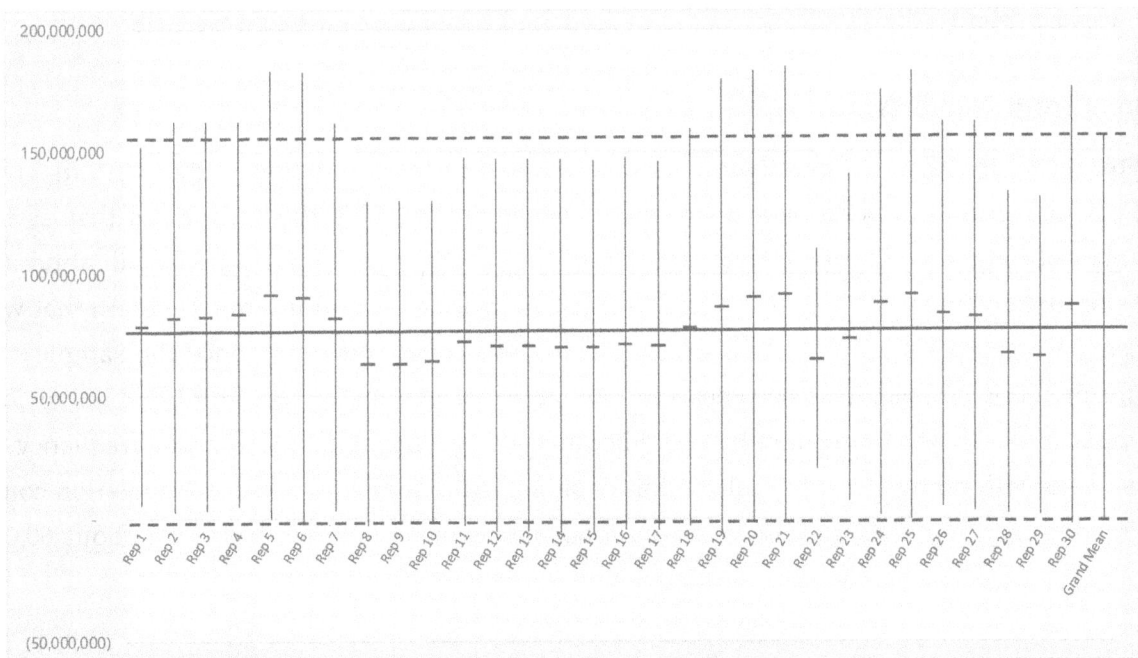

Figure 5-7 shows the magic of the CLT. When the number of samples is increased, the distribution of the sample means falls into a normal distribution. The normal distribution outcome happens regardless of the distribution of the underlying data. As to the variance of all the samples, that becomes slightly harder to explain and is called the *standard error* which is the standard deviation of the sample mean(s) divided by the square root of the sample size (n). It is expressed mathematically as:

$$\sigma_{\bar{x}} = \frac{s}{\sqrt{n}}$$

For this example, it calculates as 1,745,985 or 2.2% of the mean and represents the dispersion of the underlying distribution.

Figure 5-7. Histogram of means shown in Figure 5-6.

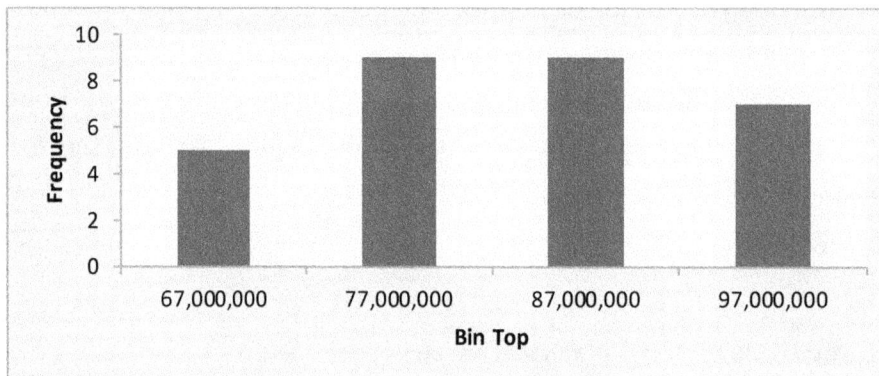

Law of Large Numbers

If you recall, the Law of Large Numbers was invoked in the discussion of the Venn diagram shown in Figure 1-2, pertaining to international air travel. This theorem states that as the number of items in a sample increases, the resulting average for those samples will approach the population average.[30] This result is also expected from the Central Limit Theorem that was based on a sample size of 30. Law of Large numbers comes into play once the sample size count of 100 is crossed. This becomes very important when one moves into big data, or into outputs from Monte Carlo simulations,[31] because the events with the highest frequency of occurrence will dominate in the calculation of an average. That is why the extrapolation could be made regarding missing passengers as total passenger count was based on about 60,000

[30] See Law and Kelton (2000) page 259 for a rigorous mathematical definition.
[31] This is mainly demonstrated by the true randomness of the random number generator. The inferior ones reuse the same stream of numbers which results in the outputs being replicated. A critical review of functionality is required to assure that the right tools are being applied.

flights that give an average of about 124 passengers per flight, which happens to be the heuristic used for air travel planning.

Case Study: Large numbers and air freight

This example has a lot of data. It has 526,623 rows of data for 220 million tons of air freight delivered to the US from January, 1990, through March, 2017 (data release is lagged six months by the Department of Transportation). This freight was delivered by 473 carriers (represented by 509 carrier codes) from 1,126 foreign airports to 569 US airports. Most people do not know that Anchorage is a major air freight hub for the transshipment of electronics and small automobile parts between 44 airports (mostly in Asia) and the US as shown in Table 5-4. Comparative amounts are displayed in Figures 5-8 and 5-9 and the total for countries is shown in Figure 5-10 and in a map view in Figure 5-11. Also note that Guadalajara, Mexico is a big shipper resulting from its role as an automotive manufacturing center. The file is available at https://technicspub.com/qasa.

Figure 5-8. Pareto chart for foreign air freight tonnage arriving in Anchorage in 2016 by point of origin.

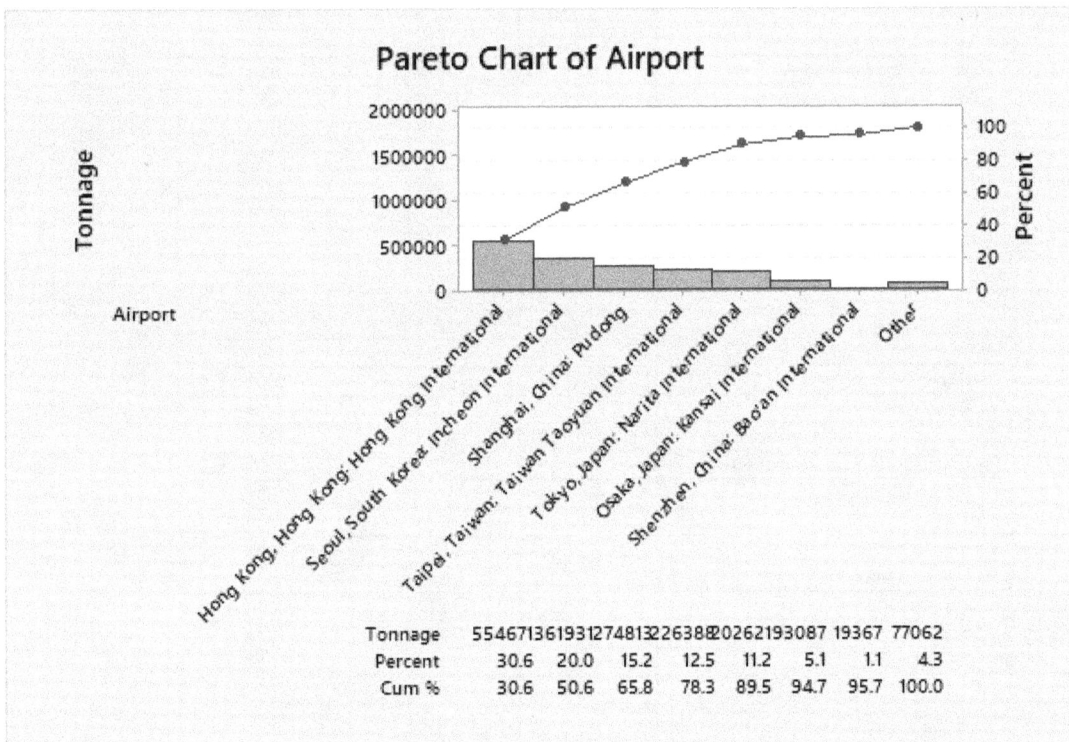

The first four airports achieve the Pareto point (where 20% of the categories account for 80% of the measurement) and six origination airports account for 95% of the tonnage. This is an example of the effects of the Law of Large Numbers. If the tonnage carried dropped 1% for Hong Kong, that would represent 5,547 tons, which is bigger than the total amount carried by 31 of the carriers. It also would be 0.31% of the total 1.8 million tons carried. If Calgary saw a

1% increase, it would only be 55 tons or 0.003% of the total. Changes for the contributions to the total for the small origination locations is just noise.

Table 5-5 lists the 29 carriers that delivered the freight with the Pareto point achieved by seven (or 24%). Comparative amounts are displayed in Figures 5-12 and 5-13. While not as drastic as the results for the origination cities, the critical few are more influential than the trivial many due to the Law of Large Numbers.

Table 5-4. Pareto analysis of foreign air freight tonnage arriving in Anchorage in 2016 by point of origin.

Foreign Airport	Sum of Total	% of Total	Cummulative %		
Grand Total	1,809,940				
Hong Kong, Hong Kong: Hong Kong International	554,671	31%	31%		
Seoul, South Korea: Incheon International	361,931	20%	51%		
Shanghai, China: Pudong	274,813	15%	66%		
Taipei, Taiwan: Taiwan Taoyuan International	226,388	13%	78%	Pareto Point 4 of 44	0.09
Tokyo, Japan: Narita International	202,621	11%	90%		
Osaka, Japan: Kansai International	93,087	5%	95%		
Shenzhen, China: Bao'an International	19,367	1.1%	96%		
Guadalajara, Mexico: Miguel Hidalgo Y Costilla Internation	17,009	0.9%	97%		
Toronto, Canada: Lester B. Pearson International	12,149	0.7%	97%		
Zhengzhou, China: Xinzheng	12,051	0.7%	98%		
Nagoya, Japan: Chubu Centrair International	12,043	0.7%	99%		
Vancouver, Canada: Vancouver International	6,886	0.4%	99%		
Halifax, Canada: Halifax Stanfield International	5,875	0.3%	99%		
Calgary, Canada: Calgary International	5,513	0.3%	100%		
Guangzhou, China: Baiyun International	1,349	0.1%	100%		
Chongqing, China: Jiangbei	850	0.05%	100%		
Edmonton, Canada: Edmonton International	423	0.02%	100%		
Petropavlovsk-Kamchatsky, Russia: Yelizovo	385	0.02%	100%		
Tianjin, China: Binhai	353	0.02%	100%		
Jeju, South Korea: Jeju International	329	0.02%	100%		
Hanoi, Vietnam: Noibai International	254	0.01%	100%		
Bangkok, Thailand: Suvarnabhumi International	216	0.01%	100%		
Sapporo, Japan: New Chitose	173	0.01%	100%		
Manila, Philippines: Ninoy Aquino International	130	0.01%	100%		
Dhaka, Bangladesh: Hazrat Shahjalal International	121	0.01%	100%		
Guatemala City, Guatemala: La Aurora International	121	0.01%	100%		
Seoul, South Korea: Gimpo International	98	0.01%	100%		
Hohhot, China: Baita	91	0.01%	100%		
Lapu-Lapu City, Philippines: Mactan-Cebu International	83	0.005%	100%		
Monterrey, Mexico: General Mariano Escobedo Internationa	82	0.005%	100%		
Frankfurt, Germany: Frankfurt Main	78	0.004%	100%		
Jakarta, Indonesia: Soekarno-Hatta International	73	0.004%	100%		
Cologne, Germany: Koln Bonn	70	0.004%	100%		
Ulan Bator, Mongolia: Chinggis Khaan International	62	0.003%	100%		
Misawa, Japan: Misawa AB	54	0.003%	100%		
Osan, South Korea: Osan AB	46	0.003%	100%		
Singapore, Singapore: Singapore Changi International	37	0.002%	100%		
Port-au-Prince, Haiti: Toussaint Louverture International	26	0.001%	100%		
Bayan Lepas, Malaysia: Penang International	16	0.001%	100%		
Tokyo, Japan: Tokyo International	9	0.0005%	100%		
Phnom Penh, Cambodia: Phnom Penh International	3	0.0002%	100%		
Beijing, China: Capital International	2	0.0001%	100%		
Keflavik/Reykjavik, Iceland: Keflavik International	1	0.0001%	100%		
Kuala Lumpur, Malaysia: Kuala Lumpur International	1	0.0001%	100%		

Figure 5-9. Sunburst chart for foreign air freight tonnage arriving in Anchorage in 2016 by point of origin.

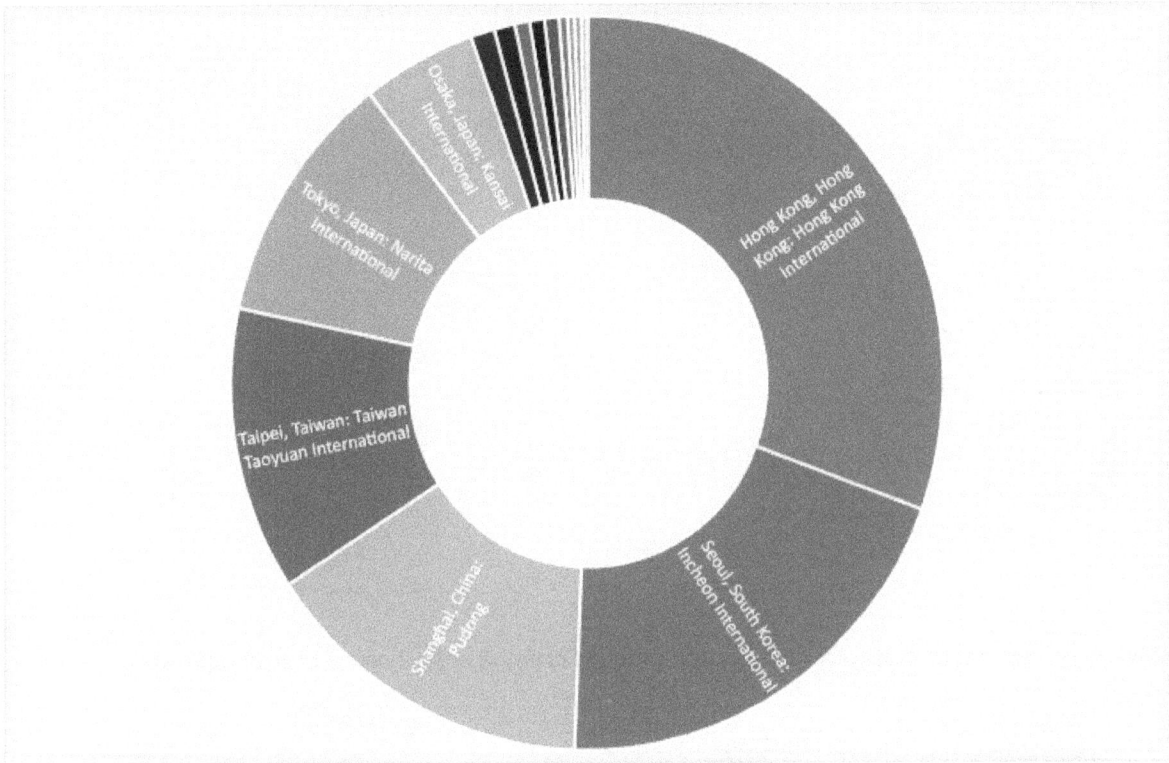

Figure 5-10. Pie chart of foreign air freight tonnage arriving in Anchorage in 2016 by country of origin.

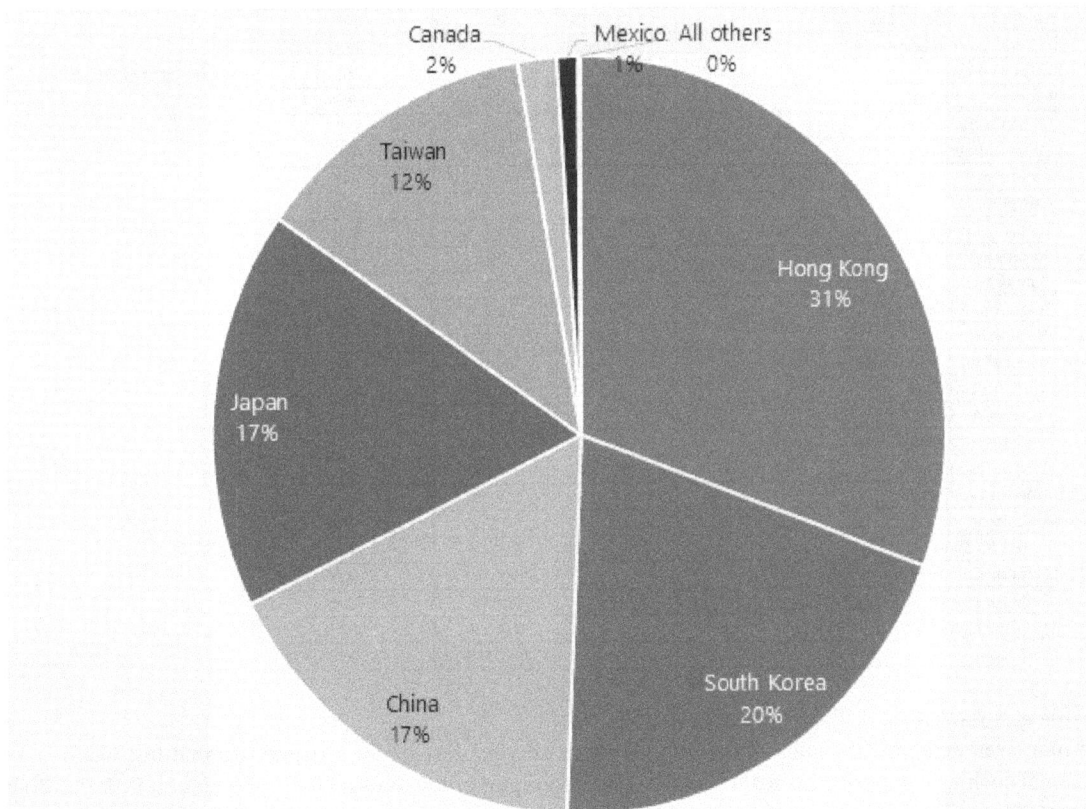

Figure 5-11. Data from Figure 5-10 in a map view.[32]

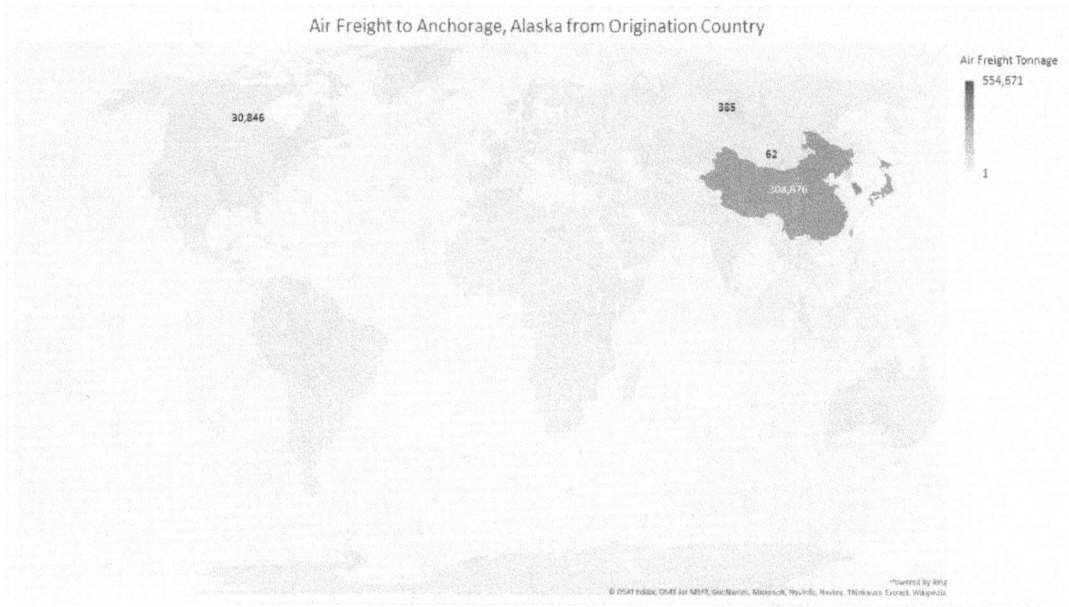

Air Freight to Anchorage, Alaska from Origination Country

Table 5-5. Pareto analysis of foreign air freight tonnage arriving in Anchorage in 2016 by carrier.

Carrier Name	Sum of Total	% of Total	Cummulative %		
Grand Total	**1,809,940**				
Cathay Pacific Airways Ltd.	344,136	19%	19%		
United Parcel Service	290,579	16%	35%		
Korean Air Lines Co. Ltd.	223,675	12%	47%		
Polar Air Cargo Airways	186,027	10%	58%		
Cal Air International Ltd.	145,008	8%	66%		
Federal Express Corporation	129,835	7%	73%		
Atlas Air Inc.	125,547	7%	80%	Pareto Point 7 of 29	0.24
Nippon Cargo Airlines	112,905	6%	86%		
China Cargo Airline	65,162	4%	90%		
China Airlines Ltd.	37,098	2%	92%		
Air China	34,758	2%	94%		
Kalitta Air LLC	24,800	1%	95%		
Britt Airways Inc.	23,947	1%	96%		
Southern Air Inc.	22,247	1%	98%		
Qantas Airways Ltd.	20,935	1%	99%		
Singapore Airlines Ltd.	16,832	1%	100%		
Western Global	3,029	0.2%	100%		
National Air Cargo Group Inc d/ba National Airlir	1,414	0.1%	100%		
Ozark Air Lines Inc.	972	0.1%	100%		
Viscount Air Service Inc.	687	0.04%	100%		
Air Transport International	78	0.004%	100%		
Condor Flugdienst	78	0.004%	100%		
American Airlines Inc.	43	0.002%	100%		
Air Canada	41	0.002%	100%		
United Air Lines Inc.	35	0.002%	100%		
Antonov Company	26	0.001%	100%		
Delta Air Lines Inc.	24	0.001%	100%		
Airborne Express Inc.	21	0.001%	100%		
Icelandair	1	0.0001%	100%		

[32] This map was generated in Excel. It has problems since very few of the expected formatting features are not functional such as color selection for the shading and zooming of the map region. Tableau and JMP Pro do better, but ArcGIS is truly needed to display this information.

Figure 5-12. Pareto chart for foreign air freight tonnage arriving in Anchorage in 2016 by carrier.

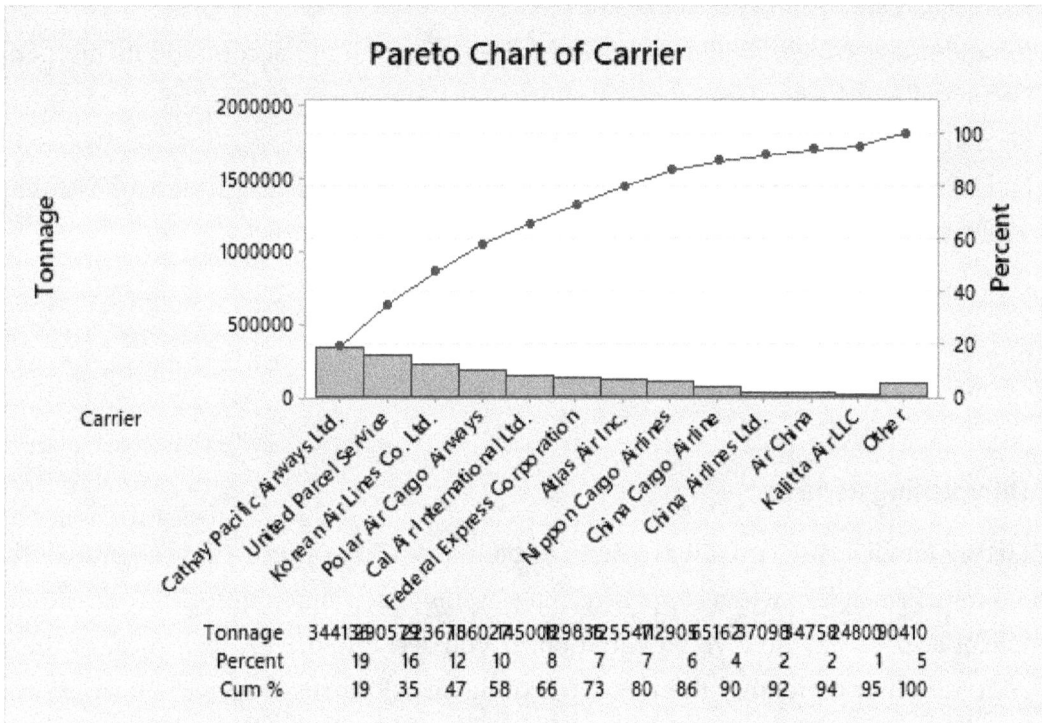

Figure 5-13. Sunburst chart for foreign air freight tonnage arriving in Anchorage in 2016 by carrier.

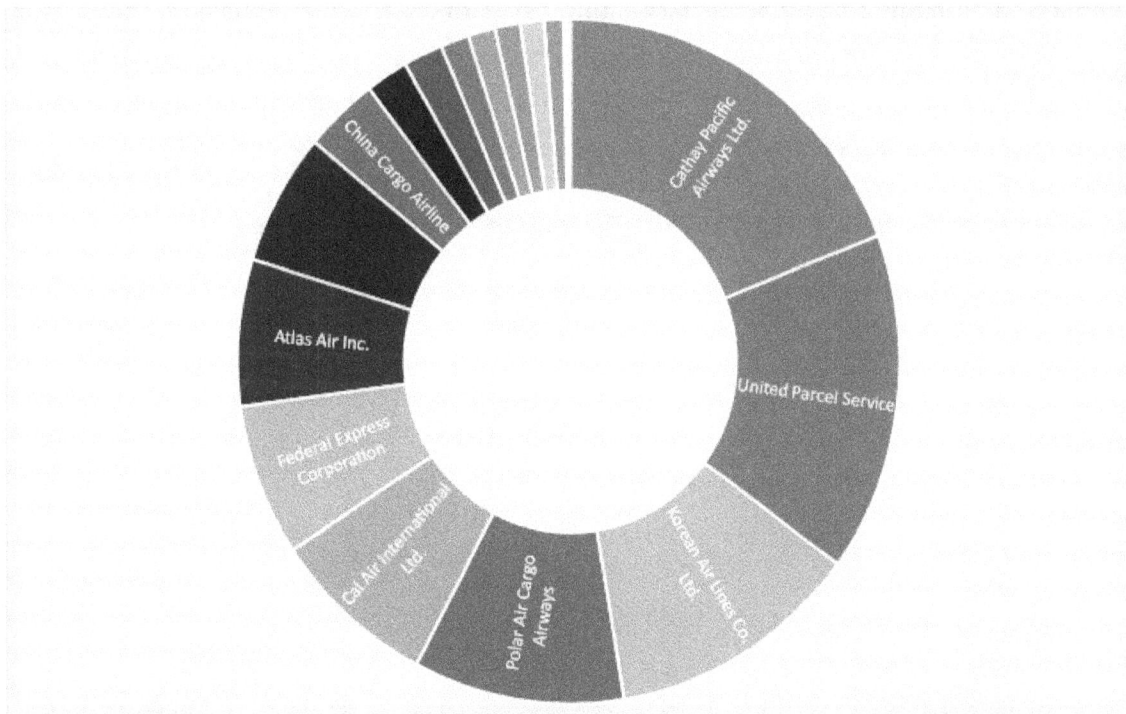

This discussion is just the tip of the iceberg. There are several other questions that can be investigated, such as:

- What difference does it make if freight-only airlines and airlines carrying both passengers and freight are separated?
- If looking at airlines, should a backward pass be applied to accommodate mergers, like Delta and Northwest, United and Continental, American and US Air?
- Should data be split on scheduled versus charter flights?
- Is the same story generated if the analysis is done on originating country? This requires a new field (column) for country for the 1,126 foreign airports.
- Are there differences in the data before and after 9/11/2001?

Have fun exploring for the answers!

Law of Diminishing Returns

When starting an analysis, you always need to be mindful if you must investigate all the way back to First Principles when applying new tools and techniques. All statistical and mathematical approaches involve an equation. If you see a 1/sqrt(n) (reciprocal of the square root of n), pause and remember the Law of Diminishing Returns. Figure 5-14 shows a classic inverse exponential relationship that asymptotically approaches, but never reaches, zero. Once the values fall below the resolution of the measuring system itself, the cross over point can be interpreted as the upper limit of large numbers for that case. Decision making must accommodate uncertainty after that limit is crossed. For the data in Figure 1-1, the error of 4.3% corresponds to a sample size of about 550. The value used for the extrapolation was about five times that value and thus carries a strong degree of certainty.

Figure 5-14. Example of effect of sample size on surety of results.

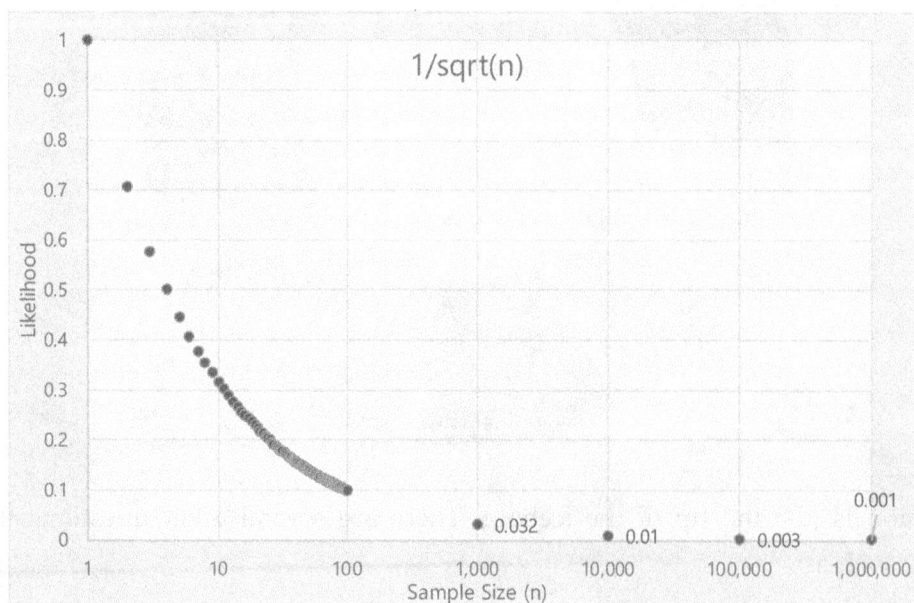

Sampling Theory

When you have mastered the lessons from the four preceding topics, you start to realize that you do not need complete holdings for every parameter for every record to make an informed decision. You can tolerate some levels of uncertainty. You can even use that to your advantage. If you know that the precision of measurement is rough (like ± 5 to 10%), you do not need to gather every bit of data. You can devise a sampling plan that provides a "good enough" answer quickly and cheaply because the CLT is on your side. Obtaining a high-fidelity answer is not what management wants if the input had low fidelity. It leads to a false sense of certainty!

Speaking of false, that brings up the concepts of false positives and false negatives (formally called *Type I* and *Type II errors*, respectively) as shown in Table 5-6. There is a quantitative tool, the *Power Equation*, that can be used to help make a rational decision as to the intrinsic risks of sampling. This is not a straight-forward topic and always leads to conflict between the statistician, process owner, and accountant.

Table 5-6. Demonstration of logic related to error types.

		Sample	
		True	False
Reality	True	Valid	Error – Type II False Negative
	False	Error – Type I False Positive	Valid

The Power Equation is stated as:

$$\text{Power} = f(\text{Effect size, Sample Size, Alpha value, Beta})$$

When any three variables are known, the fourth can be calculated. All statistics textbooks discuss these calculations in detail, while Minitab provides a handy calculator. A search of the internet will also return numerous calculators. A general discussion is provided below.

Effect size relates to how big a difference can be measured and if it is a meaningful difference. It represents the voice of the process. This is always the first criteria to judge, because a no-go decision is required if impacts will not add value. This factor has been lost in the latest big data projects with some of them hopelessly doomed because the magnitude of the noise squelches the signal itself.

Sample size (n) refers to the number of items to be measured. This term results in expenses being accrued because labor is involved and sometimes product is destroyed. In the medical world, such as drug testing, sample size is called *trial size*. There are several ways to draw a representative sample with the most prevalent being random sampling and sampling **with** replacement, such as putting the bead back in the box so that the proportion of the population is always the same. Both have their place.

An alpha value (α) is the probability of a false positive or incorrectly rejecting the null hypothesis that there is no difference in the average values. An alpha of 0.05 corresponds to a 95% confidence level. The alpha value is the control to which a p-value is compared to conclude that differences are significant. When the consequences of a failure increase, then the alpha value must decrease. Without an increase in sample size, the likelihood of a false positive is minimized, while the likelihood of a false negative is increased. It is a coupled equation where tradeoffs must be considered rationally.

The beta value (β) is the chance of a false negative and can be expressed as:

$$Power = 1 - Beta$$

Beta is not expressed very often, nor called out in specifications frequently. It is typically sacrificed during trade-off decision making. It is frequently the focus of strategic planning sessions when required. A common approach to sampling is to run a cheap and quick screen testing that generates many false positives. A second test is then run that has more power so that valid answers are maximized.

The bottom line in designing a sampling plan is as much art as it is science. Other factors that have influence are representativeness of the sampling methodology, past studies, time available, importance to stakeholders, and experience of data scientists. When in doubt, just remember that no one ever complains about too much data.

On a related note, and for binomial situations, the chance of an error can be calculated as:

$$e = \sqrt{\frac{Z^2 * p * (1-p)}{n}}$$

where e is the magnitude of the error, Z is the critical value from a standard distribution (1.96 used for 95% coverage), p is the probability of occurrence, and n is the sampling size. So, if you want to be 99% confident that there is less than a 1% chance of an error, you need a sample size of 16,512. When you apply a Million Row Data Audit, you can say that you are 99% confident with a 0.1% chance of error. That is a very small margin of error.

Do you care about the few or the many?

The portion of the distribution makes a difference in designing a sampling plan. The few can be in either the high or low tails and the many is in the center hump of the normal distribution. Sometimes this question is phrased as two-tailed or one-tailed problems, with the two-tailed test assessing the central value.

I remember helping IT with a customer complaint. A new and obnoxious manager complained to the Plant Manager that the wait times at the help desk were excessive. I was sent to help craft the response and the IT manager had a bunch of data and shared with me the mean (32 seconds) and the median (29 seconds) for the 1,300 calls in the past two years and wanted to know what graphs to use. I advised her that the center hump did not matter because we only cared about the upper tail. I asked Brenda, "What is your committed service time that you are staffed to support?" I knew the answer but let her tell me it was two minutes. My next question was how many failures were there and the answer was one, along with a story that it was on a special day when they launched an updated version of email and had expected much longer times and had sent plant-wide warnings to expect disruptions. I said you do not need a graph for Mike (the plant manager who had been trained to be a Six Sigma Black Belt). All that is needed when you go to tomorrow's staff meeting is that story you just told me. You have good news, so you can strut. Someone will ask what the average is, which you can answer. Then Mike will start paying compliments to you and your staff and will send a message to Roger to just shut-up. That is how it played out. Roger continued to make waves without ever checking the data and was demoted shortly thereafter. It helps to collect as much data as possible at the point of collection (what I sometimes call the *atomic level*) because you will be prepared to answer questions pertaining to both the few and the many.

Surveys are a tool used to collect data from a sample of people. A common tool used is Survey Monkey. There is an abundance of psychology involved in surveys like writing unbiased questions, selecting scoring scales, including duplicates for quality control purposes, and determining an adequate sample size. The final design is an optimization exercise governed by the budget available. The quantitative tools described herein can be directly applied to the survey data.

Case Study: Election polling and error bars

Every four years, the American news cycle becomes dominated with survey data such as candidate x is favored by 53% ±3%. Discussion focuses on the six-point lead (in a two-candidate race) over candidate y. If x has a 3% chance of error, so does y. The candidates are tied from a probability point of view that has a 95% confidence interval. The effect of sample size on confidence limits is shown in Table 5-7 and Figure 5-15 for a situation when "yes" is selected two-thirds of the time. When the sample size increases above 3,000, the gains become more marginal while costs greatly increase. So when asked to set a sample size for a pool or survey, you should respond with "it depends." You need to know what the population size is, what hypothetical outcome is needed, and how large the confidence band needs to be to show a significant difference.

Table 5-7. Error bar calculations.

n	p	±95%
30	66.667%	17.2%
100	66.000%	9.3%
1,000	66.600%	2.9%
2,000	66.650%	2.1%
3,000	66.667%	1.7%
4,000	66.650%	1.5%
5,000	66.660%	1.3%
10,000	66.667%	0.9%

Figure 5-15. Polling estimate with error bars.

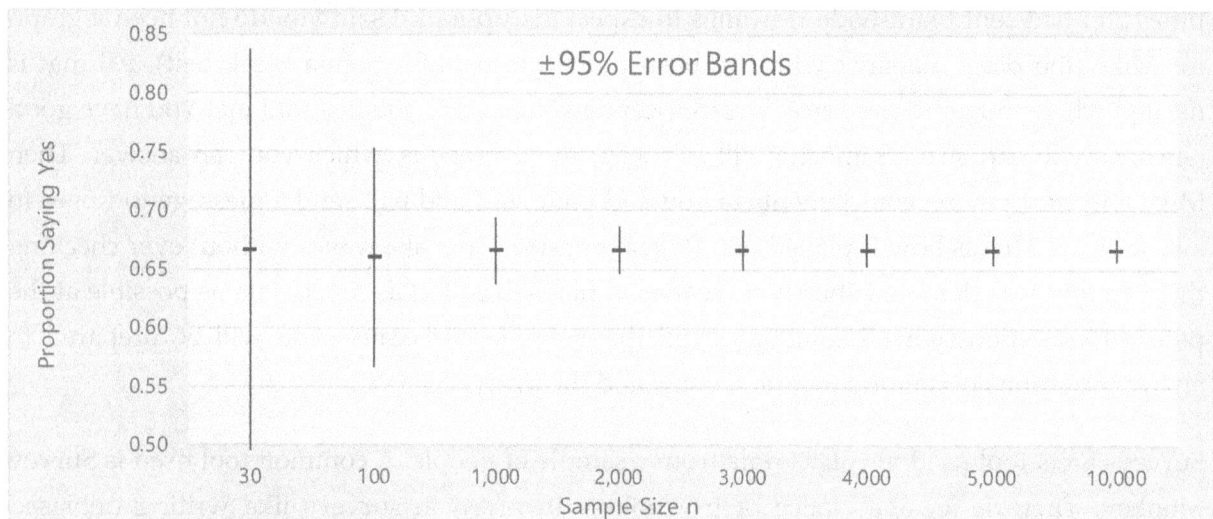

Generic outputs

Quantitative analysis always begins the same because it is necessary to review the data on hand. An example can be found in Figures C-6 through C-17 in Appendix C that are the Minitab descriptive statistics views for monthly rainfall data. Many times, I print hardcopies and spread them out on a large table. Now I find a jumbo screen in a data observatory and look at all of them at once. I look for patterns and trends. Nothing fancy is required for this step, just observation and an inquiring mind. If they look the same, they are most likely not independent and can be stratified into logical groups for analysis as shown in Figures C-19 and C-20. I follow up with inferential statistics tools to determine if the differences are significant. It also helps to talk about the data at this stage as a means of interacting with the stakeholders. The example below goes a bit further into some of the graphical tools that can be applied.

Case Study: Rank sum for largest cities

Figure 5-16 shows the 31 cities with 2016 population greater than ten million. The graphical view in Figure 5-17 summarizes the 2016 population values as well as growth rates, share of countries' population, and expected population by 2030.[33]

Figure 5-16. Location of largest cities expected by 2030.

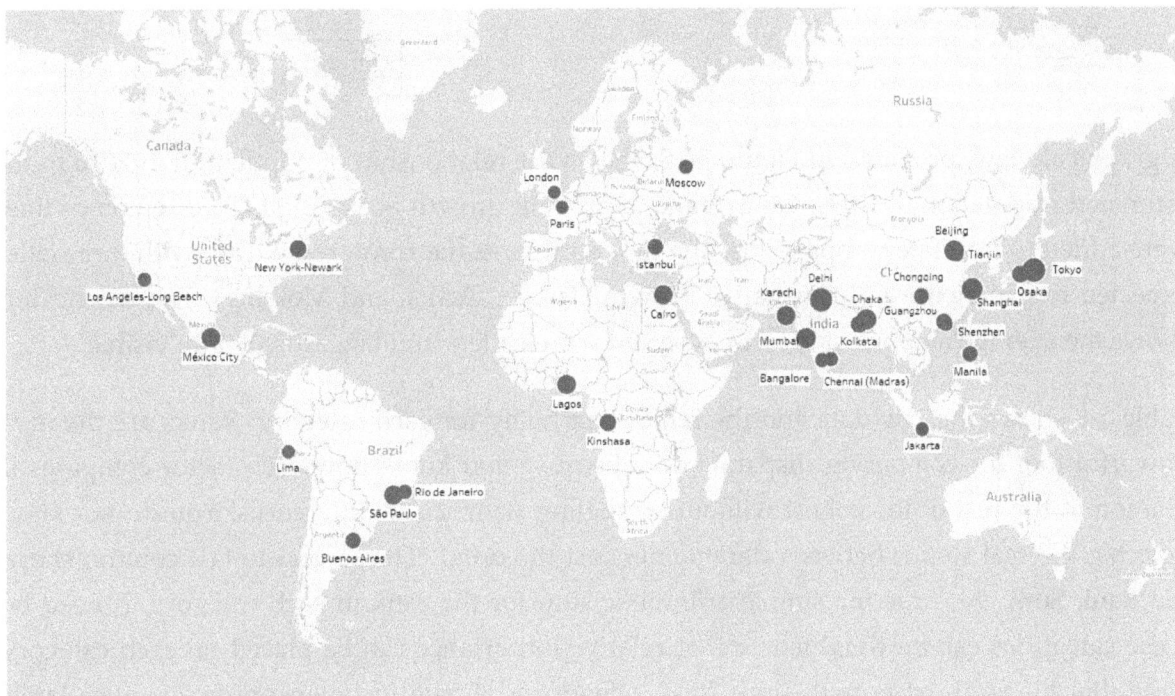

---

[33] Data from https://bit.ly/2noIXK3. An alternative source of data is from the Globalization and World Cities Research group described at https://bit.ly/2AXkyoG.

Figure 5-17. Graphical summaries from Minitab for each column.

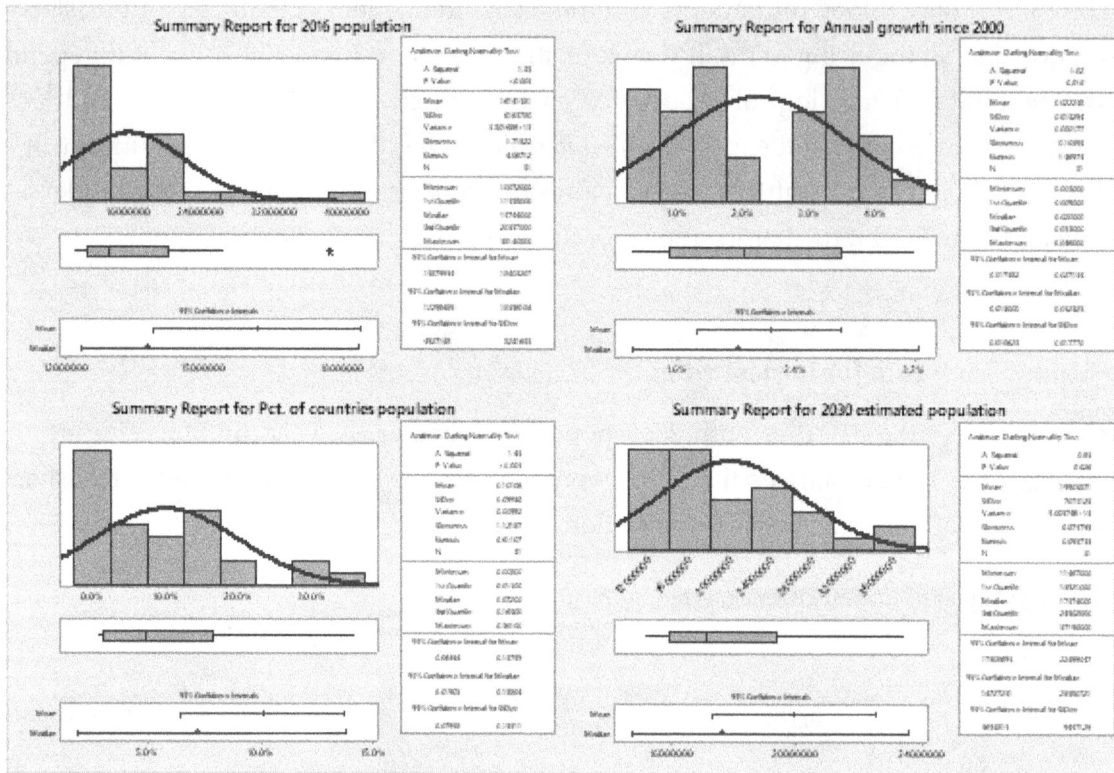

A graphical view, provided in Figure 5-18, shows the relationship of population in 2016 to the estimated population by 2030. On average 23 percent growth is expected. Two reference lines were added to facilitate comparisons. The first represents the zero-growth case with three cities expected to experience no or negative growth (Tokyo, Osaka, and Moscow). The second line shows the average amount of growth expected with leaders and laggards plainly visible.

Table 5-8 shows the raw data and the rank of each city for each category. Ranks are the most powerful tool for comparing disparate data because you know how each factor compares in relation to the rest of the cohort without struggling with scalar differences. You do not know how big the real step is between data points, just the order. The second to last column shows the Rank Sum, which is the simple arithmetic sum for the rank in each category. If need be, these categories can be weighted so that relative importance can be placed on each category. Note that the 'Annual growth since 2000' category is shown in reverse order because larger values are more desirable for this analysis. The last column is the Final Rank, which is shown in reverse order, and represents that the record with the lowest rank sum is the overall best record. While ranking appears simple, it provides powerful support for decision making.

Figure 5-18. Expected population differences between 2016 and 2030.

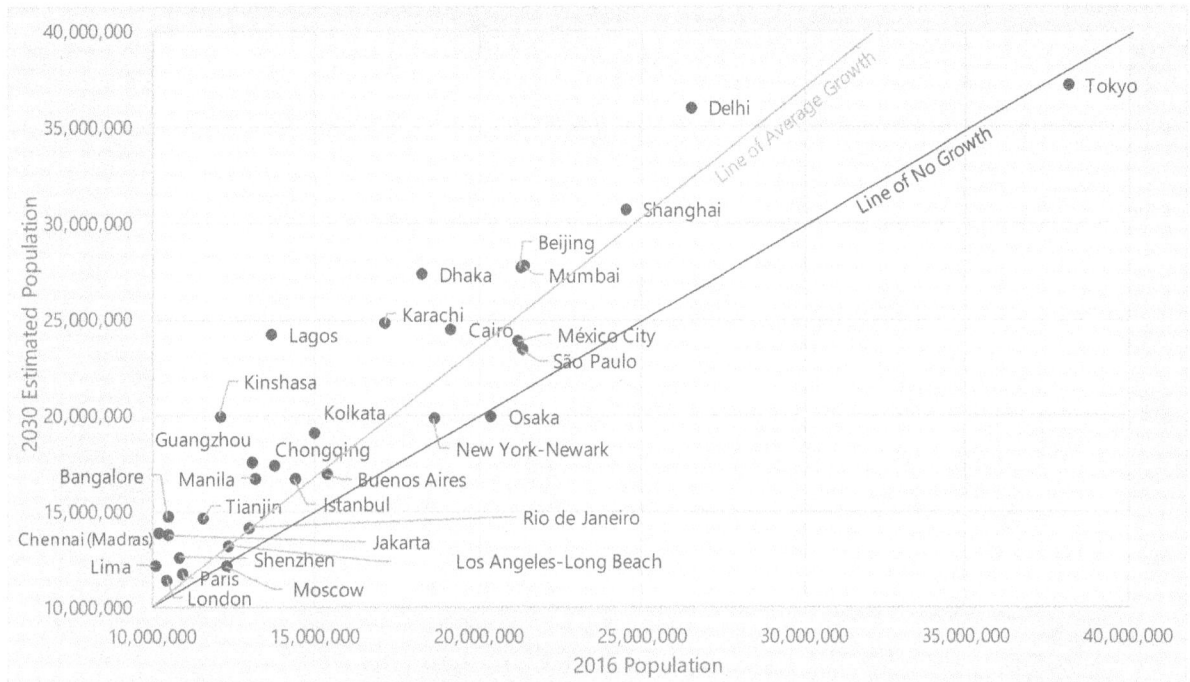

Table 5-8. Detailed data and ranks for cities greater than 10,000,000 inhabitants.

City	Country	2016 population	Annual growth since 2000	Pct. of countries population	2030 estimated population	Rank 2016 population	Rank Annual growth since 2000	Rank Pct. of countries population	Rank 2030 estimated population	Rank Sum	Final Rank
Tokyo	Japan	38,140,000	0.6%	30.1%	37,190,000	1	28	3	1	33	1
Dhaka	Bangladesh	18,237,000	3.6%	11.2%	27,374,000	11	5	12	6	34	2
Shanghai	China	24,484,000	3.5%	1.7%	30,751,000	3	7	22	3	35	3
Delhi	India	26,454,000	3.2%	2.0%	36,060,000	2	11	21	2	36	4
Beijing	China	21,240,000	4.6%	1.5%	27,706,000	6	1	24	5	36	4
Cairo	Egypt	19,128,000	2.1%	22.2%	24,502,000	9	15	4	8	36	4
Karachi	Pakistan	17,121,000	3.3%	9.0%	24,838,000	12	10	14	7	43	7
Lagos	Nigeria	13,661,000	3.9%	7.2%	24,239,000	17	3	16	9	45	8
Kinshasa	Dem. Rep. Congo	12,071,000	4.2%	16.5%	19,996,000	23	2	8	12	45	8
México City	Mexico	21,157,000	0.9%	16.7%	23,865,000	7	24	6	10	47	10
Mumbai	India	21,357,000	1.7%	1.6%	27,797,000	4	17	23	4	48	11
São Paulo	Brazil	21,297,000	1.4%	10.4%	23,444,000	5	19	13	11	48	11
Istanbul	Turkey	14,365,000	3.1%	18.5%	16,694,000	15	12	5	20	52	13
Buenos Aires	Argentina	15,334,000	1.3%	36.1%	16,956,000	13	21	1	18	53	14
Osaka	Japan	20,337,000	0.5%	16.1%	19,976,000	8	29	10	13	60	15
Manila	Philippines	13,131,000	1.7%	12.7%	16,756,000	18	17	11	19	65	16
Chongqing	China	13,744,000	3.5%	1.0%	17,380,000	16	7	26	17	66	17
Guangzhou	China	13,070,000	3.6%	0.9%	17,574,000	19	5	27	16	67	18
New York-Newark	USA	18,604,000	0.3%	5.7%	19,885,000	10	30	18	14	72	19
Lima	Peru	10,072,000	2.0%	31.9%	12,221,000	31	16	2	28	77	20
Kolkata	India	14,980,000	0.9%	1.2%	19,092,000	14	24	25	15	78	21
Bangalore	India	10,456,000	3.9%	0.8%	14,762,000	28	3	28	21	80	22
Tianjin	China	11,558,000	3.4%	0.8%	14,655,000	24	9	28	22	83	23
Rio de Janeiro	Brazil	12,981,000	0.9%	6.3%	14,174,000	20	24	17	23	84	24
Moscow	Russian Federation	12,260,000	1.3%	8.7%	12,200,000	22	21	15	29	87	25
Paris	France	10,925,000	0.7%	16.7%	11,803,000	25	27	6	30	88	26
Jakarta	Indonesia	10,483,000	1.4%	4.1%	13,812,000	27	19	19	25	90	27
London	United Kingdom	10,434,000	1.2%	16.3%	11,467,000	29	23	9	31	92	28
Shenzhen	China	10,828,000	3.1%	0.8%	12,673,000	26	12	28	27	93	29
Chennai (Madras)	India	10,163,000	2.9%	0.8%	13,921,000	30	14	28	24	96	30
Los Angeles-Long Beach	USA	12,317,000	0.3%	3.8%	13,257,000	21	30	20	26	97	31
	Average	16,141,581	2.23%	10.11%	19,903,871						

Figure 5-19 is a graphical view from Tableau that displays all four categories simultaneously. In addition to the x and y axis, the size of the mark is scaled to a category, and the color of the mark represents the appropriate bin relating to stratification of the data.

Figure 5-19. View of four attributes simultaneously in Tableau.

Rankings for 31 Largest Cities

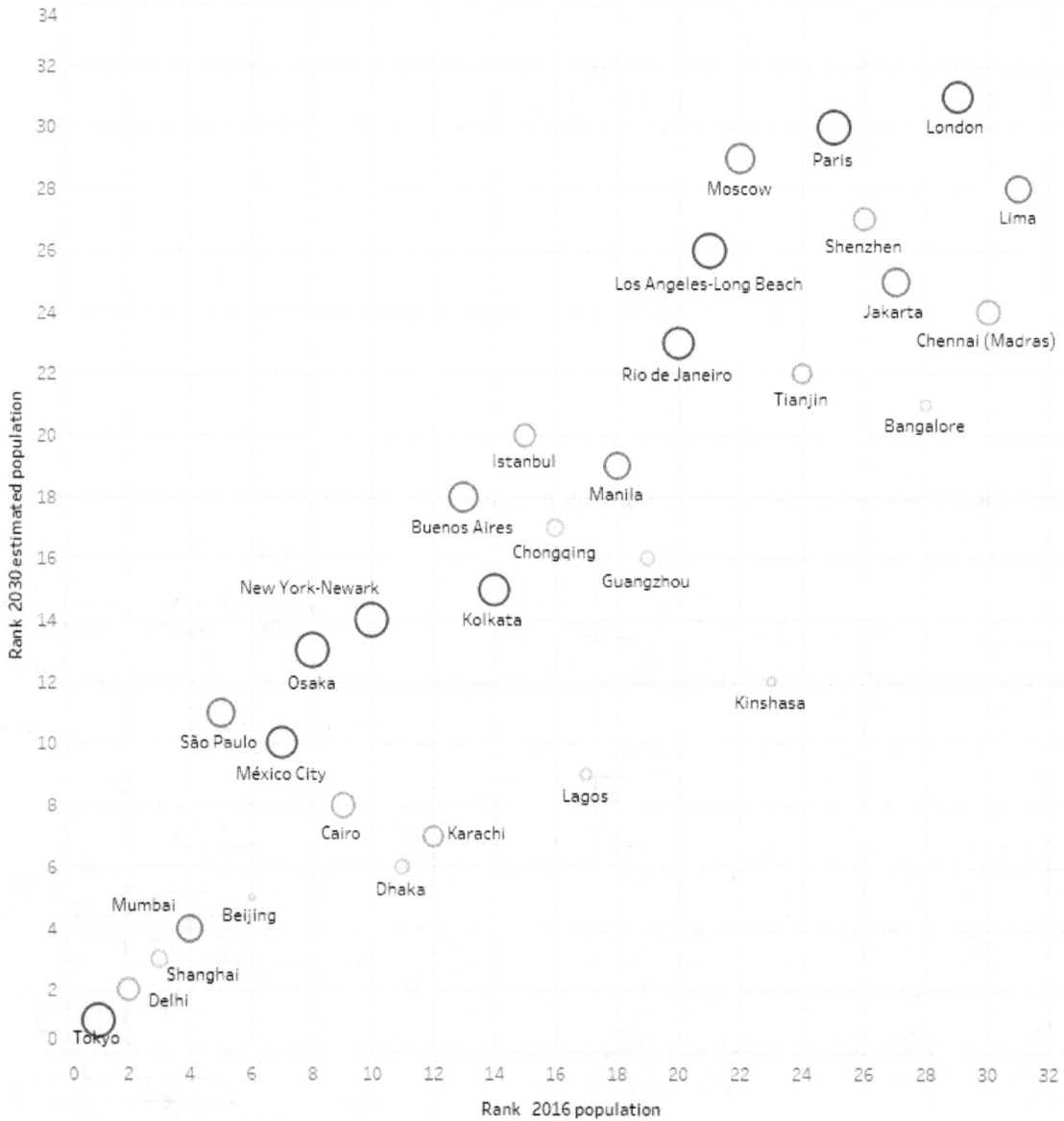

Tableau would even show a fifth dimension if it was categorical by using a different shape for each category. You can pack a lot of information into a graph when using five factors! There is an art to displaying data that comes from complex and complicated systems that have large variations in the scale of data being presented. Experience is the best teacher.

Behavior over time

Lots of systems data represent conditions that change as a function of time. The formal field of Systems Dynamics uses a tool called *behavior over time graphs* that captures trends in data. These graphs also allow interactions to be viewed to determine if changes in inputs result in direct or indirect proportional changes to outputs, as well as the magnitude of change recorded. This type of analysis assigns causality from the input parameters to the output. A quick glance at Figure 5-20 can allow the incorrect conclusion to be formed that increase sales of ice cream results in more shark attacks. While the data are strongly correlated, logic tells us this is not causation. Also, Shark Week on the Discovery Channel would have had an episode about this if it was true.

Figure 5-20. Eating ice cream promotes shark attacks!

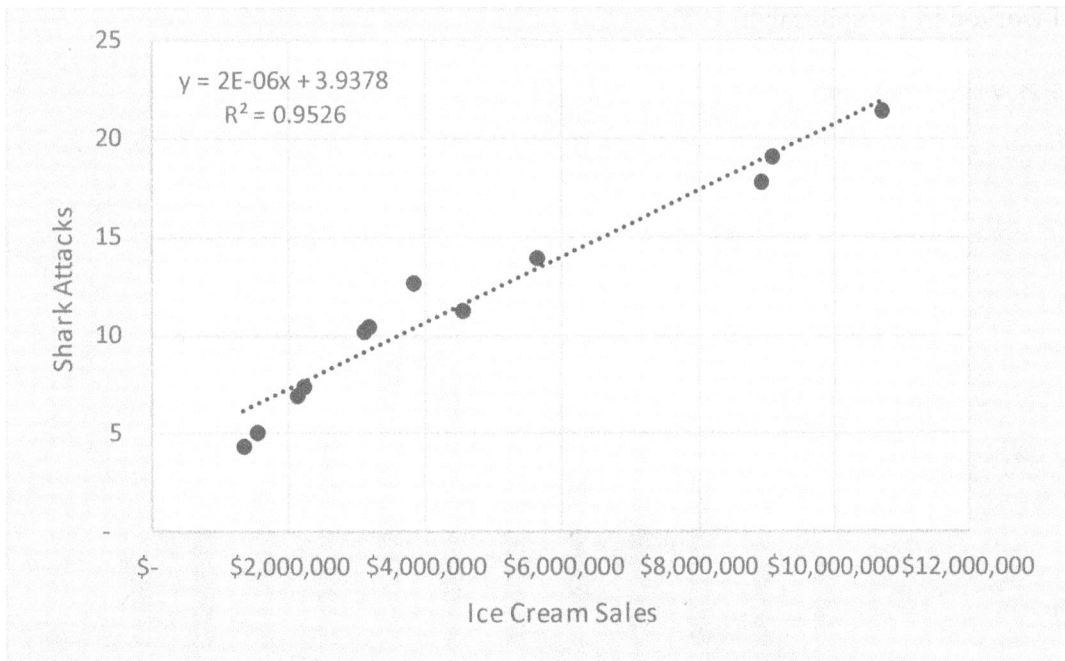

There is confusion regarding which are the dependent and independent variables. From a logical point of view, one must realize that:

- Sharks bite people.
- People buy ice cream.
- Increase the people and both shark attacks and ice cream sales increase.

That means the number of people is the independent variable and that shark attacks and ice cream sales are dependent variables. There is a correlation between the independent and dependent variables, because that is the causation mechanism. The regression line shown in Figure 5-20 is a simple mathematical line of best fit, with no correlation possible since both variables are not logically connected. They behave in a covariate or confounding manner and can be used as surrogates for each other with the r-squared value implying there is a 95% match between them. Figures 5-21, 5-22, and 5-23 display the data over time for visitor count, ice cream sales, and shark attacks, respectively. All three graphs show a summer peak.

Figures 5-24 and 5-25 show the positive correlation between shark attacks and visitor counts, and ice cream sales and visitor count, respectively, for these events are the logical causations. The regression line (which is discussed in detail in Chapter 6) goes through the paired-values of independent and dependent variables with the r-squared value showing that 95% and 99.8%, respectively, of the relationships can be explained by the independent variable, visitor count, and that the equation for the line can be used to predict outcomes for data within the range displayed. The use of regression models for predictive purposes is a cornerstone of data analytics modeling. It is not new. I was running a regression app on my HP-41C calculator when I worked in the soils lab in 1981.

Figure 5-21. Visitors by month.

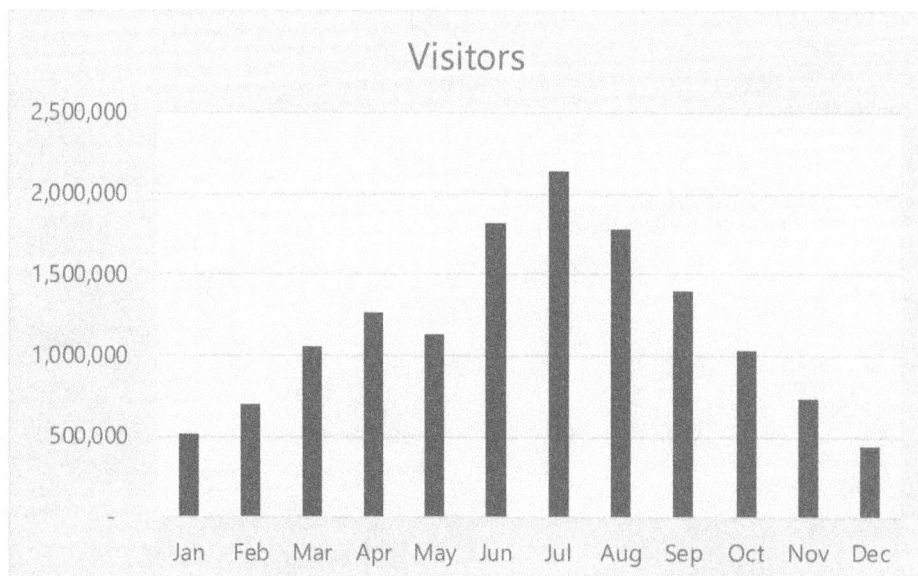

Figure 5-22. Ice cream sales by month.

Figure 5-23. Shark attacks by month.

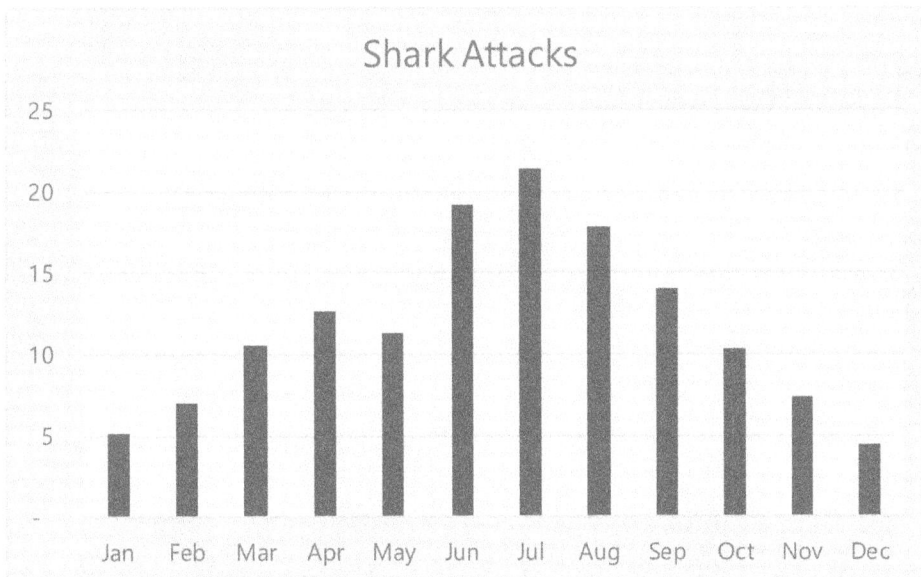

If you do not see a suite of graphs like this, then consider that causality is not proven. If there are 50 variables being discussed, expect a lengthy report that looks at each parameter individually, and then in combination. The determination of causation requires a disciplined and a lengthy analysis. One must have a full understanding of the processes active within the system so that independent and dependent variables can be determined before models can be built.

Figure 5-24. Shark attacks versus visitors.

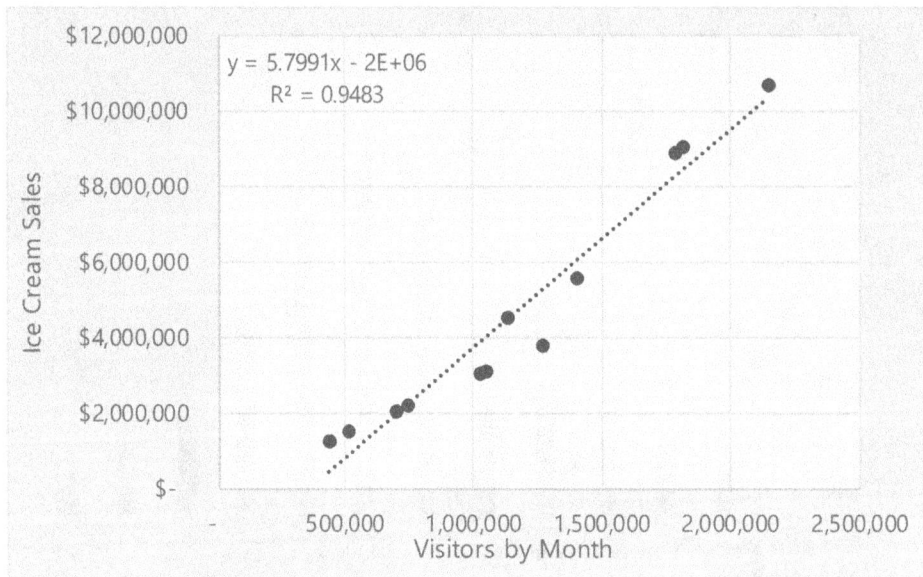

Figure 5-25. Ice cream sales versus visitors.

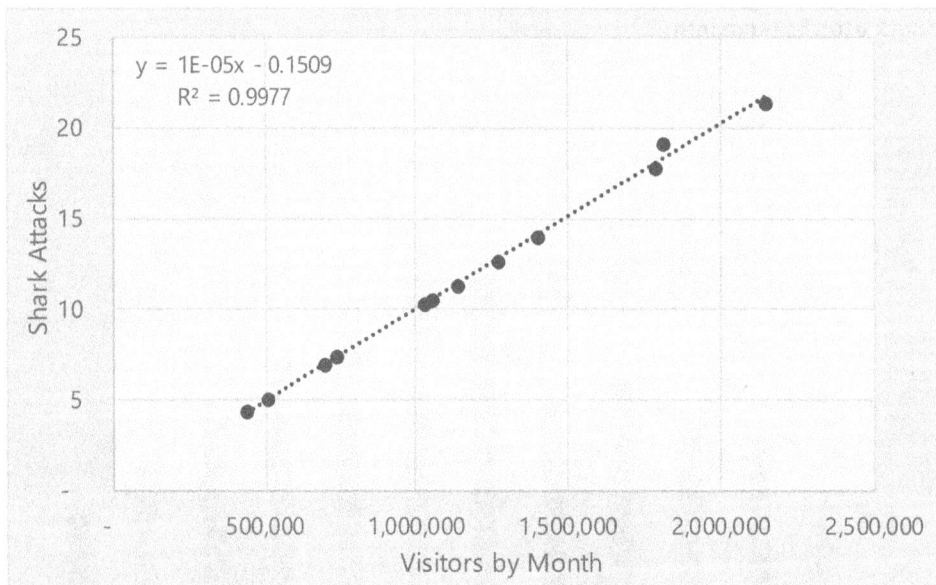

Quantitative data comes in many nuanced forms. Each form is best presented by traditional means, but sometimes advanced visualizations are needed for complex situations, cutting edge applications, or savvy audiences. Once these forms are understood, then quantitative analysis can be performed using the techniques described in the next chapter.

CHAPTER 6

Fundamental Quantitative Analysis for Systems

I have worked for many managers with a wide range of quantitative abilities. I have learned from them all. One of the managers took a moment to sit with me when he first arrived to describe features that would make his job easier. He first described how full his calendar is every day, and how utterly random the subject of each meeting is when looked at in total. He said he can sometimes spend up to the first 15 minutes of a meeting trying to shift mental gears, but it is easier when he has some graphs that serve as a focusing aid. He requested a few components be included in a presentation:

- How many items are there? Tens, hundreds, thousands?
- How long is the processing time? Hours, days, months, years?
- What direction is good? Is the goal to increase or decrease the count?
- How do these numbers compare to where we were last week, last month, and last year?

These requests were easy to accommodate, and I was effective in spreading the request to other analysts. It communicated his approach to decision making – steady and disciplined. It also served as a good foundation for doing quantitative analysis.

Being a data scientist is like being an offensive lineman: no one notices you until you do something wrong! There are fundamental aspects that must always be followed such as being disciplined in approach and rigorous in execution. Many people play fast and loose with data. The world of data science needs to be more structured so that truthful and useful solutions are achieved. Sometimes extenuating situations can scuttle an endeavor, but if the analysis was fundamentally sound, the analysis can be extended to solve related issues.

This chapter is broken into three parts: a discussion of exploratory tools, an introduction to the basic tool kit for quality analysis, and three technical sections that highlight fundamentals for descriptive statistics, inferential statistics, and modern tools for the analysis of big data. There are many good textbooks that provide in-depth knowledge for how to apply these tools. There are also many software packages to aid with this analysis that include extensive training modules to support mastery. The intent herein is to demonstrate how and when quantitative analysis approaches can be used.

Exploratory data analysis

The most exciting phrase to hear in science, the one that heralds new discoveries, is not "Eureka!" (I found it!), but "That's funny…"

Isaac Asimov[34]

As with all scientific endeavors, performing an analysis opens the door to the unknown. This is also true for data analysis. The writings of John Tukey are fascinating. His article with Martin Wilks is a must read to understand the fundamentals of data science circa 1965. Warning, we still have not mastered the key points that they described! It is truly fascinating to read what the great thinkers were expecting to happen by the computer revolution. They would be shocked about how much time ends up being wasted on the web versus actually doing work. Tukey coined the term *Exploratory Data Analysis* in his book title from 1977 that provides the following gems of wisdom:

> *Exploratory data analysis can never be the whole story, but nothing else can serve as the foundation stone – as the first step…Unless exploratory data analysis uncovers indications, usually quantitative ones, there is likely to be nothing for confirmatory data analysis to consider. (page 3)*

He was a world-class data scientist before the computer age. He figured out how to make powerful visualizations using typewriters and found patterns and trends using only an inquiring mind. One of his most powerful tools was the box-and-whisker plot, shown in Figure 4-5. Much has been written on the analysis of measures of central tendency (including in this book), but he was focused on analyzing variance in 1987, shown in Table 6-1. When analytics tools are rolled out, most of these techniques are not addressed. It is obvious that they are still rafts in the iceberg-filled sea. There is much yet to be learned. As big data are a new development, expect there to be many new tools unveiled each year for the foreseeable future.

---

[34] https://quoteinvestigator.com/2015/03/02/eureka-funny/.

The first big targets for exploration are granularity and decomposition. Data must be chunked to be processed when it is collected over time or space with the finer divisions taking longer to process. Data can be decomposed along many lines and hierarchies as dictated by the system itself. Software tools exist to help with these initial steps. It is an easy analysis because one just follows the Voice of the Data and determines which fields are the most important just by following the natural course.

Table 6-1. Tukey's approaches to analyzing variance. [35]

Tool or Technique
Classical Analysis of Variance
Aggregation
Half-Normal Plotting
Horizontalized Plotting
Scission into Bouquets of Contrasts
Pretrimming by Nomination
Post-trimming by Election
Nominated Bouquets
2nd Order Trimming (Super-Election)
Reformulating a Response
Rethinking a Scission
Refactoring an Analysis

The strategy is also called *unpeeling the onion* because each layer is unfolded, straightened, and reviewed in order to learn what is unique about that layer. The process slows when additional layers are investigated, because those discoveries must be joined with the knowledge acquired from the previous layers. Once this decomposition is finished, the data scientist is buried in an avalanche of knowledge that must be presented and prioritized for action. The output of this exploration is frequently named a *multi-vari study* and is often presented as an appendix to the study. It usually works best to present this appendix individually to each individual stakeholder so that free and open discussions can occur.

What about missing data points?

They are missing. There are approaches to accommodate them that border on magic and voodoo, but the bottom line is you do not know what you do not know. The only correct way to handle missing data points is to show that the confidence level and error bars are slightly

[35] Johnson, E.G. and Tukey, J.W. (1987). Graphical Exploratory Analysis of Variance Illustrated on a Splitting of the Johnson and Tsao Data. In *Design, Data & Analysis*. C.L. Malloe (Ed.), pp. 171-244.

larger so that decision making must be made with more uncertainty. If a data point falls below the detection limit, a two-phased approach must be taken in that the count of measurable data (an attribute) is first reported, and then it is reported as a percentage of completeness. Finally, descriptive statistics can be applied to the subset that is reported of having measured values. This is often seen on Election Night when the talking heads report that candidate X has a 10-point lead with 67% of the precincts reporting. Inferences can be made about the missing data, but a winner cannot be named until those votes are counted. However, if you are in the realm of large numbers, then a few missing data points will not be as influential. If there is a chance that data will be lost, then sampling frequency should be increased as a hedge. The tools of big data make this easier than it was in the old days when people used stubby yellow pencils and Big Chief tablets.

Descriptive statistics

Descriptive statistics are used most often during the quantitative analysis of a system. They are like knives to a chef. There are many kinds of software to use depending on where the data resides and the intensity of the analysis to be performed. There are four parameters that are captured when the descriptive statistics are generated: central tendency, variation, distribution, and ranking. The technical term is *univariate analysis*.

Central tendency

There are three values that are used to describe the center of a data set. The *mean* or *average* is calculated by the sum divided by the total number. The *median* is the mid-point with one half of the values above and below this number. The *mode* is the most frequently occurring number. These three values can vary for the same data set. Sometimes the values that correspond to the 25th and 75th percentiles are captured to bound the center half of the data.

Variation

What is the spread of a data set? How much does this vary? The first measure is the *range;* simply the high value minus the low value. Then there is the *standard deviation*, which is the square root of the variance. The units of standard deviation are the same as those that were measured. *Variance* is calculated by squaring the value that results when the sample value is subtracted from the mean. Squaring results in only positive values and was easier to process in the days of the slide rule than working with absolute values. The result is an estimate of the average amount of difference between the samples and the mean. Another factor, *covariance*, expresses the joint variability of two random variables and reports both magnitude and direction of the relationships. Other terms, such as *skewness* and *kurtosis*, are beyond this discussion. It is commonly assumed that variation is constant over the spread of data. If not, the term *heteroskedasticity* is applied and requires the application of more rigorous statistical methods. It is a good word to use at cocktail parties to determine if you have had too much to drink.

Distribution

The quartiles mentioned above are also a crude representation of how the data itself would display on a number line. If the data are cut into deciles (slices representing 10% of the range), a more exacting view is given. If $1/10^{th}$ of the data points fall in each decile, then the data are uniformly distributed. In reality, data tends to fall into a variety of patterns with the normal distribution being most common when dealing with biological systems. This is commonly called a *bell curve* with the peak of the bell corresponding to the average, the median being equal to the mean, and the distribution falls along symmetrical curves. An interesting feature of the normal distribution is that 68% of the data falls within one standard deviation, 95% falls within two standard deviations, and 99.7% falls within three standard deviations. This feature is used for many inferential tests. Other common distributions are:

- uniform (data are selected with equal probability over a range),
- triangular (three-point approximation of a normal),
- exponential (for arrival times and other data that is all greater than zero; the mean is equal to the standard deviation),
- beta (a way to present a normal distribution that is truncated at zero; created when data from multiple exponential distributions are added together), and
- lognormal (when data from multiple normal distributions are added together and the tail becomes drawn out; common in geologic systems).

Once the characteristic distribution of a population is known, then only key descriptive parameters need to be obtained for further processing of the data.

Ranking

This is simply arranging the data in numerical order from smallest to largest and assigning a count number to each value. Sometimes it is more appropriate to use largest to smallest order.

Ranks allow relative comparison of data relating to a closed system and can be used to track trends over time. The most common example is college sports teams such as football and basketball. Those with the highest pre-season ranks are expected to compete for the championship and are broadcast during the prime viewing times. When upsets occur, and a team drops in rank, coverage and analysis are extensive. Ranks can be applied to many situations and it is the extensive discussion when shifts occur that truly add to the knowledge of the system.

Case Study: Descriptive tools for census data from 3,221 US counties

For the fifty states, there are 3,143 counties. Washington D.C. is included as one county but is legally a territory. When the territory of Puerto Rico is included with the states, the count increases to 3,221. Overall, the 2010 census recorded the location of 312,471,327 people. Figure 6-1 shows the totals for the 50 states and Puerto Rico using equal area blocks. This is the lower bound of big data, in my opinion, as the skills of a data scientist are pushed by the varieties of values encountered. First, are the names with 31 counties named Washington, 26 named Jefferson, and 12 named Adams. Then there is the range of population from 82 people in Loving County, Texas, to 9,818,605 people in Los Angeles County, California. That is five orders of magnitude difference that must be accommodated in various displays. Then there are the number of counties ranging from one for the District of Columbia to 254 for Texas. However, Alaska has boroughs, Louisiana has parishes, and Puerto Rico has municipalities just to make things interesting. "Ungainly" is an effective way to describe this data set.

Figure 6-1. Block map view of population for US states based on the 2010 Census results.

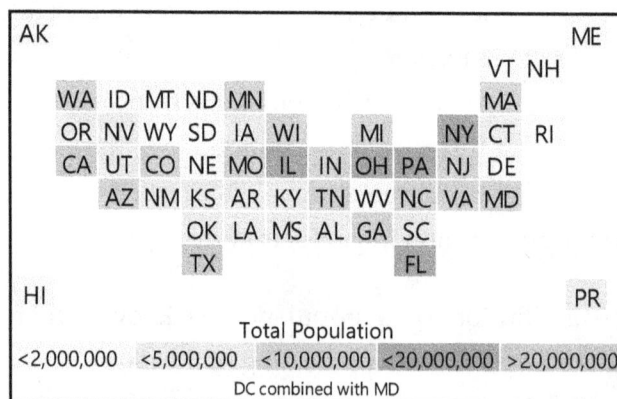

Three views of the data are shown below, but a dynamic table is best. The census data file can be found at https://technicspub.com/qasa with data filters and a pivot table already activated to allow the reviewer to "slice and dice" the data when needed. That these two Excel data mining tools are needed to access the data is the reason I consider data for a list of counties to be big data. A spreadsheet with 3,000 rows of credit and debit transactions can be easily processed without the skills of a data scientist, using these tools. Figure 6-2 is a stacked bar chart that has 52 states and territories shown across the page. Each county is represented by a rectangle with Los Angeles county in California being plainly visible while most of the 254 counties in Texas are not discernable (the large blocks are Bexar, Dallas, Houston, Tarrant, and Travis counties). An alternative approach is to show multiple graphs split by regions or grouped by key parameters.

Figure 6-2. 2010 census count by county and state from Tableau.

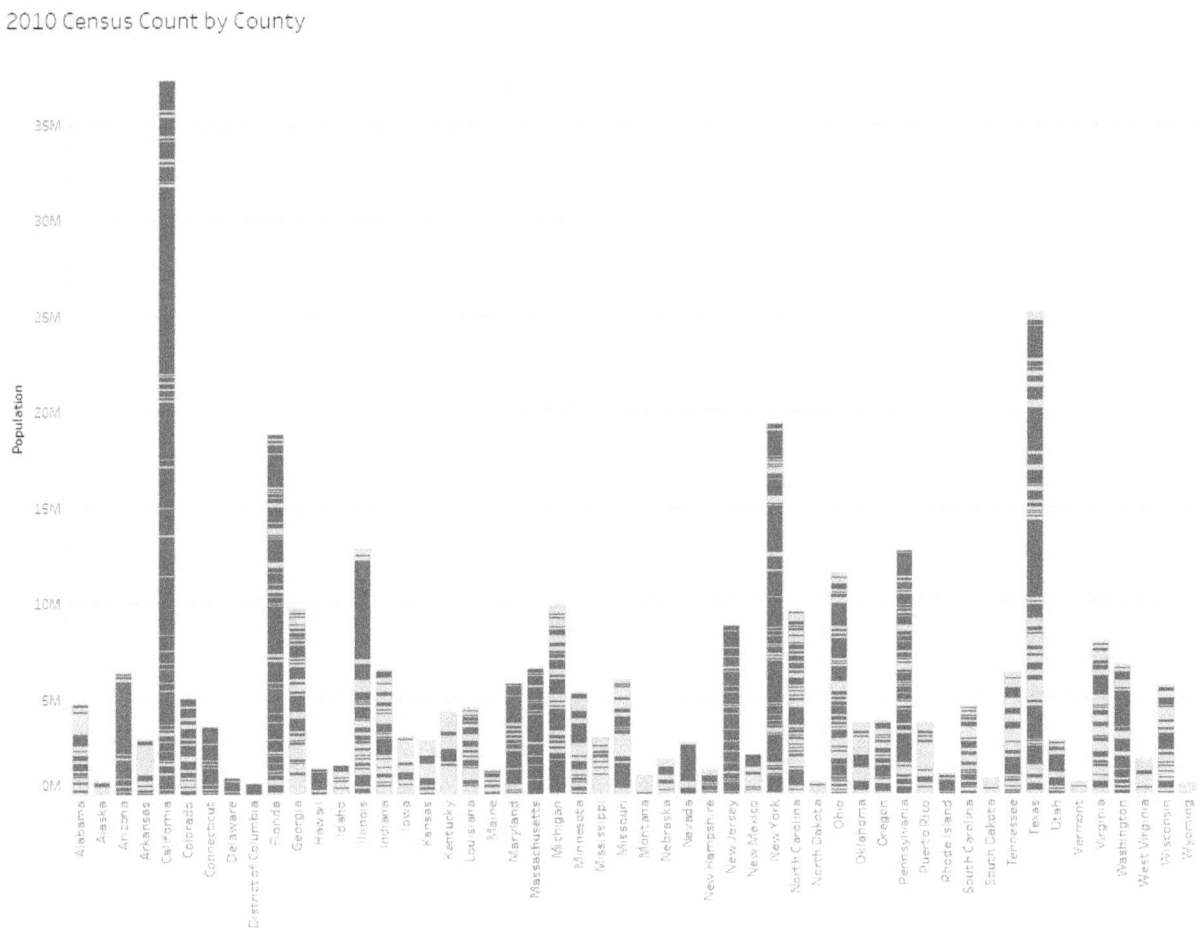

2010 Census Count by County

Figure 6-3 is a map view of the population values by county for the 48 contiguous states with a range of ten colors selected. For a meeting, I would print this on 11x17 paper size to handout (sometimes called a *placemat*). There are many challenges. The areal extent of each county is deceiving with San Bernardino, California being the largest and New York County, New York (the island of Manhattan) being one of the smallest. The 1.7 million people in Manhattan are not discernable on this map. If Alaska and Hawaii are needed for display purposes, the professional mapping software ArcGIS must be used so that multiple panels can be combined because Tableau only allows a single slice of the globe.

Figure 6-3. US census count by county from Tableau.

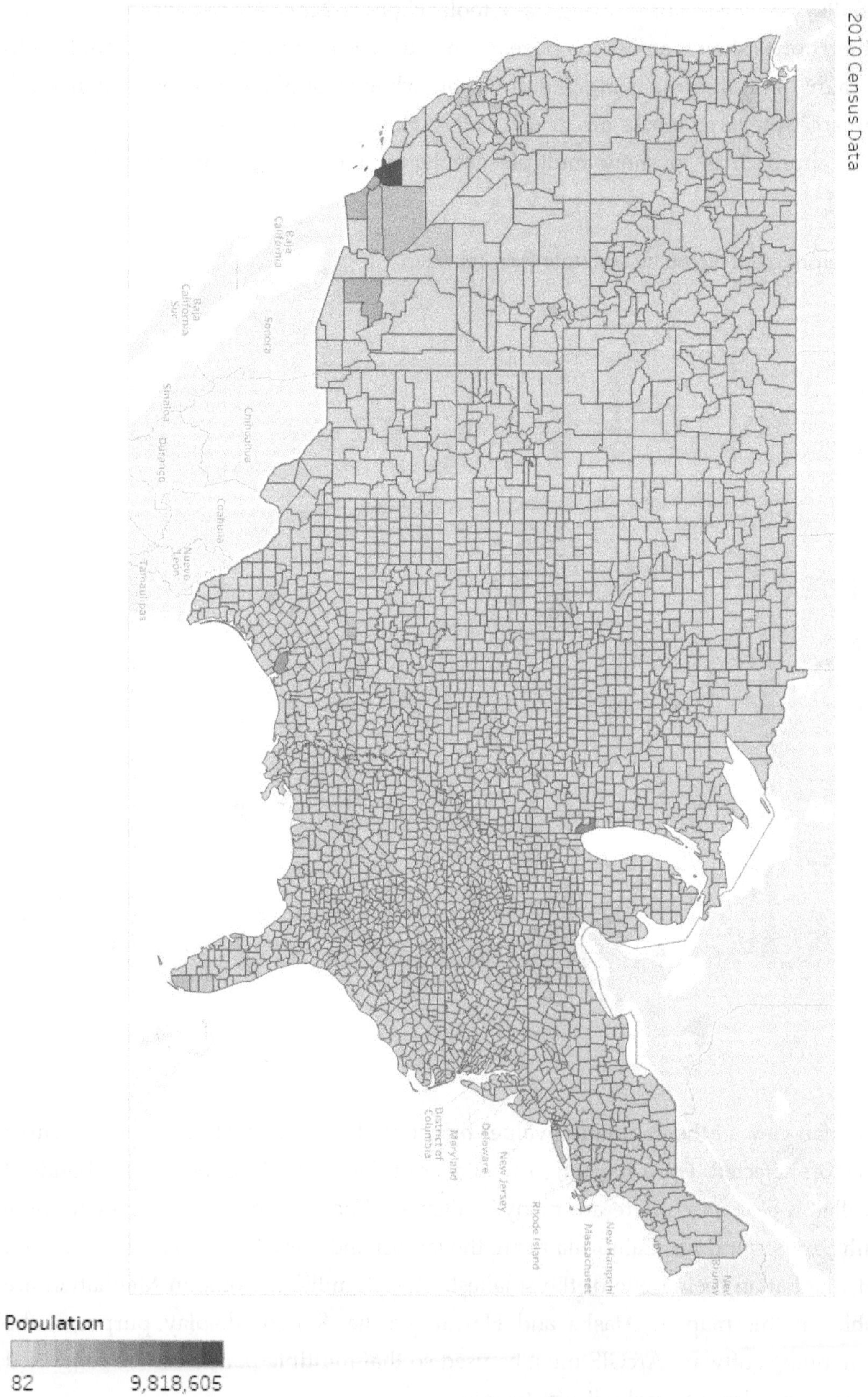

Population
82   9,818,605

Figure 6-4 is the standard panel from Minitab that summaries the descriptive statistics for the population by county. This is a standardized way of looking at data. It forces a disciplined review that is repeatable over time and between data scientists. While I might save a few minutes by not doing this review for every column of data with which I work, I have learned that it is in my best interest to fully engage with every column in every data set I process.

The first glance at the graph indicates that the data are exponentially distributed. This non-normality is supported by the low p-value from the Anderson-Darling test that states that data are significantly different than would be expected for normally distributed data (note that an α of 0.10 is used for assessing this test with a normal distribution being accepted for values greater than 0.1). The mean of 97,011 is almost four times the median of 26,076, and even larger than the third quartile of 65,982. The standard deviation of 309,299 results in a Coefficient of Variation (CV) of 318%. The data table also contains the ranks with 678 counties (21%) accounting for 80% of the population per the Pareto principle. In order to return to the original order after sorting by rank, two alphabetical sorts are needed with the first being by county and the second by state. For display purposes, these data should be stratified to separate the critical few from the trivial many by imposing a hierarchy or looking for natural breaks.

Figure 6-4. Descriptive statistics for US counties census counts from Minitab.

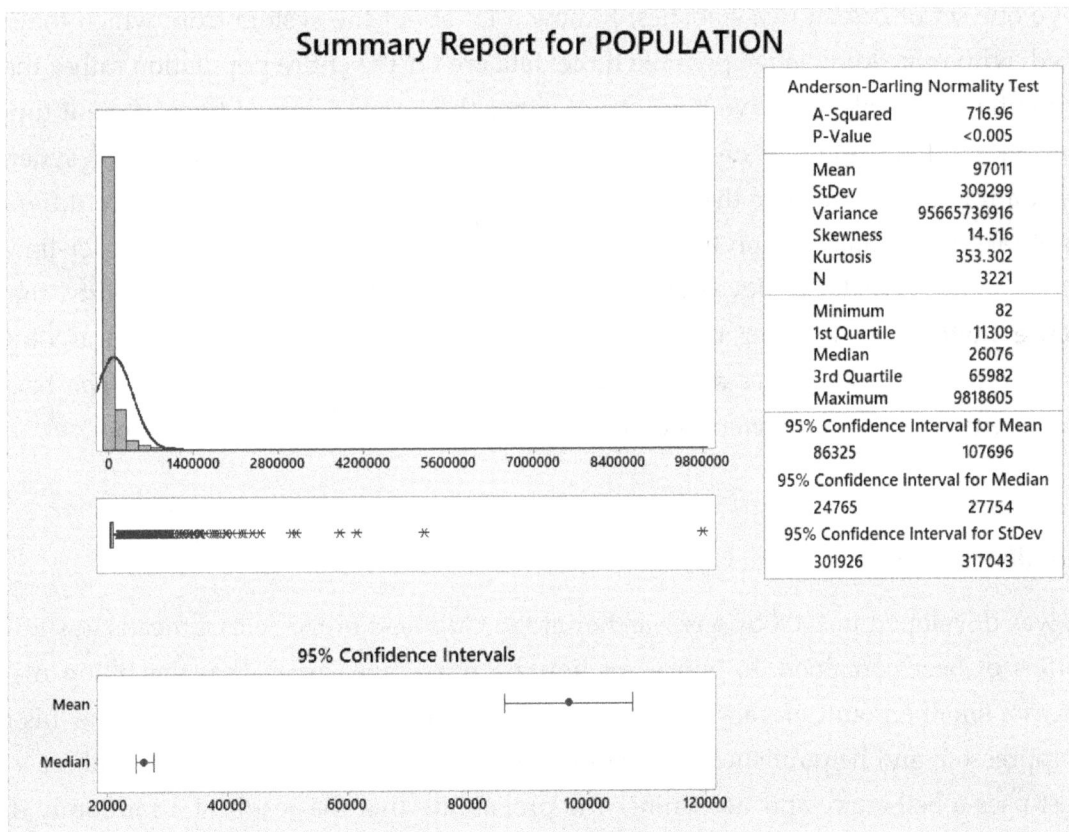

Then, if you drill down to the source data, there are over 100,000 census blocks for the counties. This data set is a challenge to a CIO. Everyone wants to download their own copy, to make changes and highlights, and to keep different versions. The files become quite large when they include pivot tables. The CIO starts to twitch when it comes to having to backup all those files. But then, the Census Bureau sometimes corrects errors, or consolidates census blocks, or adds new blocks. Ideally, every user would redownload all data and regenerate all their custom views. The CIO starts twitching more when he then must backup all these new files, while retaining the old ones. To fix this problem, all the census data back to 1790 is being moved to "the Cloud" and kept as one authenticated data warehouse. Users will generate scripts that generate their custom views and that will extract the current data from the data warehouse every time. The users need to record all access dates and append data export documentation to their reports. No data are retained, so nothing requires backup, which results in a smiling CIO. The bandwidth used is less than watching a movie, so system performance is not impacted.

Inferential statistics

If I have one set of descriptive statistics, I know a lot about the system from which they were obtained, with more knowledge possible if the data are for the entire population rather than for a representative sample. If I have two sets, of either the same system at two different times, or two instances of the same system, I will know four times as much because of synergistic factors. I am going to compare the two sets to see if they are similar or if they are different. It does not take much imagination to realize some comparisons will result in a border-line case. Statistical tools have been developed that allow us to quantify the probability that the difference matters, or to infer that the difference is significant using the phraseology of statistics. When the differences are insignificant, that variation is called *noise*. The basics of using inferential statistics to determine the level of significance are presented below.

Normal distributions

A test was developed in 1908 by a researcher at the Guinness brewery as a means to show that all bottles of beer contained 16 ounces of liquid, on average, given that the filling machine produces a small amount of variation from bottle to bottle. For proprietary reasons, his name was suppressed, and he published his methodology as a "student" of mathematical reasoning. The test uses a bell curve and determines the probability that the result of a randomly drawn sample is significantly different than the test value. He called the decision point a t-value to describe this difference. Hence the name of Student's t-test. Modern methods call this value a *p-value*, because it is easy to explain that the variation from the 16-ounce specification will only happen less than 1% of the time. When this happens, it is called a *significant difference*. For the consumer, this difference is inconsequential since consumption is measured by counting the number of bottles drunk.

Returning to the rainfall data from the case study in Chapter 5 allows us an opportunity to test an assumption. The data from the first 13 years was at a location slightly different than the location where the data were collected for the next 125 years. This comparison is made using a Student's two-sample t-test as shown in Table 6-2. The language of statisticians is very structured. They describe the case when there are no differences between data sets as the null hypothesis (H_o) and the case when there is a difference as the alternative hypothesis (H_1).

Table 6-2. Student two-sample t-test on rainfall measurement locations.

Two-Sample T-Test and CI: TOTAL, Location

Method

μ_1: mean of TOTAL when Location = A

μ_2: mean of TOTAL when Location = B

Difference: $\mu_1 - \mu_2$

Equal variances are not assumed for this analysis.

Descriptive Statistics: TOTAL

Location	N	Mean	StDev	SE Mean
A	13	22.06	7.21	2.0
B	125	20.31	5.45	0.49

Estimation for Difference

Difference	95% CI for Difference
1.75	(-2.70, 6.19)

Test

Null hypothesis	$H_0: \mu_1 - \mu_2 = 0$
Alternative hypothesis	$H_1: \mu_1 - \mu_2 \neq 0$

T-Value	DF	P-Value
0.85	13	0.411

The test requires a formal statement of the objective: there is a significant difference of the data from the two locations. This is called the *alternative hypothesis*. The null hypothesis then becomes that the difference between locations is not significant. The t-value corresponds to the chance that the result seen was due to random probability. The result is that there is no significant difference for the total as shown by the p-value of 0.411 (which is greater than the α value of 0.05) meaning the alternative case is rejected. Statisticians never say that the null case is accepted, just that the particular alternative is rejected. The user then makes the final decision and accepts the risk in the event it turns out poorly. These comparisons were also made for each month and no significant differences were observed for any of them.

Figure 6-5 is a powerful comparative view of the data from both locations that allows visual comparison of the mean, median, 25th and 75th percentiles, as well as approximations of what the spread across ± three standard deviations would be. It is safe to conclude that the data from the first 13 years at one location can be combined with the next 125 years since there is much overlap in the data. Note that the length of these whiskers reflects uncertainty because of the sample size with the 125-year view representing a case of the Law of Large Numbers. However, this analysis assumed the data were distributed normally. A different, nonparametric, test is needed because the lowest value possible is a zero which would truncate a normal distribution. A Mann-Whitney test is a distribution-free test that is a more exacting test of significance and is explained below. When comparing data like this, it is helpful to apply a PGA suite of filters:

- Do the differences make "p"ractical sense?
- Are the differences supported "g"raphically?
- Do statistical tools provide "a"nalytical proof?

Figure 6-5. Boxplot view of the comparison by location results.

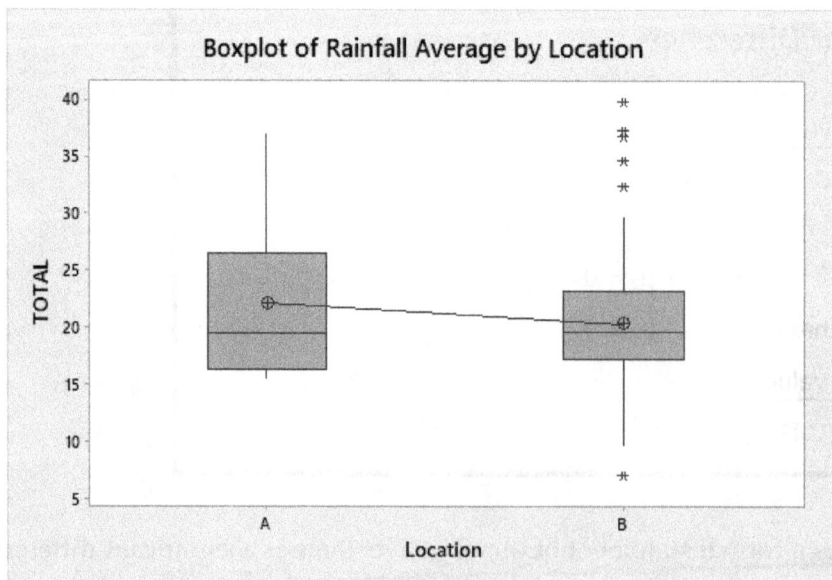

An alternative way to look at this is to analysis the components of variation themselves, as shown in the following equation:

$$\sigma^2_{Total} = \frac{\sigma^2_{Location\ A} + \sigma^2_{Location\ B}}{2} + \sigma^2_{Storm} + \varepsilon$$

where σ^2 represents variance and ε represents the error in measurement itself, or the absolute limit of resolution, due to the limitations of the rain gauge itself. The exact solution to this equation is beyond the scope of this book and is a very complicated subject. Personally, I have a rain gauge on the back fence that aids me in determining how much rain fell on my lawn because the official gauge at the airport is 15 miles away.

Nonparametric methods

If the data are not normally distributed, or if we do not know what the distribution is, a suite of robust tests is available. These distribution-free tests are termed *nonparametric tests*. The two-sample comparison test is named the *Mann-Whitney test* after the two researchers who discovered it. Instead of comparing the means, it uses the medians. The output for the rainfall example is shown in Table 6-3 with a p-value of 0.713. This test even accommodates situations when the medians are the same (tied), but the distributions vary. This level of rigor does not apply to this comparison.

Figure 6-5 is still a valid comparison and highlights the stronger similarities of the medians than the means that was reflected by the p-values. In Minitab, I usually run both tests with the graph being a default option from the Student's test. I do not always present them to the customer but retain them with the project documentation if the same conclusion is reached. Again, the level of detail I present depends on the customer's maturity with statistics.

Other nonparametric methods exist. Sometimes results are reported as a percentage, such as a failure rate by lot. Comparison is made using the Test of Two Proportions and is a straight probability-based approach. Other times, results are expressed as the frequency that a result will occur compared to its expected value. This is the Chi-squared test (χ^2). A simple example is to assess the randomness of the random number generator in Excel. If the data are truly random, then we would expect to see 10% of the results in bins that are $1/10^{th}$ of the total range (0 to 1 for Excel-generated random numbers). Table 6.4 shows the results for a sample of 100 numbers and that the 0.6's have 22 instances of occurrence.

Table 6-3. Mann-Whitney test applied to two locations of rainfall data.

Mann-Whitney: Location A, Location B
Method

η_1: median of Location A

η_2: median of Location B

Difference: $\eta_1 - \eta_2$

Descriptive Statistics

Sample	N	Median
Location A	13	19.40
Location B	125	19.55

Estimation for Difference

Difference	CI for Difference	Achieved Confidence
0.59	(-2.42, 4.49)	95.01%

Test

Null hypothesis	H_0: $\eta_1 - \eta_2 = 0$
Alternative hypothesis	H_1: $\eta_1 - \eta_2 \neq 0$

Method	W-Value	P-Value
Not adjusted for ties	954.50	0.713
Adjusted for ties	954.50	0.713

Table 6-4. Count of random numbers by bin from Excel =rand() function with Chi-squared analysis (after Conover, 1999).

Bin	Expected	Observed	O-E	O-E squared	O-E squared / E
0	10	6	-4	16	1.6
0.1	10	9	-1	1	0.1
0.2	10	13	3	9	0.9
0.3	10	4	-6	36	3.6
0.4	10	7	-3	9	0.9
0.5	10	7	-3	9	0.9
0.6	10	22	12	144	14.4
0.7	10	7	-3	9	0.9
0.8	10	11	1	1	0.1
0.9	10	14	4	16	1.6
				Sum =	25
				Test Statistic (0.95) =	16.92
					Reject Ho
					Difference is significant

Is this random? The chi-squared test involves calculating the sum of the squared differences between observed and expected values, divided by the expected value, and comparing to the test statistic from a chi-squared distribution with n-1 degrees of freedom. These values are also shown in Table 6-4 with the conclusion that there is a significant difference between the expected and observed data. Thus, we can conclude that the numbers Excel generates are not truly random for this small sample.

Figure 6-6 is an N-N view showing the cumulative value across bins versus the expected uniform growth expected. Usually there is variation across the expected line, but not in this instance. Expectations do not need to be uniformly distributed and a pie chart is a useful tool for showing the expected distribution for a more complicated situation. Restaurants, especially fast food ones, periodically perform this type of analysis on their menu choices. A similar graph, the *Receiver Operating Characteristic Curve*, is used to assess the functionality of models derived from machine learning tools.

The chi-squared test can also be used for a 2x2, or contingency, table such as the one shown in Table 6-5. This is an example based on two questions that have a binomial outcome (yes or no) with the expectation being 80% of respondents will answer yes for the first question and 75% will answer yes to the second question.

This scenario was previously discussed for Figure 2-2. The data file can be found at https://technicspub.com/qasa. The file is dynamic, so feel free to explore what happens if the affirmative is a 50/50 situation, or if the sample size is increased or decreased by an order of magnitude. Again, the null hypothesis is rejected with the observed results significantly different than those that were expected.

Figure 6-6. Comparison of expected and observed random numbers.

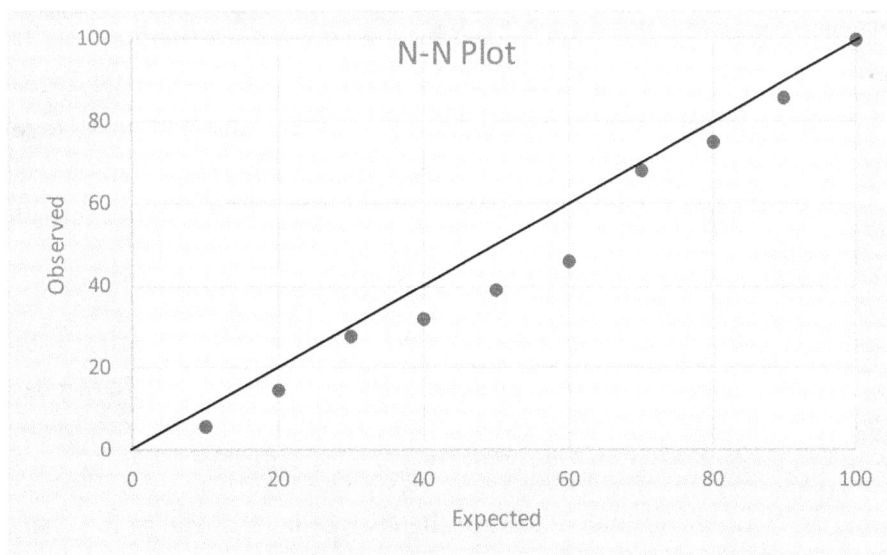

Table 6-5. 2x2 Chi-Squared analysis.

Sample Size=			3,000	

Expected			Question 2	
			Yes	No
			75%	25%
Question	Yes	80%	1,800	600
1	No	20%	450	150

Observed			Question 2	
			Yes	No
			74.7%	25.3%
Question	Yes	72.2%	1,617	548
1	No	27.8%	623	212

(O-E)^2/E			Question 2	
			Yes	No
Question	Yes		19	4
1	No		67	25

		Sum =	115
	Test Statistic (0.95) =		7.815
			Reject Ho
			Difference is significant

Correlation and regression

Recall from Chapter 5 that correlation is simply a description of the interaction between two or more variables as one of them is changed. A measure is assigned to describe both the strength and direction of the correlation and symbolized by the Pearson Correlation Coefficient, r. An r of 0.95 represents a very strong correlation that is directly proportional between the two variables – as one increases, so does the other. An r of -0.67 represents a weak correlation that is indirectly proportional between the variables – as one increases, the other decreases. An r of 0.00 represents no relationship, while an r of 0.5 represents a relationship that is equivalent to the odds of a coin flip.

This concept is taken a step further with the term r-squared (r^2). By squaring r, the directional sign is lost. By squaring a decimal, the result is even smaller as: 0.95 becomes 0.9025; -0.67 becomes 0.4489; 0.5 becomes 0.25. This term is named the *Coefficient of Determination* and it represents the degree of influence the independent variable (X) has on the observed value (Y). For the first case, one would say that 90.25% of the change in Y is due to the change in X with the remaining 9.75% of the variation due to either unanalyzed variables or noise. When multiple inputs are evaluated, a General Linear Model is generated.

Returning to the case study from Chapter 5, the data for all teams in the Bundesliga 2016-2017 season are presented in Figures 6-7 and 6-8 for wins versus goals difference and wins versus points, respectively. Also included is a trend line that best fits the data. This line was generated using a least squares approach that minimizes the overall deviation from each data point to the line of best fit. These distances are called *residuals*. The equation for the line is listed in the upper left corner of the graph, along with the r-squared value.

This equation can also be used for predictive purposes, by application in the reverse direction by solving for X. When reversed, you have a simple model. The higher the r-squared value, the better the fit. Stated another way, the equation in Figure 6-7 explains 83% of the influence wins have over goal difference, while the equation in Figure 6-8 explains 96% of the influence wins have over points. A multivariate regression could also be performed that addresses wins, ties, and goal difference. The resulting r-squared value may or may not be improved by this refinement.

Figure 6-7. Relationship between wins and goal difference for Bundesliga teams in the 2016-2017 season.

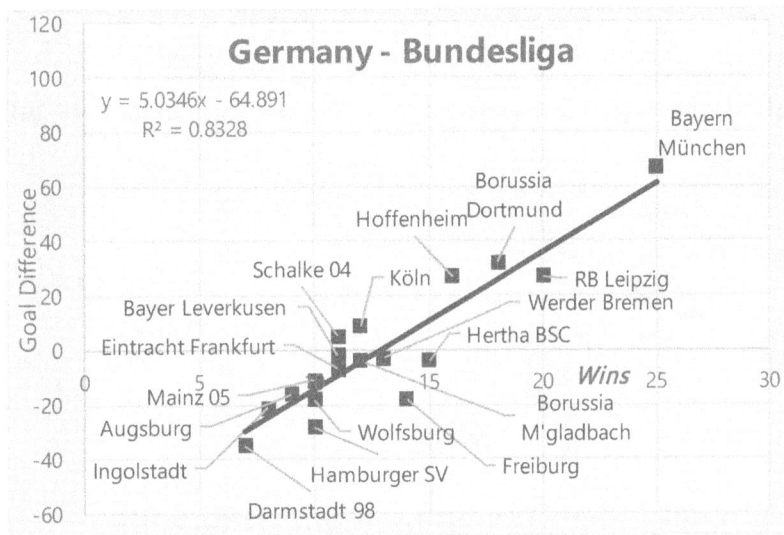

Figure 6-9 is comprised of eight graphs for eight more soccer leagues showing the relationship between wins and points.[36] The Mexican soccer league is unique because they play two distinct and independent halves for a year. I combined the results and the low r-squared is an artifact.

---

[36] Note that the range of the y axis varies between graphs. When direct comparison is important, when samples from the same population are being compared, then the range of the y axis should be the same for all graphs. Since each country's soccer league is independent, direct statistical comparisons cannot be made. However, for the European teams, the champions of each league from one year face off against each other the next year in the UEFA Cup championship matches.

The US and Canadian MLS group differs because teams are not relegated to a lower division at the end of each season. This is because the teams' owners represent a cartel in control of a closed system. When there is an association of leagues that practice relegation/promotion, each league becomes an open system and drives players performance to value winning throughout the season. The analysis performed determines if there are significant differences between MLS and the other major leagues.

Figure 6-8. Relationship between wins and points for Bundesliga teams in the 2016-2017 season.

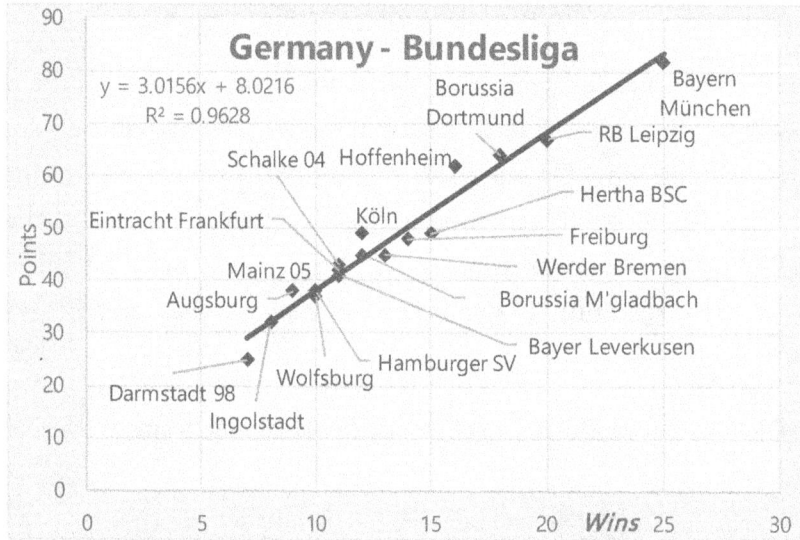

Figure 6-9. Relationship between wins and points for major soccer leagues in the 2016-2017 season.

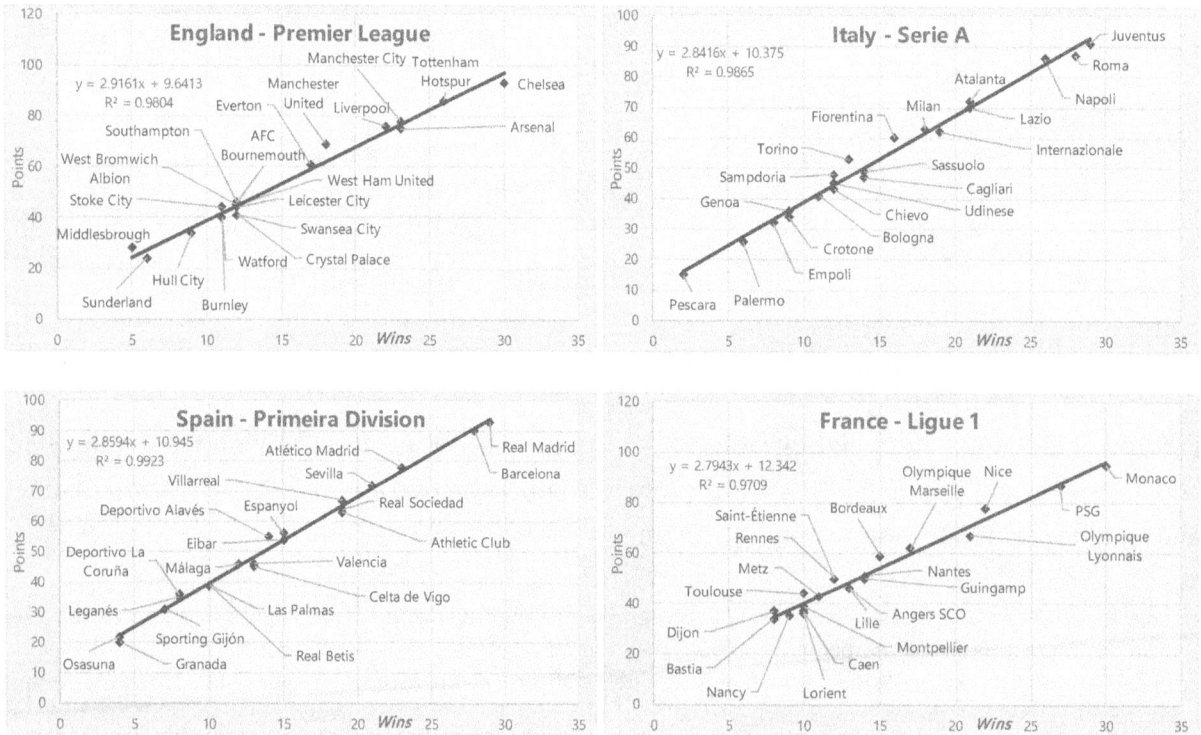

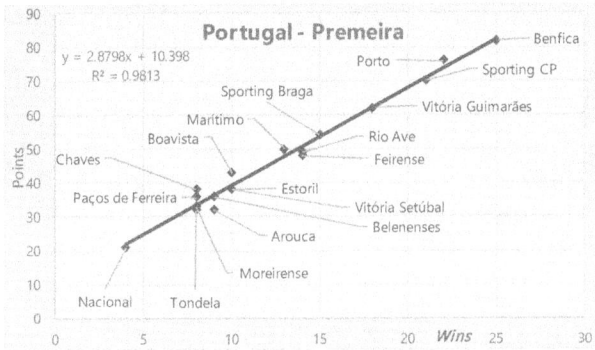

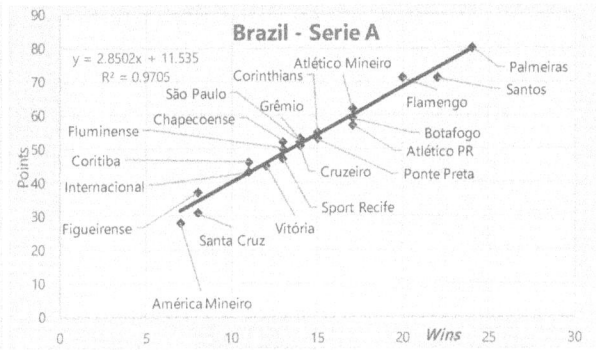

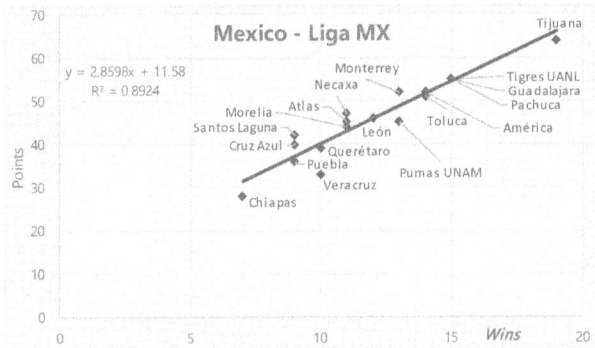

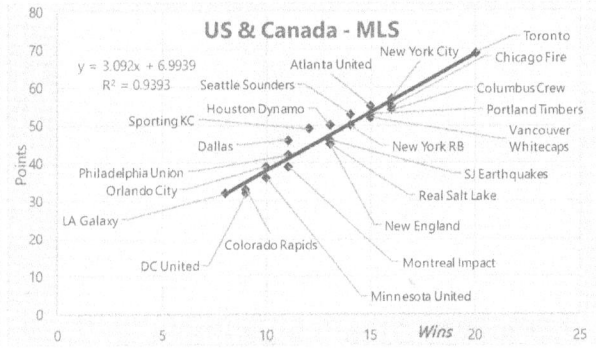

This analysis was performed using the metadata provided from the regression analysis shown in Table 6-6 that includes slope and intercept from the line equation as well as the r-squared value. In addition, the spread from highest to lowest value in each league was also assessed. A simple inferential statistics test was applied – the Z-score test. First, a mean and standard deviation were generated for each parameter. A Z-score was generated by subtracting the mean from the value and dividing by the standard deviation (X – X-bar / s).

The resultant value states how far the value is from the mean in terms of standard deviations. Assuming the data are normally distributed, then one can expect 95% of the data to fall within two standard deviations and 99.7% of the data to fall within three standard deviations. The only value that showed a significant difference beyond the 95% confidence interval was the low r-squared Z-score of -2.3 for the points regression for Liga Mexicana. This is most likely an artifact of the split season. I imagine I could collect several dissenting opinions regarding these observations from the resident subject matter experts at the local sports bar.

Another approach would be to apply multiple regression, which is when more than one input is used to make a prediction, or logistic regression that includes state values for different conditions. The ultimate approach is to run a designed experiment where all variables are controlled and the degree of variation in the outputs can be directly calculated. While this works well for optimizing fertilization rates for cotton crops, it is hard to control sporting events.

Table 6-6. Summary of regression parameters and statistical significance for major soccer leagues in the 2016-2017 season.

League	Goal Difference					Points			
	Spread	Slope	Intercept	R-squared		Spread	Slope	Intercept	R-squared
Germany - Bundesliga	102	5.04	-64.89	0.83		57	3.02	8.02	0.96
England - Premier League	92	4.40	-65.15	0.92		69	2.92	9.64	0.98
Italy - Serie A	103	4.09	-61.35	0.95		71	2.84	10.38	0.99
Spain - Primeria División	117	4.60	-66.92	0.95		73	2.86	10.95	0.99
France - Ligue 1	101	4.40	-62.91	0.91		61	2.79	12.34	0.97
Portugal - Premeira	90	4.10	-51.47	0.89		61	2.88	10.40	0.98
Brazil - Serie A	65	3.21	-45.99	0.89		52	2.85	11.5	0.97
Mexico - Liga MX	51	3.83	-46.22	0.67		36	2.86	11.58	0.89
US & Canada - Major League Socc	66	4.95	-63.89	0.68		37	3.10	6.99	0.94
Average	87.4	4.29	-58.75	0.85		57.4	2.90	10.20	0.96
Standard Deviation	21.9	0.57	8.43	0.11		13.7	0.10	1.74	0.03

	Z-Scores								
	Goal Difference					Points			
League	Spread	Slope	Intercept	R-squared		Spread	Slope	Intercept	R-squared
Germany - Bundesliga	0.7	1.3	-0.7	-0.2		0.0	1.2	-1.3	-0.1
England - Premier League	0.2	0.2	-0.8	0.6		0.8	0.2	-0.3	0.5
Italy - Serie A	0.7	-0.4	-0.3	0.9		1.0	-0.6	0.1	0.8
Spain - Primeria División	1.4	0.5	-1.0	0.9		1.1	-0.4	0.4	0.8
France - Ligue 1	0.6	0.2	-0.5	0.5		0.3	-1.1	1.2	0.2
Portugal - Premeira	0.1	-0.3	0.9	0.3		0.3	-0.2	0.1	0.5
Brazil - Serie A	-1.0	-1.9	1.5	0.3		-0.4	-0.5	0.7	0.2
Mexico - Liga MX	-1.7	-0.8	1.5	-1.7		-1.6	-0.4	0.8	-2.3
US & Canada - Major League Socc	-1.0	1.2	-0.6	-1.6		-1.5	2.0	-1.8	-0.7

Advanced statistical methods

Chapter 5 introduced two kids of data: discrete (attribute) and continuous (variable). There is a suite of statistical tests available to work with all these combinations. The Chi-squared test works when both inputs and outputs are attribute data. Correlation, regression, multiple regression (when multiple inputs are used to predict an output), and time series analysis are applied when both inputs and outputs are variable data. When inputs are variable, but outputs are attribute (yes/no, or bins), then discriminant analysis[37] and logistic regression tools are applied. When inputs are an attribute (such as a condition or a month), and output is a variable (such as yield or rainfall), then an analysis of variance (ANOVA) is applied.

---

[37] One of the key visualization tools for these analyses is a dendrogram with an understandable example provided by Lapointe and Legendre (1994) who tried to scientifically determine the best Scotch whiskey. Hopefully, this book has not driven you to that end!

These tools can be used for inferential purposes, have nonparametric versions to work with distribution free data, or can be modified for use as predictive models, which is discussed in Chapter 7. They can be found in a data scientists' toolbox with strong preferences shown for specific software packages. Given the infrequency with which some of these tools are used, data scientists like to use what they know. The labor costs associated with the learning curve to master a package from a new vendor often exceed the cost of the preferred package and take a significant amount of time. A workaround exists in that data can be named using generic codes and processed on non-sanctioned equipment. It is a trade-off between security risks, costs, and desirability of the results.

Ishikawa's seven tools

This book applies the first principles from the concept of quality as adopted by manufacturing for use with small data. One of the major champions of this concept, W. Edwards Deming, introduced the Total Quality Management paradigm which has morphed into the Six Sigma movement. His protégé, Kaoru Ishikawa, developed the seven basic visual tools of quality for Toyota in the 1950s so that the average worker could analyze and interpret data from a manufacturing process. These tools have been used worldwide by companies, managers of all levels, and employees for performing work better, faster, and cheaper.

Data scientists grasp these concepts but need to master the deployment of them and become experts in training managers and workers in their use. As big data becomes the norm, everyone involved must master these techniques to gain the benefits. The seven tools listed in Table 6-7 are based on his original intentions and broadened to more of a business intelligence context and to reflect software innovations. This list is not wholly traditional because check sheets and Pareto charts are combined, while flow charts have been added. This represents lessons I have learned from my work and Six Sigma experiences.

Table 6-7. Ishikawa's seven tools of quality.

	Quality Tool Name	Purpose[38]
1	Flow Charts	When you need to identify the actual ideal paths that any product or service follows to identify deviations.
2	Diagrams	When you need to identify, explore, and display the possible causes of a specific problem or condition.
3	Check Sheets and Pareto Charts	When you need to display the relative importance of all the problems or conditions to choose the starting point for problem solving, monitor success, or identify the basic causes of a problem.
4	Run Charts	When you need to do the simplest possible display of trends within the observation points over a specified time period.
5	Histograms	When you need to discover and display the distribution of data by bar graphing the number of units in each category.
6	Control Charts	When you need to discover how much variability in a process is due to random variation and how much is due to unique events/individual actions to determine whether a process is in statistical control.
7	Scatter Diagrams	When you need to display what happens to one variable when another variable changes in order to test a theory that the two are related.

Flow charts

These tools allow all participants to share a vision of what is being done by the process being reviewed. An example of how to implement the CRISP-DM paradigm (the structured process to build models that was shown in Figure 2-3) is shown in Figure 6-10 as a banded swim-lane chart created using Visio. Time charts from left to right. Each lane represents a team member. The round end rectangle signifies a start or stop point. The box represents a step in the process. The diamond represents a decision. Sometimes the process improvement process results in flow charts generated from three different views: supposed to be, as is, and should be. More advanced flow charts called *value chain diagrams,* add quantitative values for each box such as processing time, wait time, and elapsed time.

The primary aim is to create a shared vision of the workings of a system. In addition, time can be added to the process steps specific to the duration of the task (within a box) and between tasks (on the arrow). This later time is called *tact time.* The time values can be for the should-be, as-is, and could-be views. The values can be a specification, an average, or a distribution.

[38] From The Memory Jogger™: *A Pocket Guide of Tools for Continuous Improvement*, 1988, Goal/QPC, Methuen, MA.

Collectively, this enhanced process map is called a *Value Stream Map* and is used extensively in Six Sigma process improvement projects.

Figure 6-10. Classic banded process map from Visio.

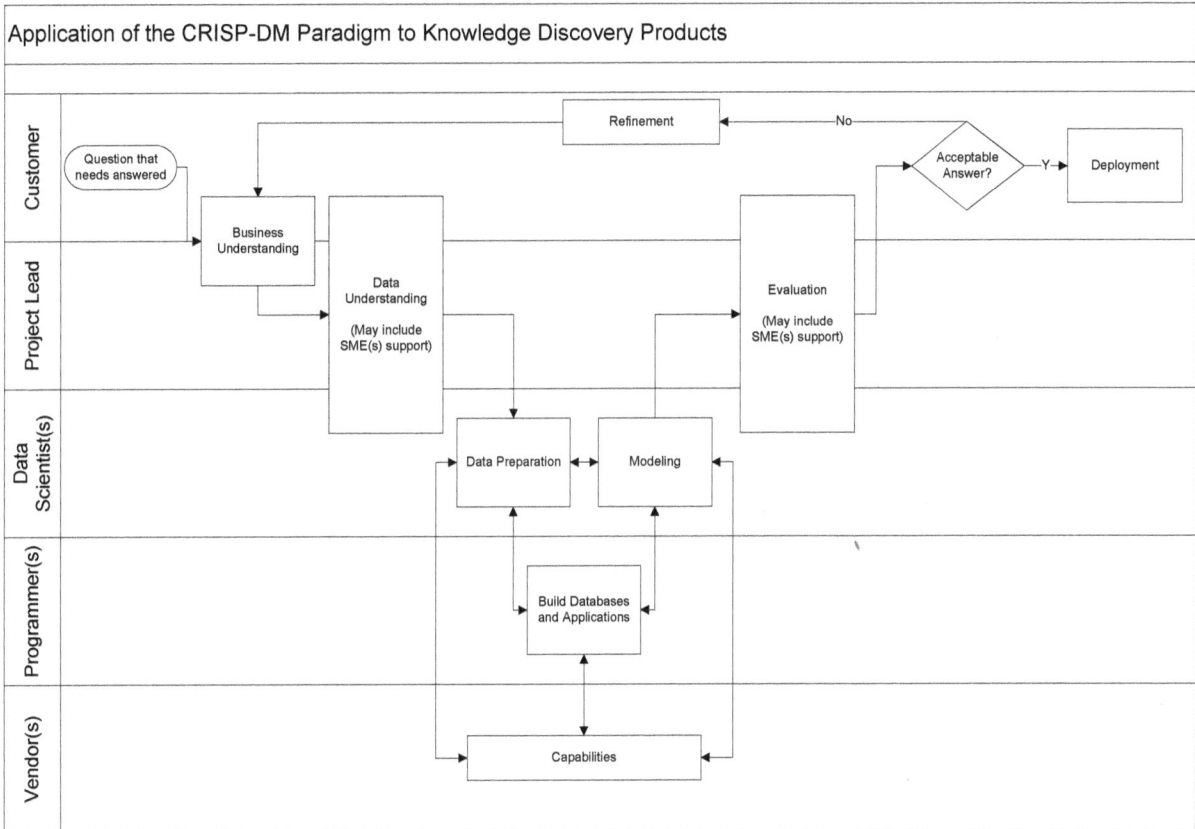

A timeline is a specialized type of flow chart that has both administrative and contextual applications in the world of big data. The example in Figure 6-11 was generated in Visio. MS Project and other scheduling tools are useful for generating Gantt charts for more complicated situations.

Figure 6-11. Simple timeline from Microsoft Project.

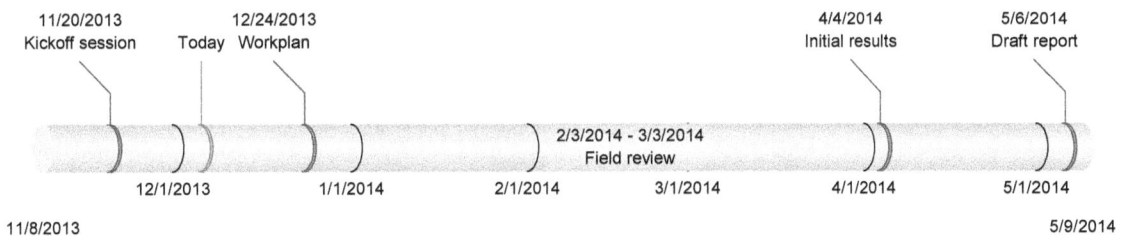

Diagrams

The old adage that "a picture is worth a thousand words" is quite applicable to the analytics workspace. Many diagrams exist for many purposes and each organization (and person in that organization) has its own preferences, taboos, and histories. Traditional views include the Venn diagram and a simple timeline. Classic quality tools are discussed below. Each application area has its standard views that must always be displayed before new views are brought online. Sometimes intermediaries must be used to transform to a complicated view.

Patience and timing are key to being successful. The master of telling stories with diagrams is Edward Tufte. The wisdom in his four books is readily available at www.tufte.com. It always helps to draw the picture at the start of a systems analysis, to vet the picture with the customer, and then to add numbers when they become available. The benefits of clear and exact communication are priceless.

Cause and effect (fishbone)

The Ishikawa fishbone diagram is a standard in the quality discipline. It is a type of cluster analysis used for qualitative analysis. The example in Figure 6-12 started with the 6-M's paradigm and assigned delay reasons to one of the primary divisions. Analysis can also be done in a more unstructured manner or using specialized "textual analytics" and self-learning categorization tools.

Figure 6-12. Fishbone diagram from Minitab.

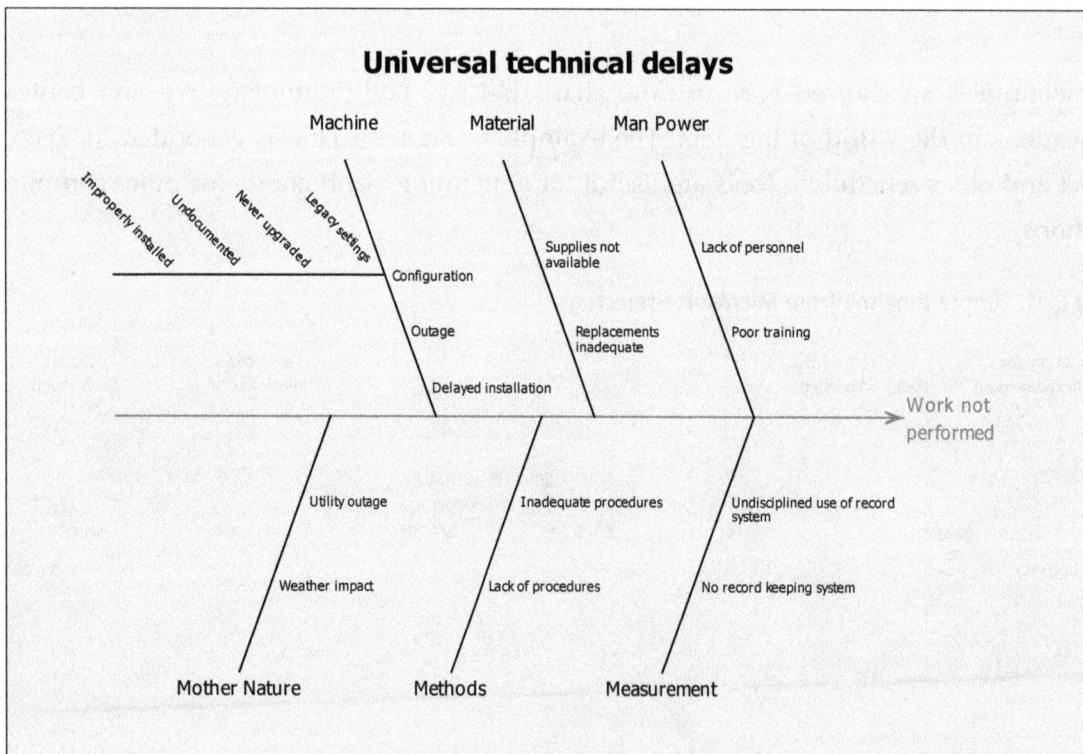

Decision trees/hierarchical breakdowns

One of the masters of analytics is the marketing world. They have been great innovators of new tools that are used for big data. Commercial sales have a strong profit motive and thus provide much of the R&D funding for software development. Fraud detection is another element that has driven many developments in analytics. The books mentioned by Davenport at the start of this training manual provide numerous examples.

Textbooks, such as the one by Berry and Linoff,[39] go into detail regarding all uses of decision trees for describing data, estimating values, and predicting results. One of the most powerful tools used is a tree diagram that shows hierarchical relationships identified by the application rules between elements and helps determine the "normal" state for the leaf categories and corresponding, frequentist probabilities.

Figure 6-13 is a static view made in Excel for a hierarchical data breakdown (classification tree) leading to discrete categories with expected probability estimates. Some methods, such as CART and CHAID, provide a dynamic interface that allows real-time arrangement of the hierarchy with calculations automatically updated.

Figure 6-13. Decomposition hierarchy.

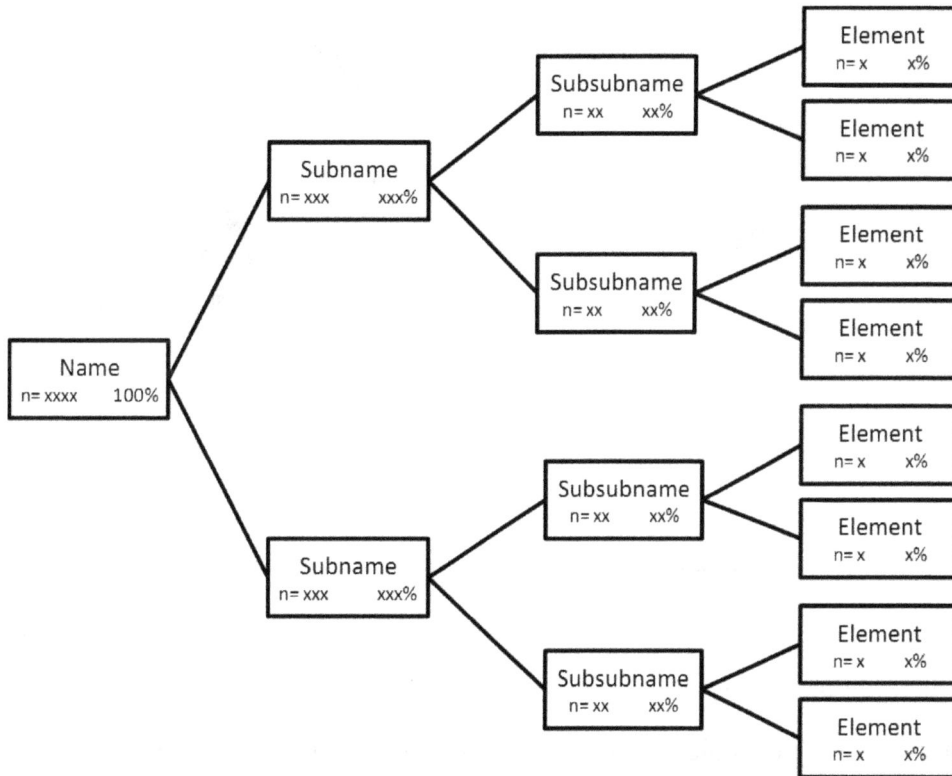

---

[39] Berry, M.J.A and Linoff, G., 1997. *Data Mining Techniques for Marketing, Sales, and Customer Support*. John Wiley and Sons, NY.

The most robust applications attempt to generate surrogate data for missing values by leveraging information contained in other fields. These approaches are used in medical studies to predict survival likelihood, as well as in marketing to determine profitable cohorts of customers. Demonstrations using a billion records in the master data table on a remote server or in "the Cloud" with sub-second response times are a *tour de force* of what can be done with "big data analytics."

Relationships

Another diagram portrays link analysis and displays relationship between different entities as shown in Figure 6-14. The diagram focuses on participants, interactions, and roles. This is the soft side of quantitative analysis and starts to uncover the biases and prejudices managers have regarding quantitative analysis. Mainly, it is an example of network analysis and it is a tool that helps ensure that logical relationships are understood. Once the logic is known, levels of quantitative data can be overlain such as frequency of contacts or duration of contacts. Other applications include transportation networks, workers and equipment, and people and website visits.

Figure 6-14. Intercommunication visualization.

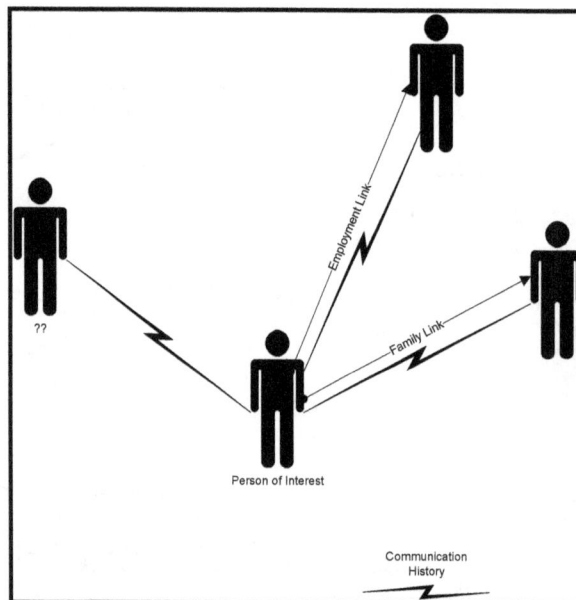

Check sheets and Pareto charts

A check sheet is a simple analog table where categories are listed on rows and a tally mark is made for each instance of the category. In the digital world, pivot tables and crops-tab reports are equivalents (See Example 6-1 for an example of a pivot table.) The categories can then be sorted from highest to lowest to make a graph called a *Pareto chart*.

Pareto charts are used to test the 80-20 rule to see how many categories (20%) contain how much of the results (80%). If the 80% sum is achieved by the top 20% of categories, then the 20% categories are named the critical few and the remaining categories are designated the trivial many. This tool is quite supportive of discovery analysis because it guides the direction for drill down. Data can be counted for a sample by simply using tally marks on a check sheet that has the categories listed. The generation of Pareto charts can be programmed into software if they are needed on a recurring basis. Most commonly, data are normalized into an Excel table and counts are determined when a pivot table is applied. The example in Figure 6-15 was created using Minitab and shows that one category accounts for 83% of the results.

Finally, the tree map in Figure 6-16 is a new tool intended to help with visualizing big data that are related to the Pareto diagrams themselves. Each small block in the tree map represents the percentage of international air passengers arriving in a city, while the entire rectangle represents the total of arriving passengers for all cities being displayed. These data are a continuation of the data from the case study in Chapter 2. When more cities are displayed, the resolution of the trivial categories becomes obscured. This example far exceeds the Pareto rule in that less than 10% of the entities account for over 90% of the results. Again, this visualization tells a story with just a glance.

Figure 6-15. Pareto chart from Minitab.

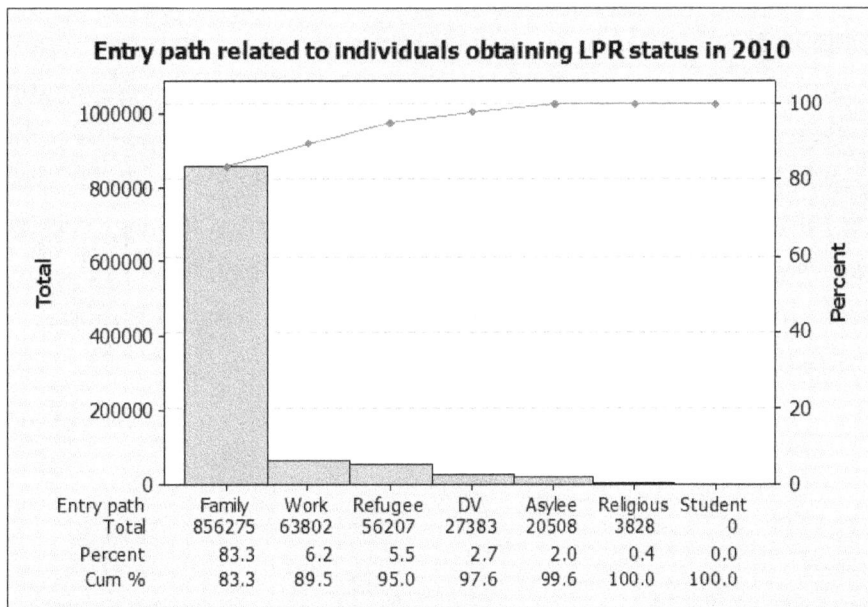

Source: US Department of Homeland Security, 2010 Yearbook of Immigration Statistics. Table 7.

Figure 6-16. Tree map for air travel data generated in Excel.

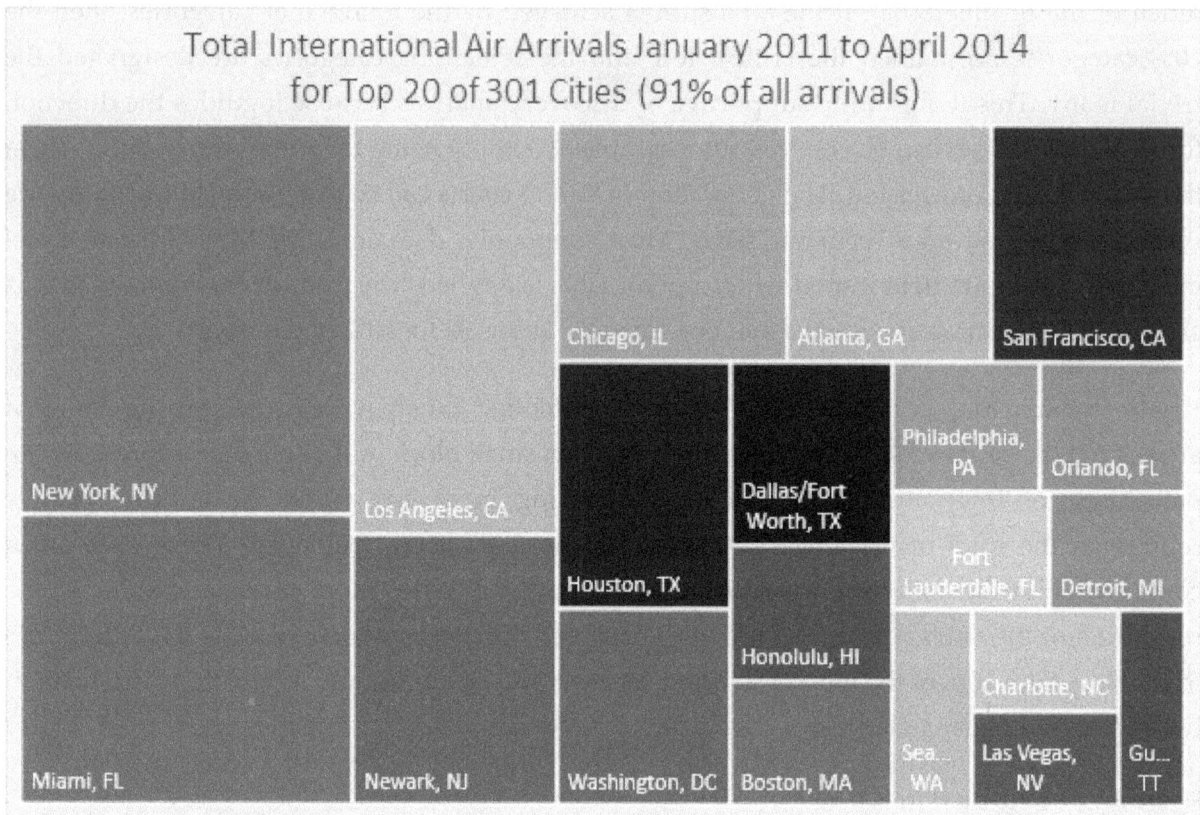

Total International Air Arrivals January 2011 to April 2014
for Top 20 of 301 Cities (91% of all arrivals)

The next example is presented as both a table (Table 6-8) and a map (Figure 6-17). The data represents the annual volume of shipping containers received at major shipping ports in the world, and first displayed in Chapter 5's case study. The table only shows the critical few locations that represent 80% of the total. The map in Figure 6-17 (generated in Excel) uses color to show the critical few versus the trivial many regarding annual tonnage received at a major port. Colors are assigned based on the spread between categories. Most of the current big data analytics software has a map output option. Results can also be exported for use in a Geographic Information System tool if more detail is needed.

Figure 6-17. Location of top 50 shipping ports with 10-year total TEU received.

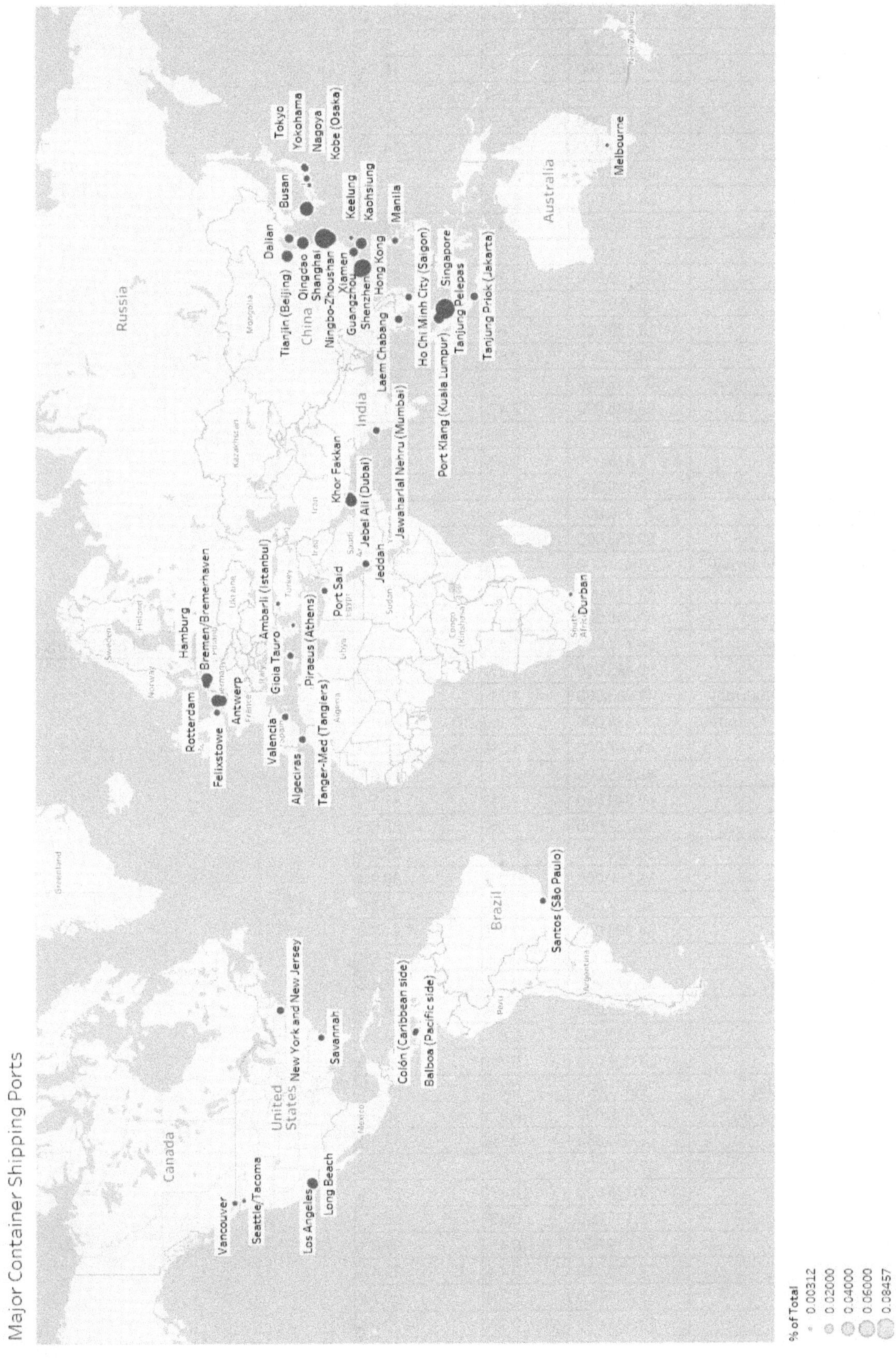

Table 6-8. Tabular view of top 50 shipping ports with 10-year total TEU received.

Port	Grand Total	% of Total	Cummulative %
Singapore	340,077,000	8.5%	8.5%
Shanghai	332,182,000	8.3%	16.7%
Hong Kong	272,894,000	6.8%	23.5%
Shenzhen	248,284,000	6.2%	29.7%
Busan	176,931,000	4.4%	34.1%
Ningbo-Zhoushan	149,296,000	3.7%	37.8%
Jebel Ali (Dubai)	138,421,000	3.4%	41.2%
Qingdao	138,193,000	3.4%	44.7%
Guangzhou	136,244,000	3.4%	48.1%
Rotterdam	129,739,000	3.2%	51.3%
Kaohsiung	116,559,000	2.9%	54.2%
Tianjin (Beijing)	113,689,000	2.8%	57.0%
Hamburg	104,225,000	2.6%	59.6%
Port Klang (Kuala Lumpur)	100,960,000	2.5%	62.1%
Antwerp	96,848,000	2.4%	64.5%
Los Angeles	94,449,000	2.3%	66.9%
Long Beach	77,626,000	1.9%	68.8%
Tanjung Pelepas	76,119,000	1.9%	70.7%
Dalian	71,998,000	1.8%	72.5%
Xiamen	69,855,000	1.7%	74.2%
New York and New Jersey	63,417,000	1.6%	75.8%
Laem Chabang	61,759,000	1.5%	77.3%
Bremen/Bremerhaven	60,749,000	1.5%	78.8%
Tanjung Priok (Jakarta)	54,727,000	1.4%	80.2% Pareto Point reached with 24 of 50 ports (0.48)
Tokyo	50,425,000	1.3%	81.5%
Ho Chi Minh City (Saigon)	47,983,000	1.2%	82.7%
Jawaharlal Nehru (Mumbai)	45,884,000	1.1%	83.8%
Jeddah	44,467,000	1.1%	84.9%
Valencia	43,845,000	1.1%	86.0%
Algeciras	42,893,000	1.1%	87.1%
Felixstowe	39,732,000	1.0%	88.0%
Manila	38,184,000	0.9%	89.0%
Gioia Tauro	36,630,000	0.9%	89.9%
Yokohama	36,102,000	0.9%	90.8%
Khor Fakkan	34,909,000	0.9%	91.7%
Port Said	34,868,000	0.9%	92.5%
Santos (São Paulo)	33,575,000	0.8%	93.4%
Savannah	32,153,000	0.8%	94.2%
Nagoya	31,241,000	0.8%	94.9%
Vancouver	27,009,000	0.7%	95.6%
Balboa (Pacific side)	26,461,000	0.7%	96.3%
Colón (Caribbean side)	22,698,000	0.6%	96.8%
Ambarli (Istanbul)	20,044,000	0.5%	97.3%
Seattle/Tacoma	17,548,000	0.4%	97.8%
Keelung	17,136,000	0.4%	98.2%
Tanger-Med (Tangiers)	16,133,000	0.4%	98.6%
Kobe (Osaka)	15,619,000	0.4%	99.0%
Piraeus (Athens)	14,619,000	0.4%	99.4%
Durban	13,407,000	0.3%	99.7%
Melbourne	12,533,000	0.3%	100.0%

Run charts

For those processes that generate transactions as a function of time, run charts provide a simple visualization of the count of items per a unit of time. The time slices are determined to fit the context of the application. The example in Figure 6-18 is a single criterion (new Lawful Permanent Residents) from 1820 through 2010 (generated in Minitab). The large year-to-year variation in this example is because of policy changes and is a common event in all government data. Therefore, review by a subject matter expert is typically required so that context can be applied to results. In addition, multiple criteria can be graphed together to allow for direct comparison, and target and goal lines can be added for evaluation needs. These are the charts where trends and cyclicity are observed, both as a direct relationship and in a lagged relationship when responses are delayed.

Figure 6-18. Simple run chart from Minitab.

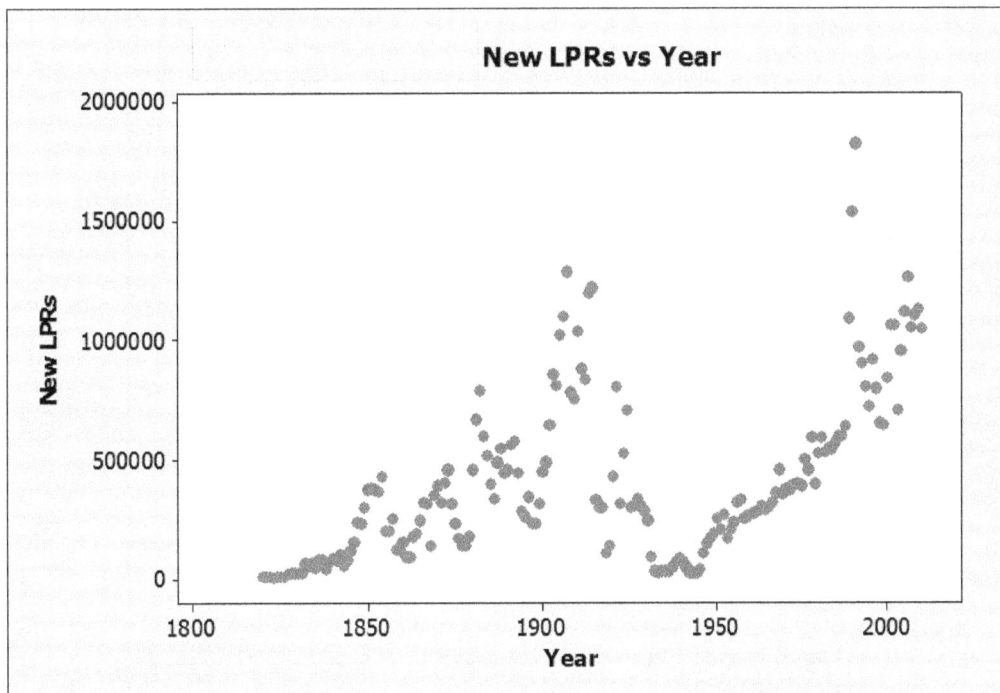

Source: US Department of Homeland Security, 2010 Yearbook of Immigration Statistics. Table 1.

Histograms

Sometimes, data are best examined by use of bins (or buckets) rather than in a continuous mode. The graphical display is reminiscent of stacks of poker chips. A smooth line can be fitted over the stacks with the most common form being a normal distribution (bell curve), most concentrated in the center. Many patterns can result for many reasons. This tool provides a rapid assessment of the nature of the data.

The example in Figure 6-19, which consists of three panels generated in Minitab, is of a bimodal cluster and shows how tools can be used to unravel a complicated situation. The first chart does not have a large center of mass as would be expected if the data had been collected from a normal distribution. Upon scrutiny, it appears that there are two humps in the pattern.

The second chart shows that, in fact, the results represent two different processes with each having a unique pattern and very little overlap between the two. The third chart extrapolates the data from the histogram into a smooth line graph representative of a normal distribution using the concept of sample and population discussed above. When two populations are comingled, the data are said to be stratified and should be analyzed in separate parts.

Figure 6-19a. Histogram view of a complete data set in Minitab.

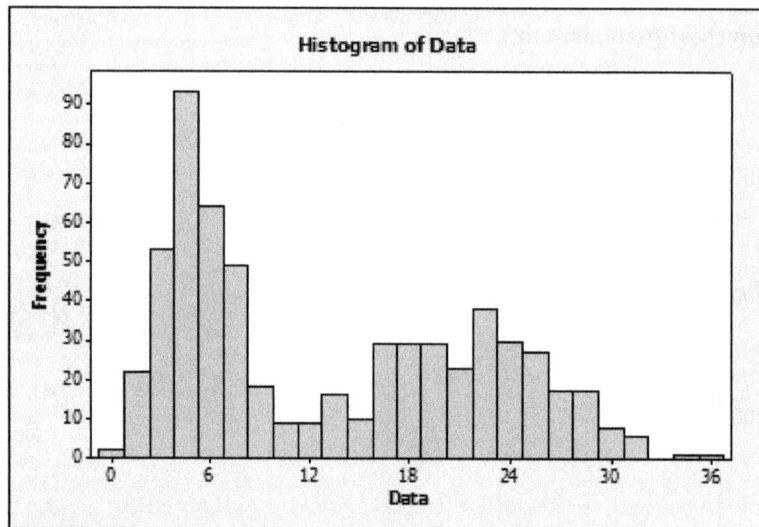

Figure 6-19b. Stratification demonstrated using a dot plot view in Minitab.

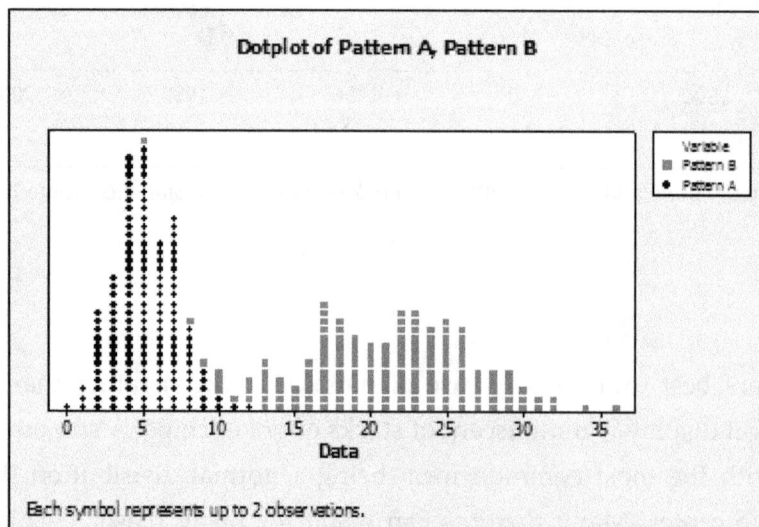

Figure 6-19c. Derived bell curves of stratified data from Minitab.

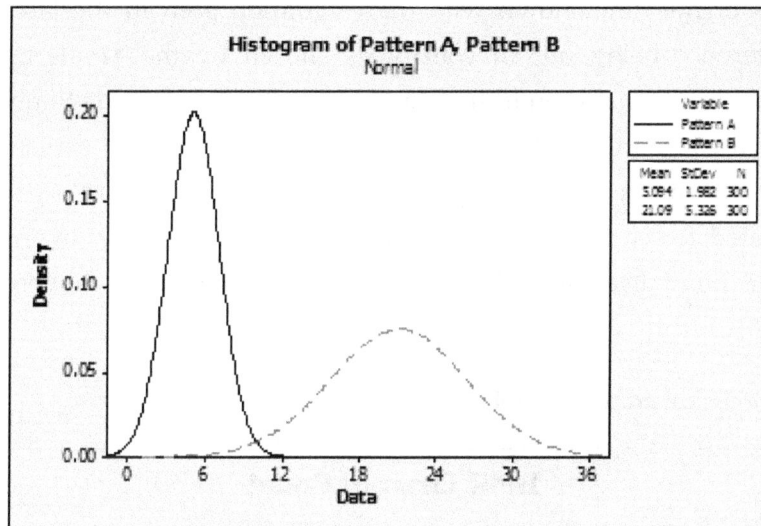

Control charts

Statistical process control charts were invented in the 1920s by AT&T (by their manufacturing subsidiary Western Electric) and represent the merger of run charts and histograms. The control limits represent the typical range of variation for a process that is expected with the changes in the graph representing the noise in the process. When a control limit has been exceeded, management is required to investigate the cause and take appropriate actions to regain control. The term for this investigation used in the Japanese's quality movement is *genchi genbutsu*, or "go and see for yourself" or more commonly called *management-by-walking-around*.

The detailed explanation of control charts is beyond the scope of this book with Montgomery (2013) providing a comprehensive treatment. The charts consist of two panels with the upper one representing variation within a process, and the lower one representing variation between measurements. The upper chart can show averages for a batch or an individual measurement for continuous data. For attribute data, charts exist for fractions (percentages) that are called p and np, and for defects (counts) that are called c and u charts. The lower panel can show a Moving Range value (difference between two consecutive samples) and range or variance that corresponds to a mean for a batch of measurements.

The example in Figure 6-20 shows that the process has produced outputs at three levels with lengthy periods of stability at each level being the normal condition. The original hypothesis was that there was an upward trend. Upon review with subject matter experts, there were changes in the processes that corresponded with each step up. This results in #6 alerts early in the time span and #1 alerts later. In addition, there are occasional large swings in output for the first three-quarters of the time shown with high variation seen in the last quarter and one week-to-week difference being out of control as shown by the #1 alert. What happened between those two weeks? That is an indication that management needs to review the situation and determine what caused control to be lost. It is possible that the explanation is a simple statistical variation, but it must be investigated and documents. Mature organizations have an explanation generated for each alert, with many software packages allowing the value to be tagged to the data point itself and displayed when the pointer is hovered over the alert number.

Figure 6-20. Classic statistical process control chart from Minitab.

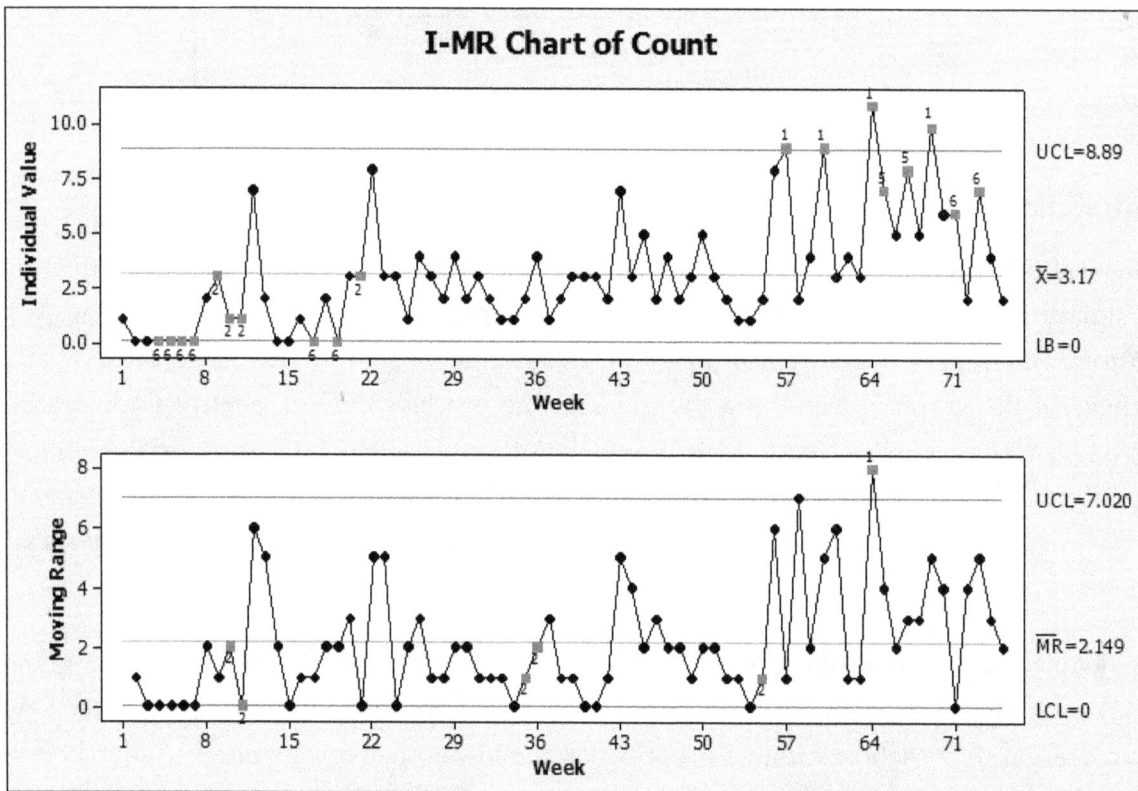

Table 6-9 is a listing of the major tests associated with control charts. These tests were derived from empirical studies and correspond to distinct patterns in the data. Test 1 is the most commonly used, with flags for tests 2, 5, and 6 displayed in Figure 6-20. The visual impact of the red ink highlights the three levels in the data. This tool does not uncover all patterns in the data, like cyclicity and coupling, but it is extremely powerful and quick. A major benefit of adopting control charts is to break the cycle of tribalism. Quantitative analysis knowledge in most companies is tribal – that is, processes and methods are not documented, but rather handed down from the journeymen to the apprentices. Only the masters can try something new, and maybe implement it if it works.

Table 6-9. The Western Electric tests for determination of special cause variation. [(Western Electric Co. (1956) and Minitab (2007)]

Identifier	Test condition
1	1 point > 3 standard deviations from center line
2	9 points in a row on same side of center line
3	6 points in a row, all increasing or all decreasing
4	14 points in a row, alternating up and down
5	2 out of 3 points > 2 standard deviations from center line (same side)
6	4 out of 5 points > 1 standard deviations from center line (same side)
7	15 points in a row within 1 standard deviation of the center line (either side)
8	8 points in a row >1 standard deviation of the center line (either side)

Control charts capture the Voice of the Data and reflect on the nature of the output. Control charts are never used to assess if specifications are being meet. This evaluation is done using a different tool, a Process Capability Chart, and is used in formal Yield Analysis studies. An example is provided in Figure 6-21 for 326 data points.

Figure 6-21. Process capability chart from Minitab for air freight data from Chapter 5's case study.

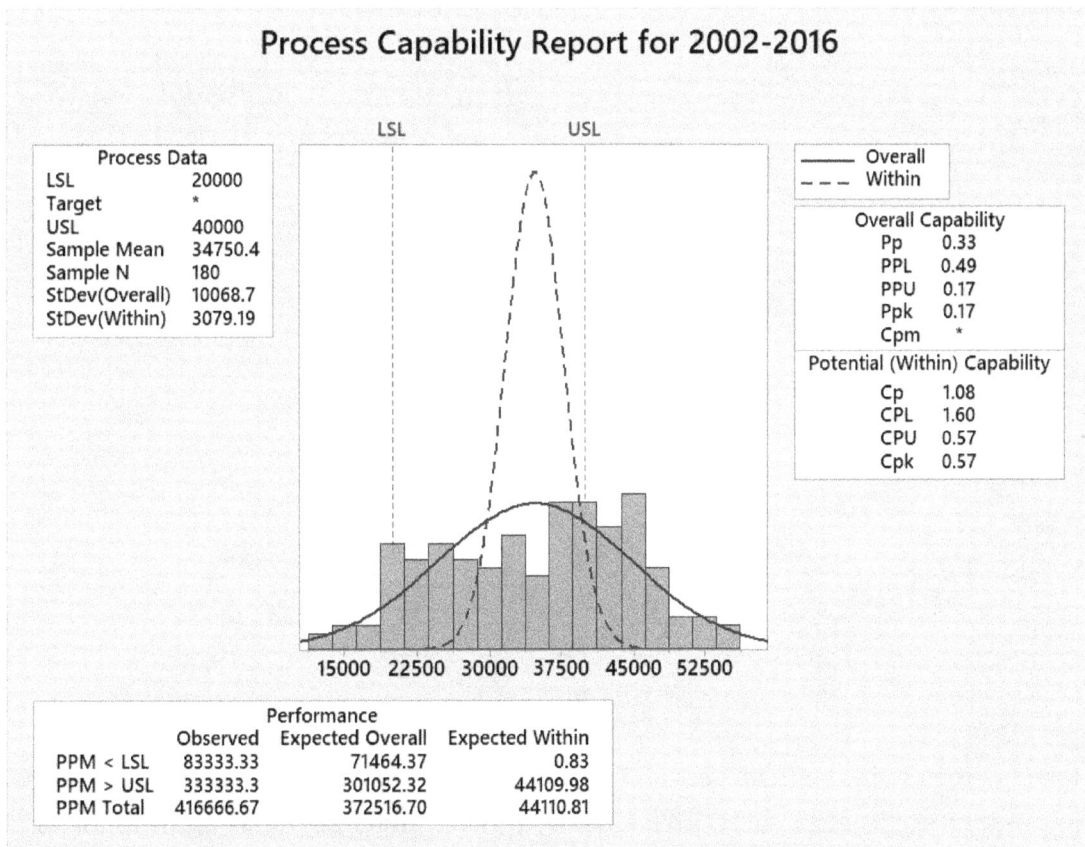

The data represents air freight tonnage, shown in Chapter 5's case study, that was shipped from Hong Kong to Anchorage from January, 1990, through March, 2017. Figure 6-22 is the same data shown on a control chart. As the control limits represent ±3 standard deviations, and if you assume the data are normally distributed, then you would expect three values out of every thousand (1-99.7%) to be out of control and require investigation. Care needs to be taken not to over load data scientists with analyzing noisy data by adopting a sampling approach (like one hour per day, or the first five minutes of every hour in a day, or one day per week).

When process changes are made, and data become tighter, then larger samples can be processed. If there are a million rows, you would expect 3,000 values to be out of control, which would be extremely hard to analyze. Once root causes are identified, go back to the step in the process where the defect occurred. Then data can be stratified with a flag and displayed on separate charts for different conditions and investigated in a more focused manner. A prominent trend is plainly visible in Figure 6-22 and supported by the number of low alerts on the left side of the graph and high alerts on the right side. This is a condition that is not supported in classic I-MR charts with a solution presented below.

Figure 6-22. Control chart view of the data in Figure 6-21.

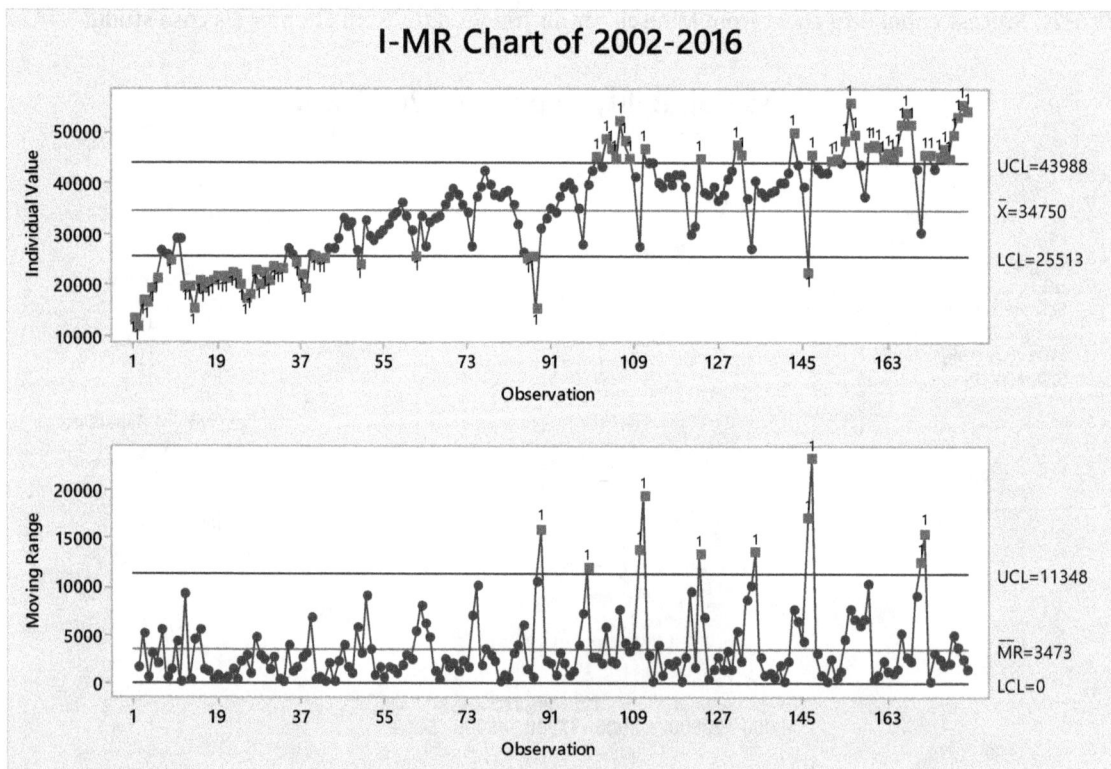

It is not easy to convince a manager who does not have a quantitative background of how to read and interpret a control chart. I have used simulated data before as an introductory tool. For a years' worth of data, I would generate 365 random numbers using a normal distribution scaled to the actual mean and standard deviation of the process being reviewed. I would then parse the month breaks and generate a series of charts. The chance of an alarm being issued when the control limits are exceeded is 0.3%, or three in a 1,000. This value corresponds to the fact that 99.7% of values fall within three standard deviations of the mean for a normal distribution. For a slice corresponding to a year (365 days), one would expect one alarm to be issued based on straight probabilities.

For a year's worth of data, one alarm would be expected and a second would not be that unusual. I found that the manner would focus on learning how to interpret the control chart itself, rather than thinking through the process. When I would present actual data, the manager had been biased to what would be expected by a random process and was able to deeply evaluate runs in the data that varied from the preconceived notation. The differences were typically idiosyncrasies in data that resulted from nuances in the process, but sometimes represented an opportunity for improvement. Finding these opportunities are the primary benefit of using statistical process control charts.

Scatter diagrams

Scatter diagrams are used to show the relationship between variables. Figure 6-23 is a specialized view, a trend chart, with time plotted on the x axis. This graph generally shows that the number of new LPRs has been increasing each year since 1945. You can state that the increase is correlated over time.

Figure 6-23. Regression plot from Minitab. (Note three high years in the late 1980s.)

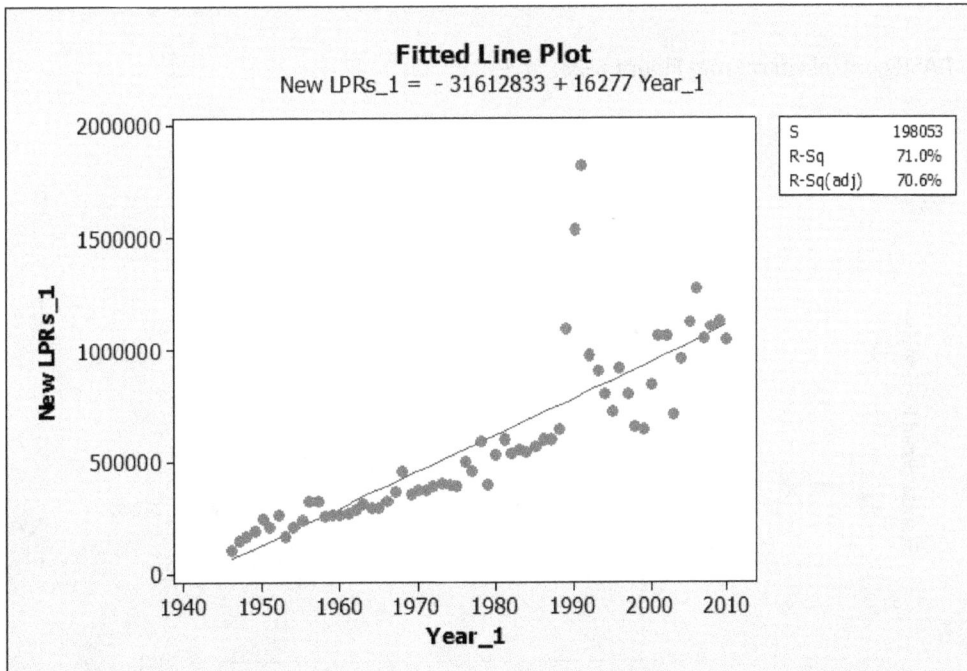

Source: US Department of Homeland Security, 2010 Yearbook of Immigration Statistics. Table 1.

The role of scatter diagrams is to show how one variable relates to another. This example also has a linear regression result included showing the line of best fit (generated in Minitab). While Ishikawa did not originally describe regression techniques, I include them because they can be added with just a right-click in Excel. The equation that defines the line explains 71% of the increase (based on the r-squared value). Three data points (1989, 1990, 1991) are the cause of the poor fit. The large gains represent a policy decision by the US Government to accept large numbers of Vietnamese refugees. Also note the increased amount of variation after the late 1980s, which is a classic example of heteroskedasticity (change in variation over time). This is a real-world example of applying quantitative tools to policy decisions. This level of analysis is required to prove that advanced tools cannot be applied to this amount of noise.

Finally, the trend line above cannot be directly plotted on an control chart but must be transformed and analyzed with a RASR (Regression Analysis Standardized Residuals) control chart to assess the high points and determine if they, indeed, are out of control. This fusion of two tools was also not presented by Ishikawa but is a logical extension because software packages can generate these outputs with just a few mouse clicks.

Figure 6-24 shows that only two out of three of the suspect data points are out of control. The first point (1989) is within the bounds of normal, but is an artifact caused by the extreme nature of the next two points. If the data had been viewed in real-time, the 1989 point would have been out of control relative to the proceeding data. The RASR chart also shows a downward trend prior to 1989, that was apparent on the scatter plot, and that requires further investigation.

Figure 6-24. RASR control chart from Figure 6-23.

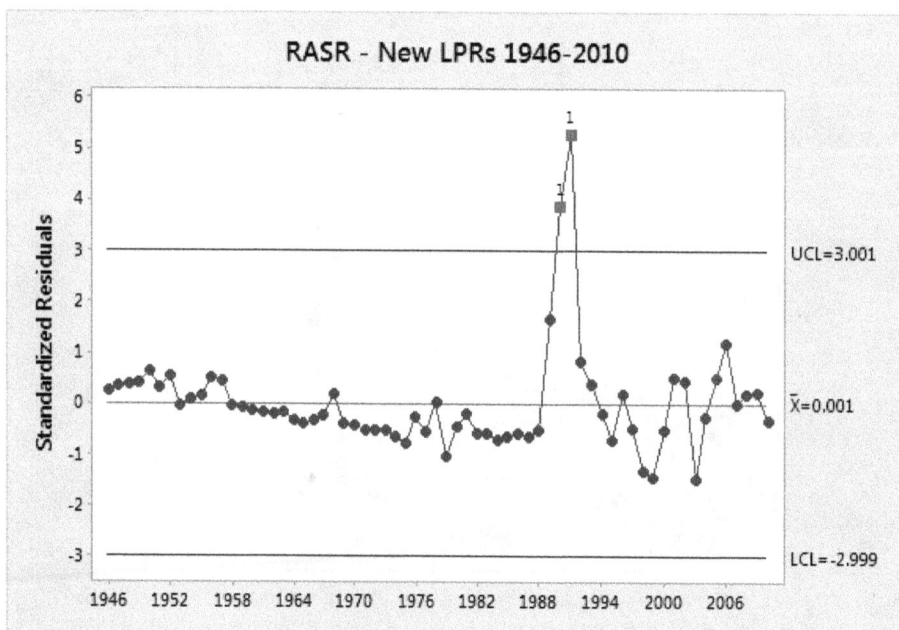

Returning to the air freight data shown in Chapter 5's case study, the I-RASR control in Figure 6-25 shows an extreme condition, because the data are showing multiple trends over time with the first ⅓ being relatively flat. As part of the regression analysis, Minitab generated four views of the residual values (the differences between the observed and predicted values corresponding to each input value) shown in Figure 6-26. The r-squared value for a linear fit is 0.912. Additional fits were attempted with a polynomial equation yielding an r-squared of 0.915 and a power equation yielding an r-squared of 0.86. The bottom panel of Figure 6-25 is a RASR chart that shows transformed data – the regression residuals are presented in a standardized form (the mean was set to 0 and the standard deviation set to 1) which eliminates the variation from the trend. The first third of the data, the flat part on the left, appears as a reverse trend in the RASR chart. Note that there is one high alert and four low alerts that are represented by noise in the top panel view.

Figure 6-25. I-RASR control chart from All Air Freight Data from Chapter 5's case study.

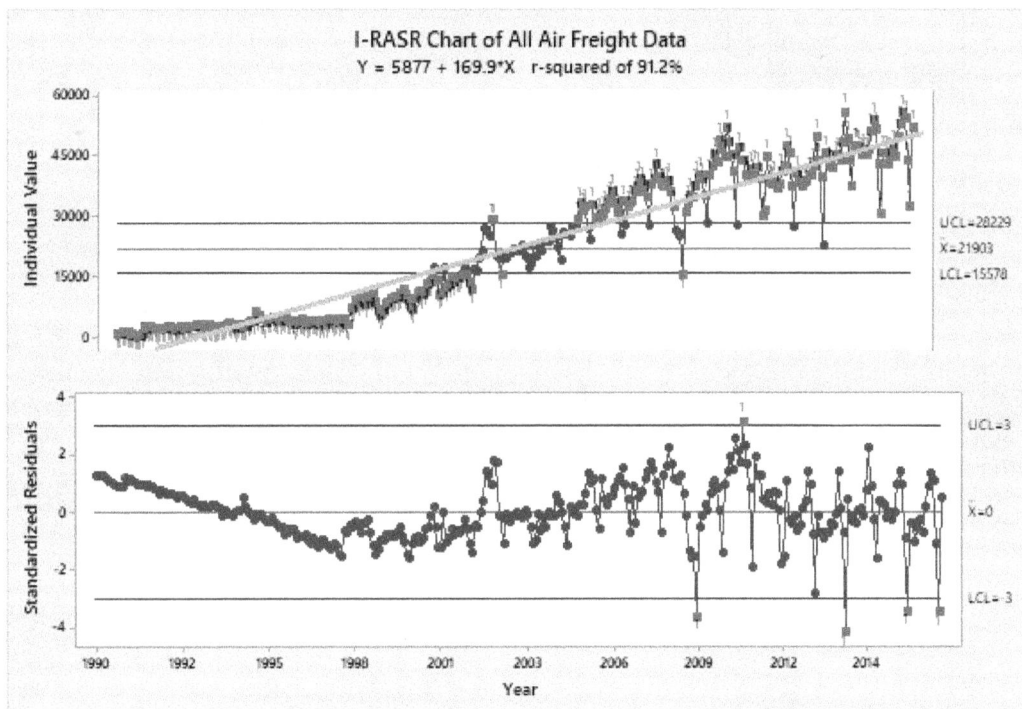

As all three attempts of fitting a line could not handle the change in trends starting in 1998, the data were split into two segments and analyzed separately, shown in Figures 6-27 and 6-28. Figure 6-27 shows that the data through 1997 has its own trend with the regression reporting an r-squared of 60%. While noisier than the entire data set, a strong correlation over time is present. It also shows that the spike in 1994 did exceed the upper control limit. Figure 6-28 shows that the data for 2008 through 2016 is less strongly correlated (r-squared of only 30%) with two points being below the lower control limit. For more on the I-RASR chart, see McGrath (2009).

Figure 6-26. Residuals plots corresponding to the regression analysis shown in Figure 6-25.

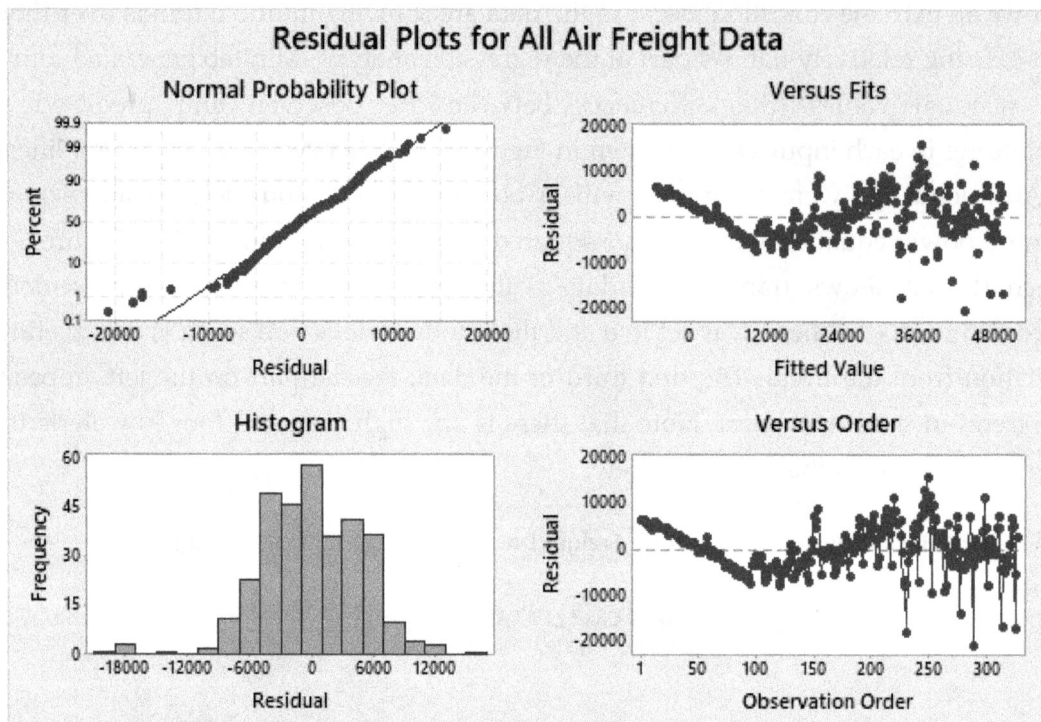

Figure 6-27. I-RASR control chart from Figure 6-25 for data from 1990 to 1997.

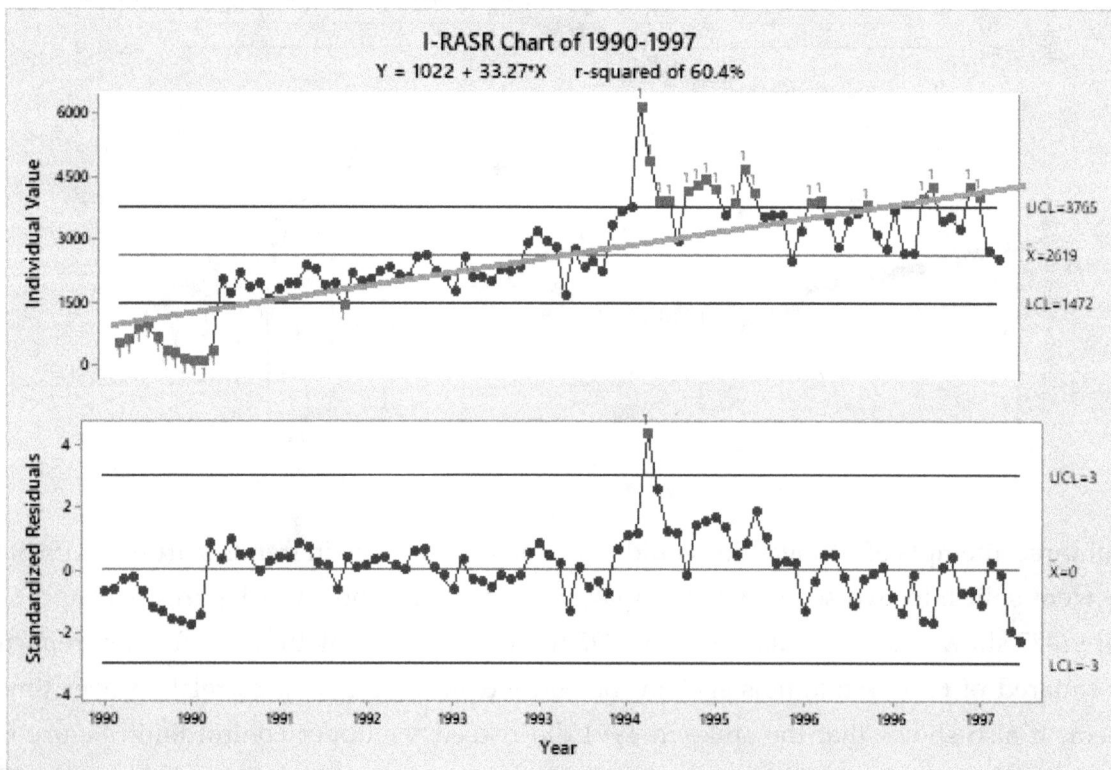

Figure 6-28. I-RASR control chart from Figure 6-25 for data from 2008 through 2016.

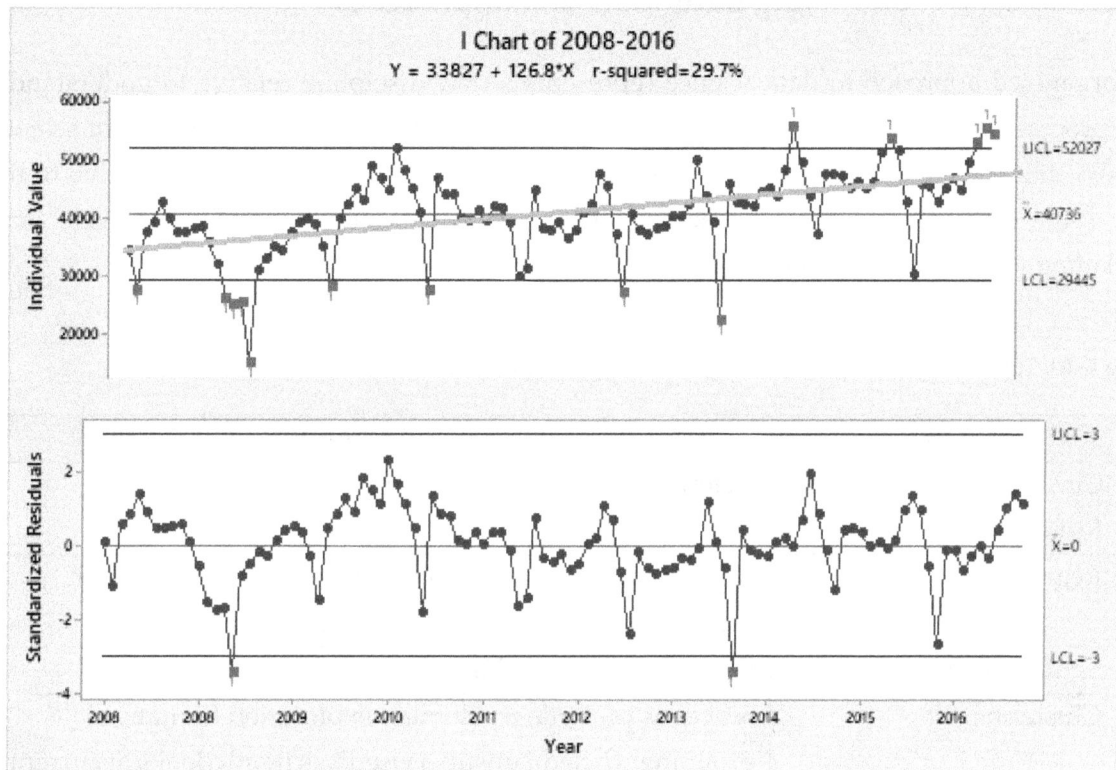

Benefits

The seven tools Ishikawa championed were developed to support manufacturing applications. The main reason they were applied was that there was a large amount of data available as a result of the applications of scientific management to prove processes were efficient. Today, almost everything in day-to-day life has been quantified, including the number of steps I take and whether they were at walking or running speed. The benefits of applying this suite of seven tools are:

- Allows for the essential analysis of any data set,
- Can be easily implemented as software tools are readily available,
- Approaches are conducive to automation for dashboard generation, and
- Abilities correspond to those taught in high school mathematics.

Big data tools and approaches

An organized approach to data science represents a new discipline relative to understanding and processing information. The components have evolved as fundamental pieces of scientific inquiry, but the advent of cloud computing has proven to be a game changer relative to their use. Table 6-10 presents the key techniques data scientists use to produce answers in the marketing world and anywhere millions of dollars are wagered on every analysis.

Table 6-10. Typical big data tools used by marketing professionals.

	Type	Function
1	Classification and Class Probability Estimation	Determines whether something will happen.
2	Regression	Determines how much something will happen. Called "supervised" because it predicts based on the past.
3	Similarity matching	Determines "likes." Foundation for 1, 2, and 4.
4	Clustering	Serves as preliminary domain exploration for natural grouping. Called "unsupervised" as it only looks at current data. Realm of machine learning.
5	Co-occurrence grouping	Identifies associations in transactions and produces statistics on frequency of the occurrence and how surprising that occurrence is.
6	Profiling	Describes behaviors in order to determine "typical" or "normal" conditions as a baseline for detecting anomalies and outliers. Degree of mismatch = suspicion score which can trigger an alarm or flag for action.
7	Link prediction	Traditional networking tools and neural network displays. Advance methods quantify the strength of the link.
8	Data reduction	Creation of subsets for model training and validation. Trade-off: loss of information versus improved insight.
9	Causal modeling	Results in creation of a predictive model using either experimental or observational approaches. Trade-off: cost to refine versus benefits of status quo.

Source: Provost and Foster, 2013, *Data Science for Business*

These tools can be applied for all data types where the focus is on the center hump of the distribution. The methodologies are not as useful if the concern is only with the high or low tails. Conversely, processing big data focuses on the strength of the center hump and is not sensitive to outliers. That is, it is robust. The simplest way to solve this problem is to stratify the data and work with cohorts that represent the low tail, the center hump, and the high tail. In order to make the splits, a detailed knowledge of the underlying processes is required, or some spatial (zip code) or economic (credit score) categories can be used as filters. These splits will not always be made perfect on the first attempt, as the technique is more art than science.

Once these fundamental tools have been mastered by the data scientists, the managers have been trained on how to read and use the outputs, and the databases updated using big data solutions, then the true benefits of analytics can be realized. These benefits are:

- Complete and thorough quantitative analysis of the system,
- Determination of the significance of differences,
- Repeatability of the analysis and subsequent conclusions, and
- Construction and deployment of models for predictive purposes.

The next chapter introduces modeling and examples that demonstrate applications to complex systems.

CHAPTER 7

Models and Prediction

One of our principles is to make everything as simple as possible, but not simpler.

Einstein

A model is defined as "a system of postulates, data, and inferences presented as a mathematical description of an entity or state of affairs."[3] It goes without saying that the model is only as good as the data that was used to build it. It also goes without saying that the model is to be used to predict what a result will be, given a specific suite of inputs, and that the results are hedged with probabilities, rather than being perfectly predictive. The goal is to develop a model that is "good enough" to support the decision that prompted the creation of the model. Software designers have a term, *minimal viable product*, to describe early versions of their output. Models are one type of such outputs. Early versions are used to train users and to ensure conformance, and with maturation of the model expected over time.

Model-based simulations are great for unravelling the short-term and long-term complexities of systems, training personnel for the outputs to expect when taking actions, and aiding in identifying opportunities for improvement, as well as any unintended consequences of those opportunities. Models are used when it is not possible to experiment with the actual system itself. Sometimes, like at NASA, full-scale physical models are built. However, it is very common for mathematical models to be built. That is the focus of this chapter. When a what-if analysis is incorporated into a mathematical model, it is called a *simulation*.

A model-based simulation is simply a tool that allows for rapid experimental observations within bounded conditions. However, models are not reality, they only provide the perception of reality (ala Plato's Cave). This is because the models and simulations represent a virtual world that is a friction-free environment, generating false impressions and conclusions separated from a reality where "real people, with their unpredictable ways, can seem difficult to contend with after one has spent a stretch in simulation."[40] Some even consider them to be

---

[40] Sherry Turkle, (2015). *Reclaiming Conversation*, p. 7.

the proverbial eierlegende Wollmilchsau (egg-producing wool-coated milk sow) because they can do everything.

Too much reliance on modeling and simulation has the potential to become a large problem. The term I use is *Normalization of Fantasy*. I derived it from the Normalization of Deviation concept that is championed by High Reliability Operations (HRO) practitioners. The poster child for HRO is the sheep that cried wolf too many times and the concept occurs when participants become calloused to false alarms. Normalization of Fantasy occurs when a data scientist is overly focused on derived outputs and does not grasp the realities of the system. A gateway is fantasy football leagues. The application of analytical tools for an uncontrollable situation with no real meaning will cause the data scientist to become calloused to the outputs from the tools. This attitude can lead to applying the same tools for critical applications with subpar interpretations.

Sorry to burst the bubble, but the Weather Channel shows how models don't work. This happens every September. It happens when they show the hurricane landfall spaghetti maps consisting of the output from different weather prediction models. Municipal emergency managers must then make evacuation decisions for expected conditions 72 hours into the future. Conservatism is favored – it is better to evacuate when only two of 12 models indicate a risk, rather than to shelter in place – but it runs the risk that the public will lose trust when the evacuation results in a false alarm event. The false positive error is a familiar example of a Type I error. Emergency managers have a hard job! This tactical approach of combining results from multiple models is termed "ensemble modeling." In its simplest form, ensemble modeling is built from classic what-if analyses that identify the boundaries of possible outcomes given a controlled amount of variation of inputs.

Models are also built when quantitative inputs are lacking, or when logical actions by people are to be assessed. This is common in fiction writing and occurs frequently in the genre of science fiction. The most noted example is *I, Robot* by Isaac Asimov from 1950 that explored how man and machine will interact in society. A classic from the manufacturing world is *The Goal* by Eli Goldratt from 1992 that describes scheduling challenges and solutions. A friend of mine never finished reading this book as the realism caused him to have heart palpations. There also is a whole profession that spends a significant amount of time evaluating what-if scenarios – our friends the attorneys. Eventually, there will be a subset of quantitative modeling dedicated to supporting the needs of the legal field.

This chapter introduces models. It will portray how each can be used and their roles in the future work environment. Examples are provided using system dynamic tools that focus on the holistic view of the system, and what behaviors are expected over time when inputs change.

Analog and digital simulation approaches

A fun analog simulation is the game of Plinko (also called *Quincunx*[41]). A disk is dropped and then bounces off an array of pins that form gateways to the next lower level. The disk comes to rest in a bin on the bottom level. The long-term outcome of playing the game is a normal distribution (bell curve). There are many other examples, but the key is that there are physical, moving parts. They are great for training and learning but are not useful for decision making.

Note that simulation can be run for both discrete and continuous processes. Take a supermarket for example. Number of customers is a discrete process – the count changes when one enters or exits the buildings (ignoring online ordering and delivery for now). The value of the merchandise in the shopping cart is a continuous process and changes regarding the count of items as well as the individual price for each item.

The engine is the random number generator. This is an algorithm in the software package that provides the input for each sequential calculation. The sequence of numbers generated is also repeatable, so that changes in runs can be evaluated because they use the same data set. There is some extreme mathematical power in how the random numbers are generated with Law and Kelton (2000) providing detailed discussions.

The heart of a simulation is the distribution assigned to each input factor. The concept of distributions was introduced in Chapter 6. There are two fundamentals that need to be explained. First is the difference between discrete and continuous modes. The pips on a dice are discrete and rolling one die results in a uniform distribution with each number between one and six having a 16.67% of occurring. When two dice are rolled, the result is a triangular distribution because seven is the most likely number to occur. An example of the continuous mode, is the gallons of gas purchased at a filling station by each car. The value can range from 0 to the upper purchase limit as displayed in the precision of the pump. I truly doubt the accuracy of the pumps that display out to the third decimal place.

When a large amount of continuous data are available, a normal distribution (bell curve) works well. When results from multiple normally distributed data sets are added together, a log normal is better, since the high end tails off. When the lower value is bounded at zero, like for waiting times, an exponential distribution is a one-sided tool (and that has the same value for the mean and the standard deviation). The simplest approach is to use a uniform distribution wherein all values across a range have the same likelihood of being selected. A triangular

---

[41] A demonstration is available from https://bit.ly/2AZBTgI.

distribution consists of three points: lowest expected, most likely, and highest expected. The sides are not required to be symmetrical.

A hybrid between uniform and triangular is the trapezoidal distribution[42] as shown in Figure 7-1. This view is of the continuous mode with a discrete mode also available. This distribution has two ranges: the most likely, represented by a uniform distribution, and the possible, represented by the low and high halves of a triangular distribution. This approach works well when trying to capture expert opinions about a factor, or when sample sizes are small.

Figure 7-1. Trapezoidal distribution generated in Excel.

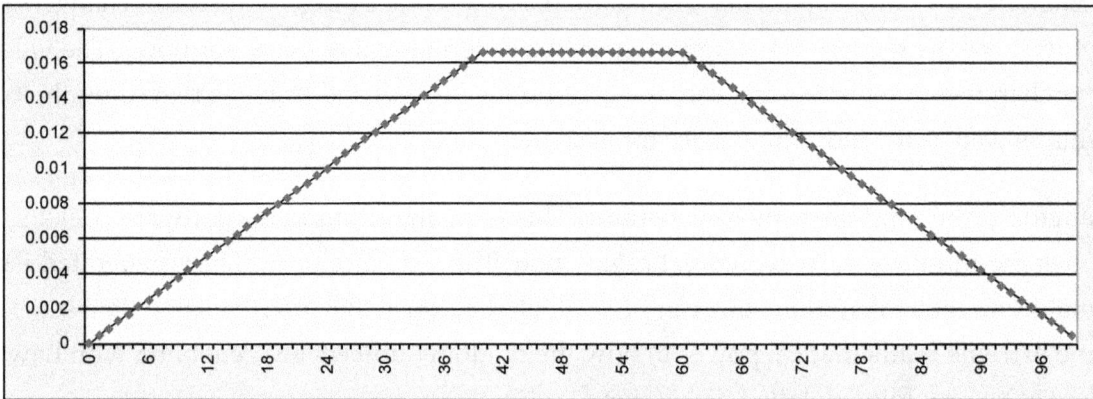

The distribution can also be modified to reflect nonlinear conditions as shown in Figure 7-2, where the uniform portion can slope, and the tails can curve to better capture what is expected.

Figure 7-2. Modified trapezoidal distribution generated in Excel.

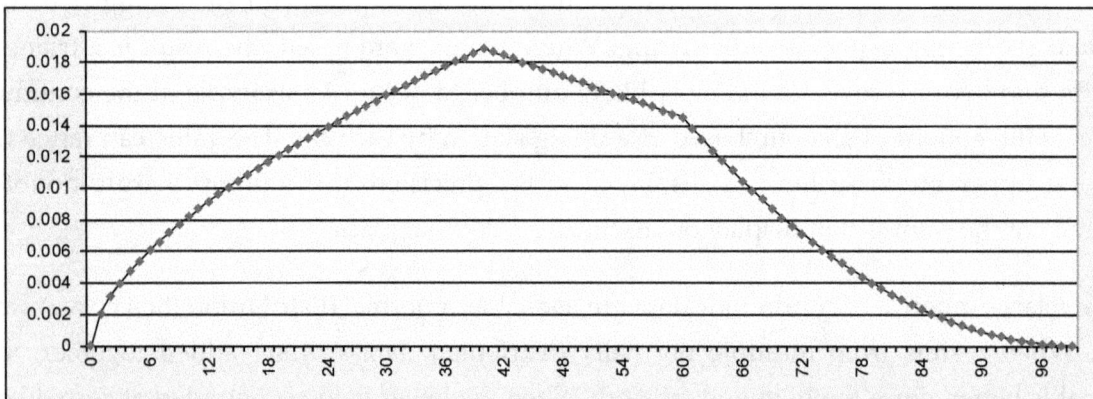

This is accomplished using a mathematical equation that contains seven different variables. It is an expedient way to build models because input data only needs to be "close enough." The

---

42 Samuel Kotz and Johan René van Dorp, 2004. *Beyond Beta: Other Continuous Families of Distributions with Bounded Support and Applications. World Scientific.* Page 150.

Central Limit Theorem will dominant simulated runs and precision can be gained by increasing the number of runs in a Monte Carlo simulation.

Building the model

A precursor to digital simulation is the building of a deterministic model, which is a mathematical construct that describes the behavior in question. In science, these models are called *laws*, such as the Ideal Gas Law, Predator-Prey ecological balance law, and the laws of Thermodynamics. In business, they are called *formulas* such as the Economic Order Quantity and Product Pricing. Sometimes the term algorithm is used. Once built, the model is then solved by entering the values for the input parameters and calculating the expected value. There are two common approaches:

- Deterministic – Enter exact values and generate a number for the output that is as accurate as the input and used for prescriptive decisions.
- Probabilistic – Enter a range for the inputs, randomly select values, and generate outputs as a normally-distributed range and used for predictive decisions. CLT predicts the normalized output.

Probabilistic (Monte Carlo) approaches are what most people think of when they hear a digital simulation is being run. Figure 7-3 is a simple example from a hapless football team.

Figure 7-3. Simple Monte Carlo simulation example using Crystal Ball.

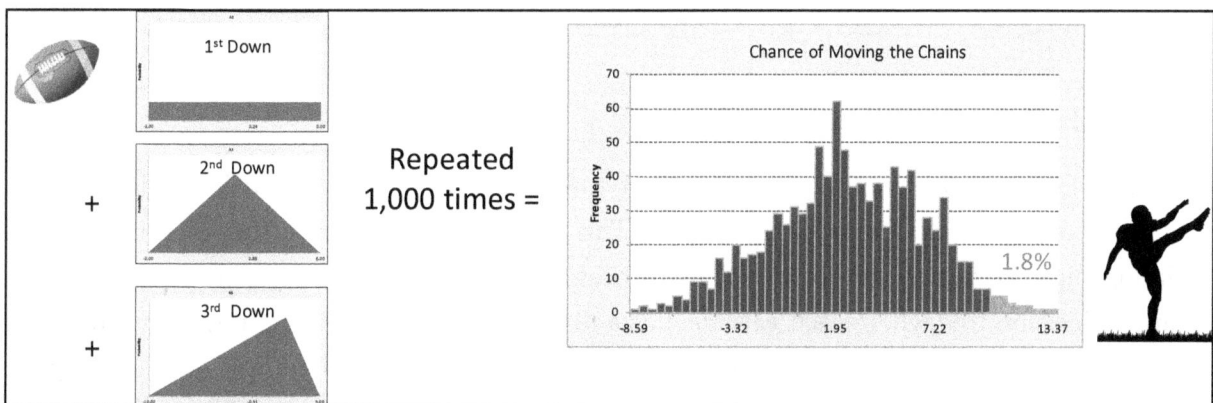

The offensive performance on first down is represented by a uniform distribution ranging from -2 to 5 yards. Performance on second down is best represented by a triangular distribution ranging from -2 to 6 with a peak at 2 yards. As expected, third down becomes a passing down represented by an asymmetric triangular distribution ranging from -10 to 5 with a peak at 3 yards. When values are randomly selected from each distribution, as if the game was actually being played, a predicted outcome is generated. When run a 1,000 times, a slightly high-biased triangular distribution is created with a peak at almost 2 yards. There is only a 1.8% likelihood of the outcome exceeding 10 yards and receiving a new set of downs.

There is quite a body of knowledge related to building models for a system. The most common is the *Generalized Linear Model*, an application of multiple regression and ANOVA statistical tools. When the model is being built, indicators can be discovered that help with calibration, such as observing correlation among select variables that represent the response variables themselves. The model is refined by the removal of predictors with variance close to zero, elimination of highly correlated values, and centering and scaling of each predictor so that responses are maximized. In terms from analog days, emphasis needs to be given to the factors that move the needle for the ranges under consideration.

For complex situations, an alternative approach can be used. This alternative is *Discrete Event Modeling*. It is used when the interactions between variables are unknown, when feedback loops occur in the process, and when relationships are nonlinear. A representation of the system is built and the processes are allowed to run for a fixed amount of time. The results are then reviewed using formal verification and validation approaches. An example is provided in Appendix D relating to traffic flow during egress from a Fourth of July fireworks display. This model was built using Arena.

The use of the CLT has ramifications because outcomes can be over normalized, and the extreme values lost, which sometimes impairs utility. There are advanced methods that try to eliminate these problems.

Advanced approaches

Bootstrap forest is an advanced approach that assesses each variable as an individual and in combination with all the other variables. This approach allows for the identification of relationships not intuitively obvious to the process experts. An analog description can be made using a decision tree where each variable has a turn being the first discriminator and the other variables are assigned to the 2nd and 3rd positions. While it sounds complicated, it is easily implemented using JMP Pro from SAS.

The Amarillo rainfall data from Chapter 5's case study is a good example of what can be done with bootstrap forest techniques. This is not simulation because the data are handled as discrete values and combined across all perturbations. Think of a slot machine with twelve wheels. The first run is for all 12 months in 1880. The second run advances the December wheel to 1881. The third run, 1882. This approach will produce two boundary values: the minimum for the driest value in all months of 0.27 inches and the maximum for the wettest values in all months of 76.99 inches.

Table 7-1 also shows the median and mean values from summing each month. Note that skewness is accentuated, but the means are almost identical. The difference is due to rounding errors. All values in between will be generated by this discrete approach which can then be shown on a histogram. A distribution can be fitted that truncates at boundaries corresponding to the actual driest and wettest years, or the actual values on the histogram can be used as discrete input values. However, this will require a significant amount of computational power as there are 138^{12} possible combinations, or 4.77×10^{25} data points. Output could also be trimmed to more reasonable bounds, say one-in-a-million, by integrating the area under the curve.

Table 7-1. Comparison of annual and summed monthly rainfall data for 138 years for Amarillo, TX.

	Minimum	Median	Mean	Maximum
Annual Precipitation	7.01	19.535	20.479	39.75
Sum of Months	0.27	16.225	20.471	76.99
Jan	0	0.415	0.597	3.17
Feb	0	0.470	0.637	2.93
Mar	0	0.565	0.922	4.14
Apr	0	1.065	1.501	6.45
May	0.04	2.460	2.888	9.81
Jun	0.01	2.520	3.073	10.73
Jul	0.04	2.215	2.684	8.02
Aug	0.15	2.770	2.994	8.07
Sep	0.03	1.795	2.029	6.42
Oct	0	1.055	1.677	7.64
Nov	0	0.430	0.785	5.09
Dec	0	0.465	0.684	4.52

Once these data are available, they can be used for simulations rather than the ubiquitous Monte Carlo approach, with its CLT-induced problems. This is a realistic, versus deterministic or probabilistic, approach. Luckily, software does not have to run every perturbation, but uses a random sampling approach to define correlations and terminates when the rate of change in the correlations becomes trivial. It makes intuitive sense to use probabilistic data to build a probabilistic model. It also shows the challenges faced by civil engineers trying to design drainage systems for urban areas. Which number is the best to use?

Even more complex is the use of *boosted tree* approaches. This is a black-box method that allows the computer to scale input variables such that they all have the same impact, like moving the needle the same way, to use an analog description. Results are then scaled back for presentation. The approach stems from classic what-if analysis, but the computer will tirelessly perform all perturbations to find the optimal settings.

Many times, organizations retreat to reductionism approaches. This happens when time is of the essence to provide an answer, or if a proof of concept is needed to justify scaling up a full-blown project. It involves abandoning the universal approach that solves every problem and trimming the data to just what is needed to solve the most pressing problem. Many of the transformations and calculations are done by hand, which takes time and introduces another set of possible errors, but results in a useable answer of a known pedigree. These studies can then be used to validate more automated approaches when those models are finalized.

Regardless of the modeling approach taken, there is a universal tactic in use that maximizes the use of the available data. That tactic is termed *jackknifing* or *folding*. This tactic involves randomly sampling the available data for, typically, three uses: training, tuning, and verifying, the model. The first pass provides an estimate of model parameters and uses ~20% of the data. The tuning pass uses the bulk of the data, ~60%, to develop parameters that cannot be directly calculated. The third pass provides information on the precision and accuracy of the model when given new data (sometimes called the *held-out block*). If the predictions made during the third pass are not acceptable, the model is said to be over-fitted to the training data. The model can then be re-generated, using less data for the training phase, to gain more predictive performance for a generalized case. Professional grade packages have these functions automated. A variation exists when smaller data sets are encountered, called *bootstrapping*, that involves random sampling **with** replacement. This basically creates a simulated response of all possible outcomes again.

What will the future work environment look like?

It will be data-centric and "the Matrix" shown in Figure 1.3 will govern all work. Everything will be in one system and accessible via common tools. The proof that this will happen is shown in Figure 7-4, that states data scientists currently spend as much time on the ETL phase (data wrangling) as the analysis phase.

Figure 7-4. State-of-the art someday after 2018.

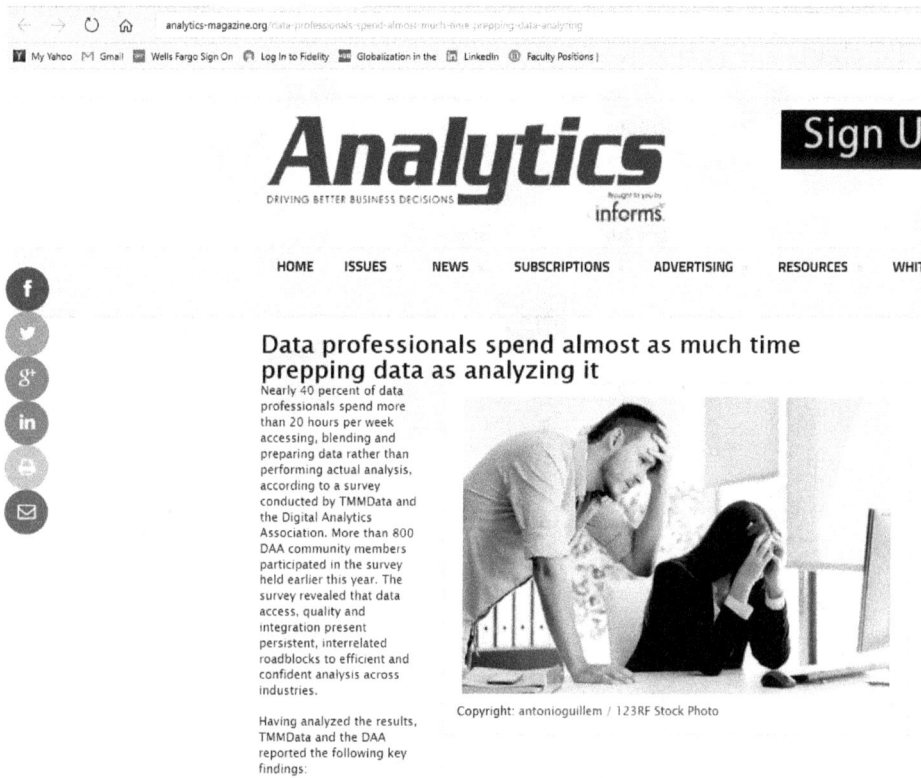

This inefficiency is not sustainable as data scientists will burn-out and the limited pool of resources will become depleted. This will happen because senior managers will become tired of receiving mistake-prone analyses, CFO's will weary of paying for the army of data scientists, and the data scientists' wrists will burn out from always experiencing "The Meditation of 10,000 Mouse Clicks" when a new class of questions is broached. Every transaction, signal, analysis, and communication will be stored in a database that is interconnected with all other systems. The key to success will be a culture of rigorous and disciplined documentation. Viva the Vogons!

Figure 7-5 provides a schematic showing what the future holds. All data will be captured as either a signal or a transaction. Until maturation is achieved, raw data will need to be transformed into a tidy format by various post-processor engines. Legacy data will always present a challenge because the quality will not be of the current caliber of compliance. When standardization is achieved, all data will eventually be generated in a tidy form. Automation will allow for many decisions to be made in mechanistic manners using deterministic algorithms, like automatic credit approval if credit score is over 840, or automatic denial if credit score is below 500. This automation also requires an investment in in-memory processing hardware to support the balancing actions. Tidy data will need to be analyzed, perhaps by a Daily Management approach as presented in Chapter 4, and summarized and displayed by day, by week, by month, as well as by year. Statistical analysis will be performed to establish a baseline as well as to determine if the difference between reporting periods is significant. SPC charts are a great aid for this and can be easily automated.

Figure 7-5. Comprehensive mental model of Quantitative Analysis for System Applications.

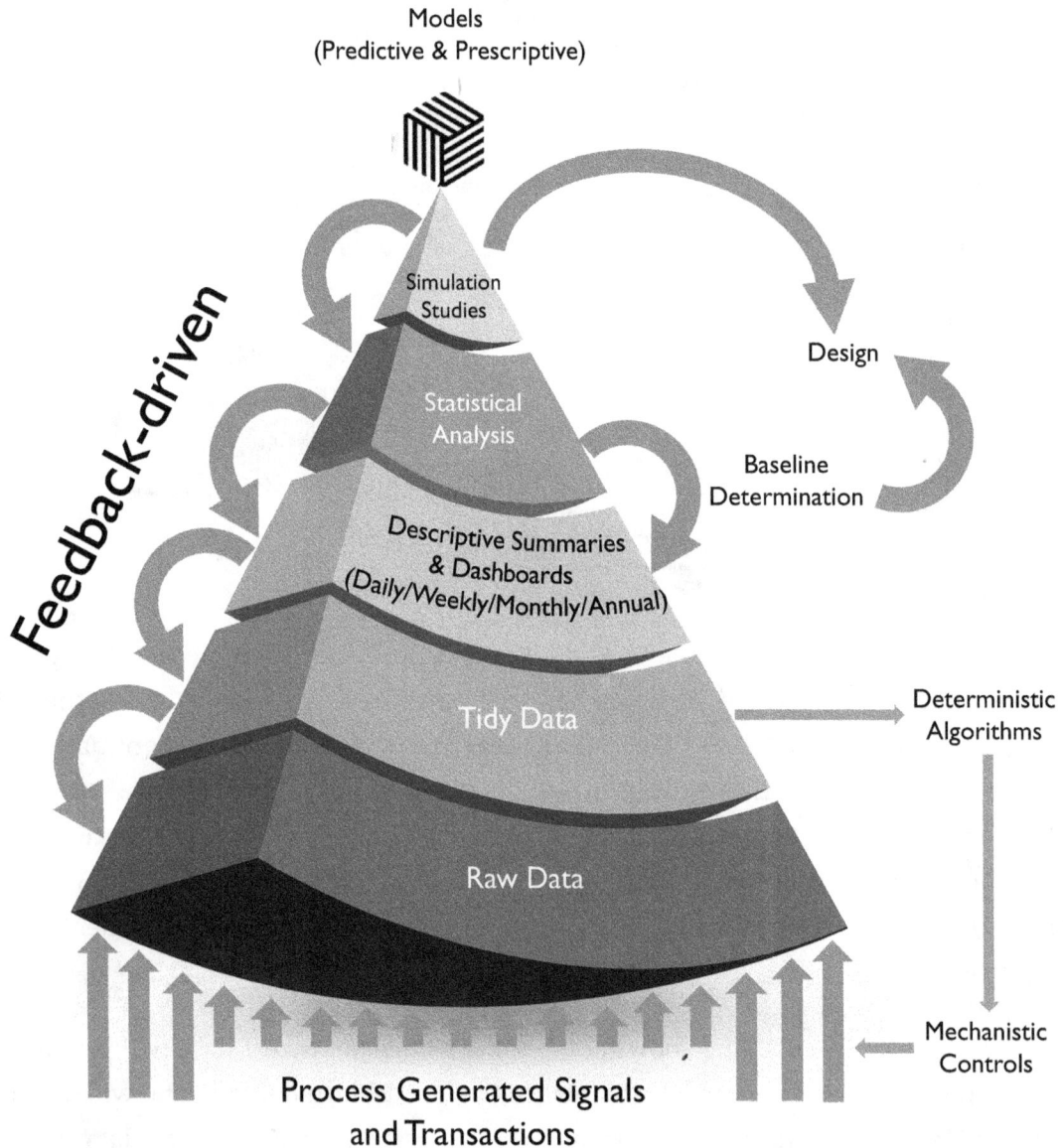

This baseline data will be used for design purposes to determine signal generation requirements, IT processing and storage requirements, action levels, and deterministic algorithms, as well as economic realities. Legacy analysis will be embedded because it will be used for verification and validation activities during the build phase. This design work will be supplemented with simulation studies where What-If scenarios can be evaluated prior to actual deployment. The final goal is to create a model that can be used for predictive and prescriptive purposes as well as *ad hoc* analysis to support problem solving challenges.

The model is not reality, but only a perception of reality. This is represented by the wire-frame element that captures only essential elements that are clearly exposed. This starkness highlights the inevitable tendencies hidden in the complexities and minutia of the system itself. Validation results will assign the amount of residual error inherent in the model, and thus the amount of uncertainty in any decision made using it. All these steps are conducted in an environment that requires feedback-driven improvements so that perfection can be pursued when maturation occurs. All metadata related to the generation and maintenance of The Matrix will be stored in a knowledge management suite (like Atlassian) so that problems can be researched in a complete manner. Resistance will be futile because all will be assimilated!

System dynamics models

In addition to the quantitative analysis of systems, there is a more formal approach that looks at the everyday aspects of life where nonlinear behaviors are observed in complex arrangements using tools and models called *System Dynamics*[43] that were invented at MIT in 1956.[44] While data are needed, the models can be run in a conceptual mode that addresses the direction of behavior over time. In order to build these models, there are some fundamental questions that need to be answered first. What is your starting point? What is the scale of interest? What are the timeframe and the time steps? Einstein's Unified Theory is an example of an extreme end point. Software helps, but simple diagrams can also be used to describe the system. The most common software tool is VENSIM.[45] Simple examples are provided below for Natural systems, Man and Nature systems, Man and Machine systems, and Human systems.

Natural systems

A classic example of a natural system is the predator-prey model, such as foxes and rabbits, and how population numbers change based on wet and dry years and the corresponding changes in birth rates for both species. There is a complex feedback loop that makes it difficult to generate an empirical solution.

---

[43] https://en.wikipedia.org/wiki/System_dynamics.

[44] http://mitsloan.mit.edu/phd/program-overview/system-dynamics.

[45] http://www.ventanasystems.co.uk/services/software/vensim/.

Another example is from pedology, the study of soils. Hans Jenny, in his book the *Factors of Soil Formation*, developed a conceptual equation to explain how soil is formed:

$$Soil = f(cl, o, r, p, t, ...)$$

where cl – regional climate; o – organisms: r – relief; p – parent material; and t – time. Jenny left the ellipsis open to indicate that there might be other variables in the function, such as the fallout of dust from volcanic eruptions. The solution of this equation becomes very complicated due to issues of scale and time steps. It also is the basis for predicting the fate and transport of pollutants.

Case Study: Rainfall and soil infiltration in Vensim

What happens when it rains? The drops fall on leaves or the soil. Some of it runs off, some of it infiltrates, and some of it pools and then evaporates. Of the portion that infiltrates into the soil, some eventually percolates to the groundwater table, with another portion absorbed by plant roots. The interactions are quite complex. Figure 7-6 is an example generated in VENSIM that also shows the movement of nutrients along with water. This is a generic view, but it is equally applicable to rain forests and deserts.

Figure 7-6. Soil infiltration model in VENSIM.

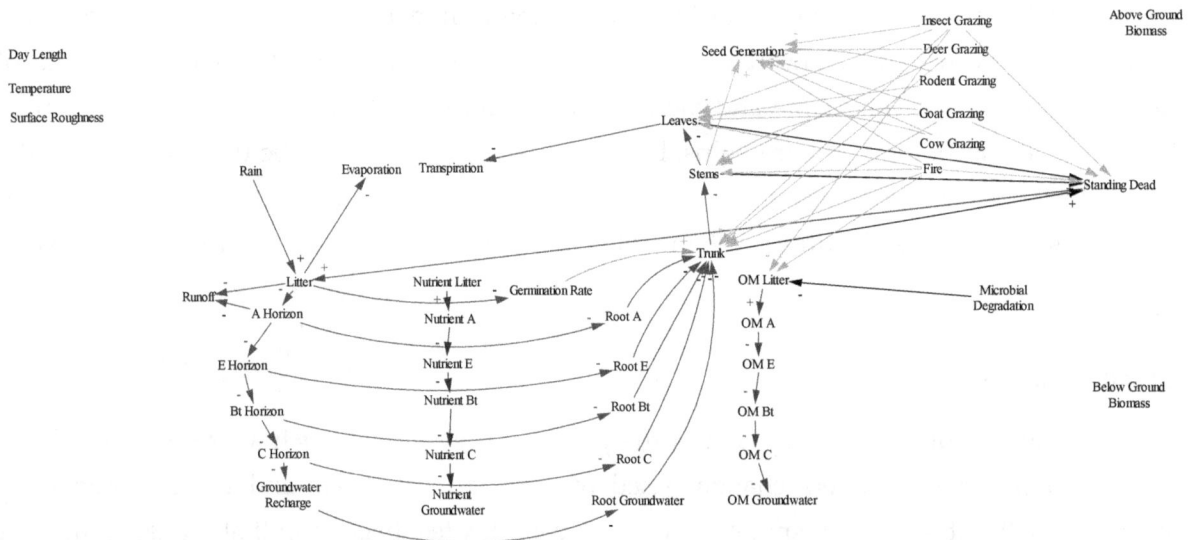

Man and nature systems

The interface between man and nature is complicated. Early in our evolution, man was controlled by nature. Over time, man has gained control of many aspects of nature with the current geologic era being called the *Anthropocene* due to the drastic changes man has caused to the system. These impacts are expected to increase because the soon-to-be ten billion inhabitants want to be fed and to surf the Internet every day. Take water for example. The case study from Chapter 5 and Appendix C discusses rainfall. Example 7-1 discussed the fate of rainfall on the soil. Civil engineers build on those models to ensure that adequate drinking water is supplied to their customers and that waste water is processed to remove hazards and preserve the resource for reuse. The solutions quickly become formulaic and are used around the world. This reuse of a concept is termed *isomorphism* in system theory terminology.

Case Study: Downtime expected value for lightning warnings

Another common challenge involves lightning. What happens when lightning hits a manufacturing facility? The electronic control systems may become fried. If the facility involves operations that are inherently risky, then extreme consequences may be realized. Many such facilities cease operations when sensors indicate that conditions are risky. From a planning standpoint, a model needs to be built that defines the expected delays that could be experienced.

Smith and Merritt (2002) and Smith (2003) proposed a methodology of how to use a model to visualize project (schedule) risks and formulate effective plans for countering them, such as establishing buffers. He proposed calculating an expected productivity loss (Le) as:

$$Le = Pl * Pi * Ls$$

where Pl = probability of lightning duration (or other disruption), Pi = probability of critical exposure step being delayed (or other impact), and Ls = length of time scheduled. The average annual time under lightning warnings (Pl) is 10.7% of the time (see the data in Figure 4-5), the probability of critical exposure (Pi) (where work cannot be performed during lightning warnings) is 5.75%, and the average schedule time (Ls) per unit is 30 days. Overall, the expected productivity loss (Le) is 0.18 days or 4.4 hours. When calculated by month, Le ranged from 0.2 to 9.9 hours. Instead of using the average, the 75th percentile is a more conservative approach for some operations, since the average means the answer is correct only half of the time.

This model can be made more dynamic by linking it to real-time weather forecasts. Real-time data will allow production control to determine if it is truly worth running a production line and paying overtime for a crew to wait out multiple thunderstorms. While workers usually like being paid time-and-a-half, they also like their free time better than being idle on the production line. This balancing of work and life can be incorporated in system dynamic models by using different layers to describe different concepts.

Man and machine systems

One of the defining characteristics of being human is our use of tools. When these tools have moving parts, they are called *machines*. Some are digital for the transactional world. Most are analog and can cause damage to life and limb when things go wrong. In the modern world, machines are associated with phenomenal amounts of data that must be processed and stored. SysML[46] is a software tool aimed at helping the engineer build detailed and complex models of machines. It is being widely deployed during the design stage for electric, hybrid, and autonomous cars as a means to define the logic that must be programmed into the vehicle.

Case Study: Global transportation

There are 65 nations involved in China's One Belt, One Road (OBOR) initiative as shown in Figure 7-7. The goal is to link Central, Southeast, and South Asia with Europe, while including critical supply chain locations. Those countries represent 4.5 billion people (62% of the world's population) and $23 trillion (30% of global GDP).

Figure 7-7. Block map for the 65 OBOR countries.

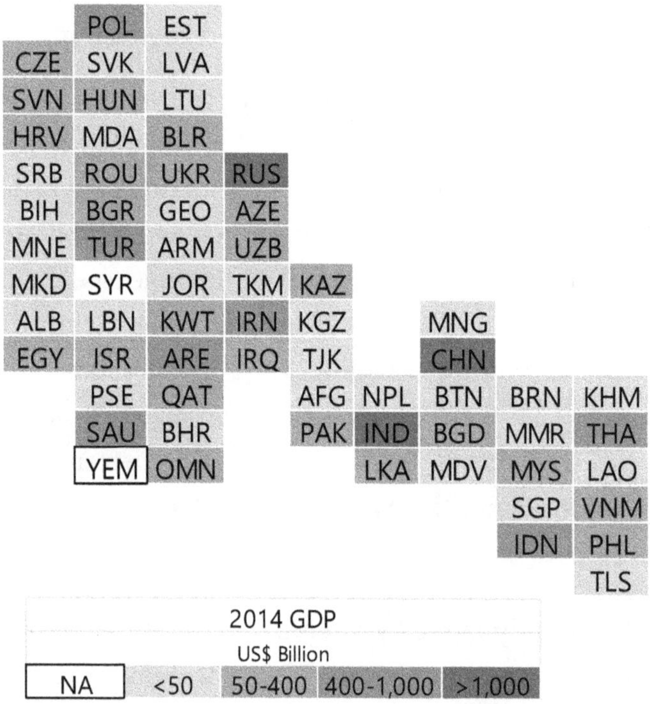

In addition, China is investing heavily to improve transport from African farms to Chinese tables. A parallel effort involves building the digital Silk Road; the fiber optic network and server farms that will handle and process all the digital information needed to make OBOR work. This initiative is truly a systems approach because some of the tricky engineering

---

[46] http://sysml.org/.

projects may never be profitable on their own (such as in Pakistan, Myanmar, and central Asia), but are needed to complete the network. By taking a holistic approach, expenses are spread system-wide with everyone benefiting in the end. There are complaints that this build-out represents an expansion of Chinese control akin to Imperialism with the counter argument being that only the Chinese banks were willing to risk their capital. Once built, all will see rewards. Much has already been written on this subject with Kaplan (2018) providing discussions from various points of view.

The eventual result will be an integrated logistics machine that delivers food, fuel, medicines, and goods to two-thirds of the world's population. Any disruptions will have profound impacts. Detailed models will be required that will be composites of many tools that address machinery, weather, demands, and policies. It will truly be represented by "the Matrix."

Figures 7-7 and 7-8 display the challenges that will be faced by such a systems model. The first iteration was on a map, but the size discrepancies between Russia and Hong Kong made for a poor view.

The next iterations involved bar and pie charts for population and GDP with poor views resulting. Rather than just using a tabular view, a cartoon approach was selected. Figure 7-7 uses equal area blocks and country codes. It can be cryptic as you must know that HRV stands for Croatia. Figure 7-8 has a balloon of matching size for each entry (which was a challenge in itself given the length of some of the country names) and color shading was selected to distinguish between groups selected by natural breaks. Both views require active mental interactions by the reader.

Human systems

Models can also be applied to a variety of human systems such as human resource programs and economics. Human Resources (HR) has always used models to evaluate new hire retention, sick leave utilization, and cost reductions resulting from wellness programs. These models take a macro-view based on probabilistic approaches that are suitable for comparing alternatives, but that cannot be applied down to the micro-view of each individual who is free to make choices.

Models of economic systems are quite complex and must account for the unknown risks driven by growth of big government and multi-national corporations (tragedy of the commons), danger of over planning in a world full of disruptive innovation, and the uncertainty of growth and governing policies themselves. They also must make meaningful connections within and between disparate systems (sometimes called *system of systems engineering*). An interesting example is called the *Big Mac Index* that is a measure of purchasing power parity because it compares a tangible asset (a hamburger) measured in one currency (US dollars) across multiple locations and multiple currencies. It is an elegant solution for reducing many variables to one index. Two economic-related examples are presented below.

Figure 7-8. Relative economic strength of the 65 One Belt-One Road countries.

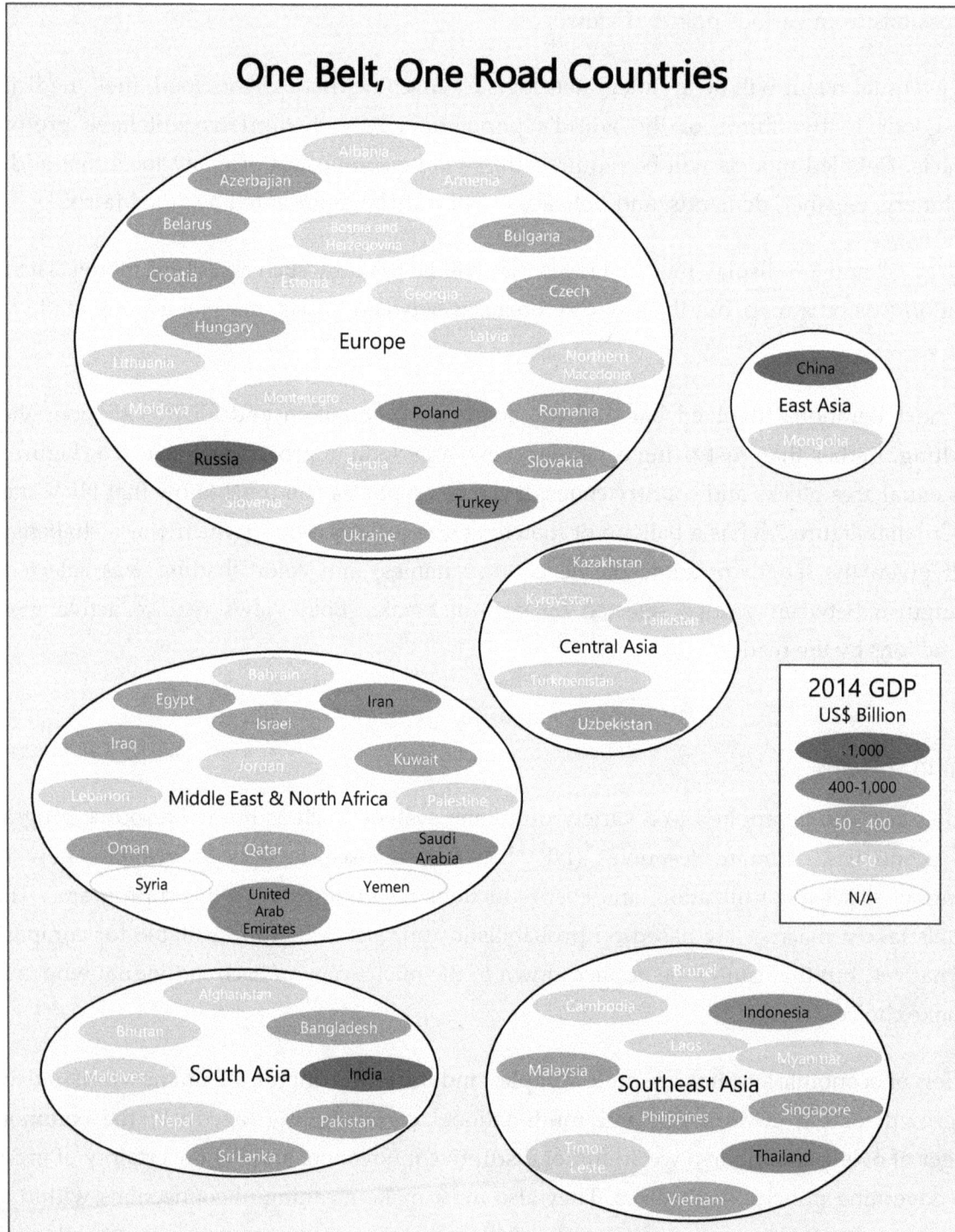

Case Study: Investing in emerging markets

One of the keys to the success of the world's economy is growth. The most growth potential lies with developing countries that are commonly called *emerging markets*. For growth to be sustained, profits need to be reinvested in building infrastructure, shown in Figure 7-9. What is not shown are the damages that occur when leadership steals funds for their own use. One of the greatest factors of complexity involves the exchange rate of currencies and the price for natural resources.

Hedging is a way to ensure success, but detailed and exacting models must be used to minimize risks. Hedging uses balancing models that place buy and sell market-futures orders tied to future dates so that monetary gains can be made in the opposite direction of the movement of the price of a commodity. The classic example of hedging occurs at the horse racing track when you buy a win-place-or-show ticket for your favorite entry. You do not get as much for a win as a pure bet would pay, but you are compensated for a poor finish.

Figure 7-9. CONOPS of investment flow to and from Emerging Markets.

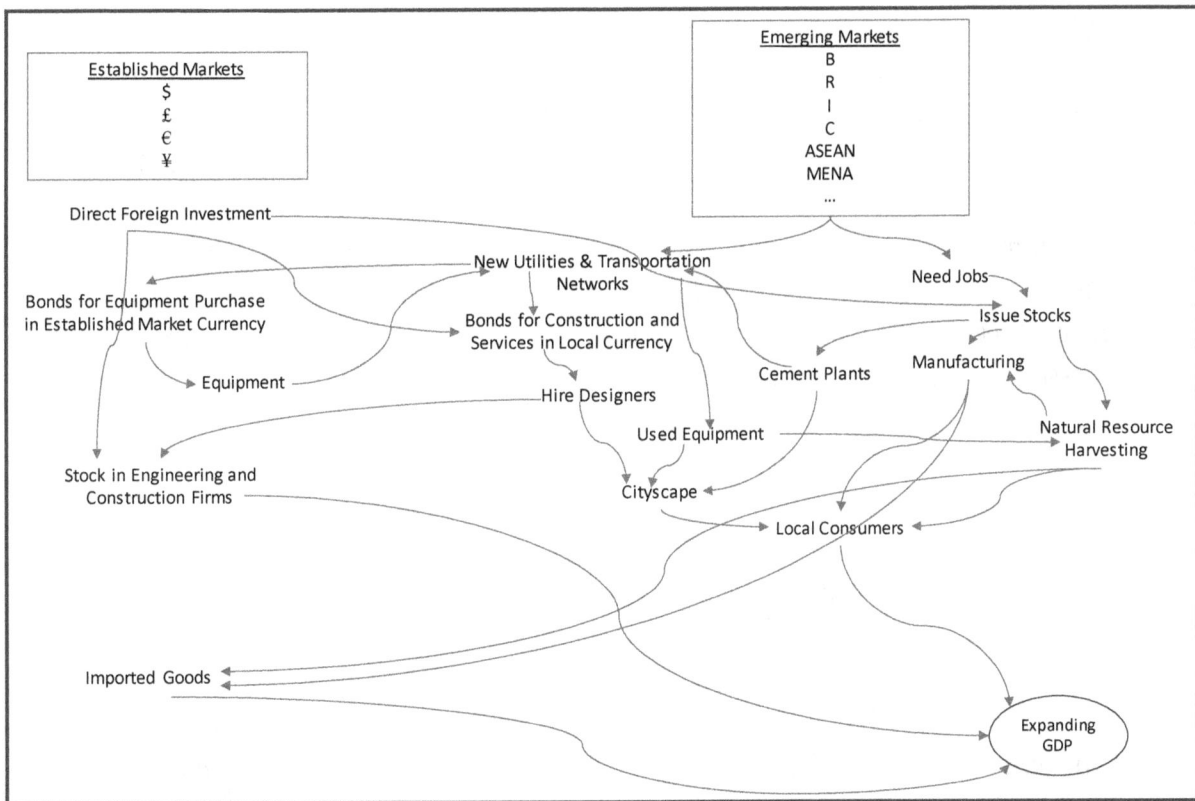

Case Study: Ricardo's Economic Model

The[47] universe of David Ricardo, set forth in his Principles of Political Economy in 1817, is dry, spare, and condensed; there is none of the life, the lively detail of Adam Smith. Here is nothing but principle, abstract principle, expounded by an intellect which is focused on something more permanent than the changing flux of daily life. This is as basic, bare, unadorned and architectural as Euclid, but, unlike a set of pure geometrical propositions, this system has human overtones: it is a tragic system.

To understand that tragedy, we must take a moment to introduce the main characters in the drama. They are not, as we have said, people: they are prototypes. Nor do these prototypes, in the everyday sense of the word, live: they follow "laws of behavior." There is none of the bustle of Adam Smith's world here; instead we watch a kind of model puppet show in which the protean aspects of the real world have been reduced to a kind of one-dimensional caricature; this is the world stripped of everything but its economic motivations.

Whom do we meet? First, there are the workers, undifferentiated units of economic energy, whose only human aspect is a hopeless addiction to what is euphemistically called "the delights of domestic society." Their incurable penchant for these delights results in every rise in wages promptly being met with an increase in population. The workers get their dry crust, as Alexander Baring put it, for without it they could not perpetuate themselves. But over the long run they are condemned by their own weakness to a life at the margin of subsistence. Like Malthus, Ricardo saw only "self-restraint" as a solution for the working masses and although he wished the workers well, he did not put too much faith in their powers of self-control.

Next we meet the capitalists. They are not Adam Smith's conniving merchants. They are a gray and uniform lot, whose entire purpose on earth is to accumulate—that is, to save their profits and to reinvest them by hiring still more men to work for them; and this they do with unvarying dependability. Perhaps Ricardo's training in the dispassionate world of international finance blinded him to the variety of motives besides money-making which motivated even a nineteenth-century industrialist; but for whatever reason, his capitalists are nothing but economic machines for self-aggrandizement. But the capitalists' lot is not an easy one. For one thing, by competing among themselves, they quickly erase any undue profits which might accrue to one lucky soul who invented a new

---

[47] This quote is from Robert L. Heilbroner (1953). *The Worldly Philosophers*. Simon and Schuster. Time Reading Program printing, 1962, pages 91-95.

process or found an unusually profitable channel of trade. For another thing, their profits depend largely on the wages they have to pay, and as we shall see, this leads them into considerable difficulties.

But so far, save for the lack of realistic detail, it is not a world too far removed from that of Adam Smith. It was when Ricardo came to the landlord that things were different.

For Ricardo saw the landlord as a unique beneficiary in the organization of society. The worker worked, and for this he was paid a wage; the capitalist ran the show, and for this he gained a profit. But the landlord benefited from the powers of the soil, and his income—rent—was not held in line by either competition or by the power of population. In fact, he gained at everyone else's expense.

We must pause for a moment to understand how Ricardo came to this conclusion, for his morbid outlook for society hangs on his definition of the landlord's rent. Rent, to Ricardo, was not just the price one paid for the use of the soil, much as interest was the price of capital, and wages the price of labor. Rent was a very special kind of return which had its origin in the demonstrable fact that not all land was equally productive.

Suppose, says Ricardo, there are two neighboring landlords. On one landlord's fields the soil is fertile, and with the labor of a hundred men and a given amount of equipment he can raise fifteen hundred bushels of grain. On the second landlord's field, the soil is less fecund; the same men and their equipment will raise only one thousand bushels. This is merely a technical fact of nature, but it has an economic consequence: grain will be cheaper, per bushel, on the fortunate landlord's estate. Obviously since both landlords must pay out the same wages and capital costs, there will be an advantage to the man who secures five hundred more bushels that his competitor.

Now it is from this difference in costs that rent springs, according to Ricardo. For if the demand is high enough to warrant tilling the soil on the less productive farm, it will certainly be a very profitable operation to raise grain on the more productive farm. Indeed, the greater the difference between the two farms, the greater will be the differential rent. If, for example, it is just barely profitable to raise grain at a cost of $2.00 a bushel on very bad land, then certainly a fortunate landowner whose rich soil produces grain at a cost of only 50 cents a bushel will gain a large rent indeed. For both farms will sell their grain in the same market, *and the owner of the better ground will pocket the difference of $1.50 in their costs.*

All this seems innocuous enough. But now let us fit it into the world that Ricardo envisaged, and its unpleasant consequences will become quite clear.

To Ricardo, the economic world was constantly tending to expand. As capitalists accumulated, they built new shops and factories. Therefore, the demand for laborers increased. This boosted wages, at least temporarily, for better pay would soon tempt the incorrigible working orders to avail themselves of those treacherous delights of domestic society and so to undo their advantage by flooding the market with still more workers. But—and here is where the world of Ricardo turns sharply away from the hopeful prospects of Adam Smith—as population expanded, it would become necessary to push the margin of cultivation out further. More mouths would demand more grain, and more grain would demand more fields. And quite naturally, the new fields put into seed would not be so productive as those already in use—for it would be a foolish farmer who had not already used the best soil available to him.

So as the growing population caused more and more land to be put into use, the cost of grain would rise. So of course, would the selling price of grain, and so too would the rents of well-situated landlords. And not only rents, but wages would rise, as well. For as grain became more to produce, the laborer would have to be paid more, just to enable him to buy his dry crust and stay alive.

And now see the tragedy. The capitalist—the man responsible for the progress of society in the first place—has been put in a double squeeze. First, the wages he has to pay are higher, since bread is dearer. Secondly, the landlords are much better off, since rents have been rising on good land, as worse and worse land has got pushed into use. And as the landlord's share in society's fruit increases, there is only one class that can get elbowed aside to make room for him—the capitalist.

What a different conclusion from Adam Smith's great pageant of progress! In Smith's world everybody gradually became better off as the division of labor increased and made the community more wealthy. In Ricardo's world, only the landlord stood to gain. The worker was forever condemned to the margin, for, poor fellow, he tended to chase after every wage rise with a flock of children and thereby to compete his wages right back town to subsistence. The capitalist, who worked and saved and invested, found that all his trouble was for nothing: his wage costs were higher, his profits smaller, and his landed opponent far richer than he. And the landlord, who did nothing but collect his rents, sat back and watched them increase.

Figure 7-10 presents four system dynamics model views of the above discussion using VENSIM. To be exact, they are causal loop diagrams that show interactions and whether the effect is balancing (-) or reinforcing (+). Each view has a discrete starting event, and all accumulate money leading to the Social Standing field. When run in dynamic mode, the rate of monetary movement will have a profound effect on the final magnitude at Social Standing.

Figure 7-10a. Base system model of Ricardo's view in VENSIM.

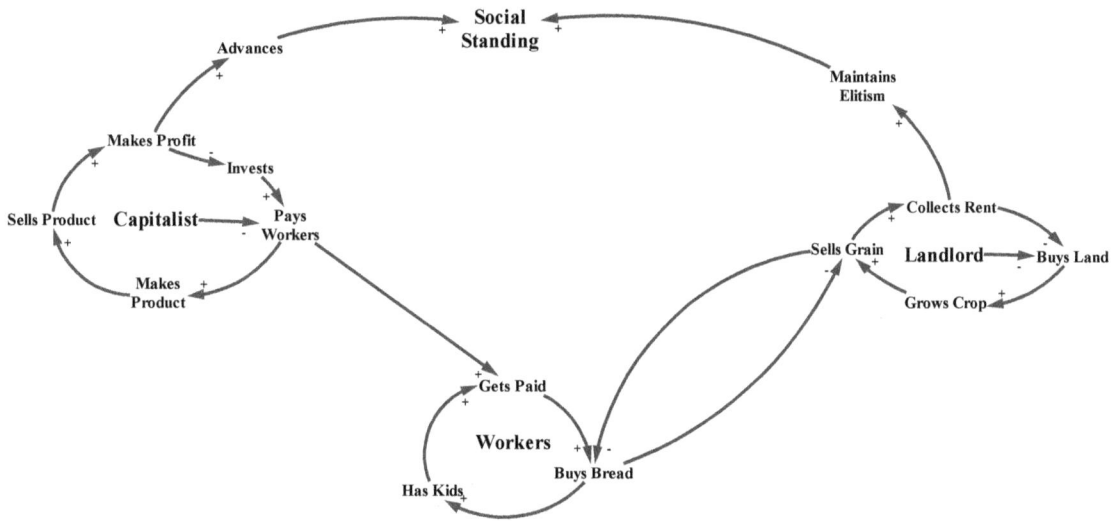

Figure 7-10b. Addition of price increase.

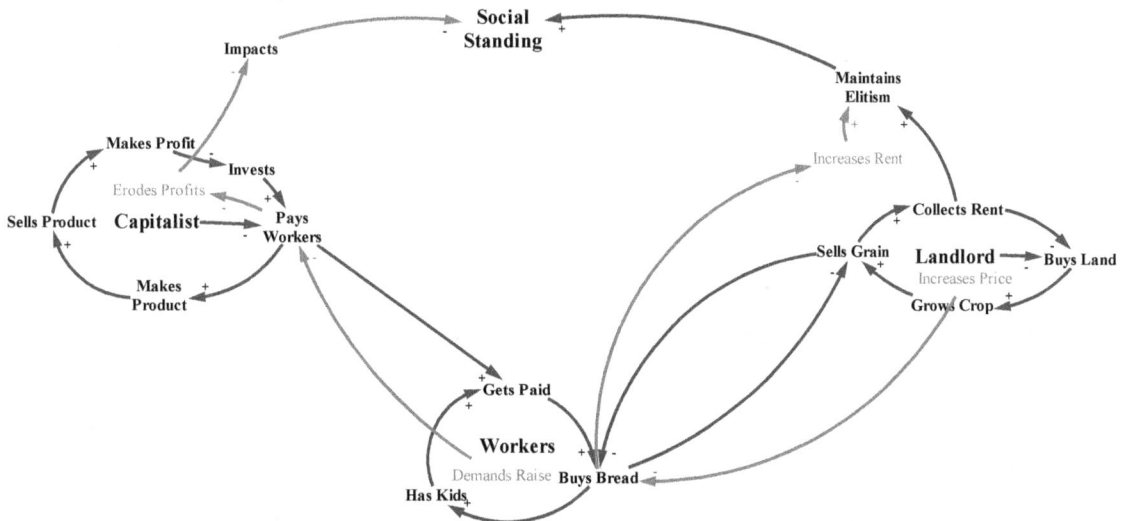

Figure 7-10c. Addition of profit sharing.

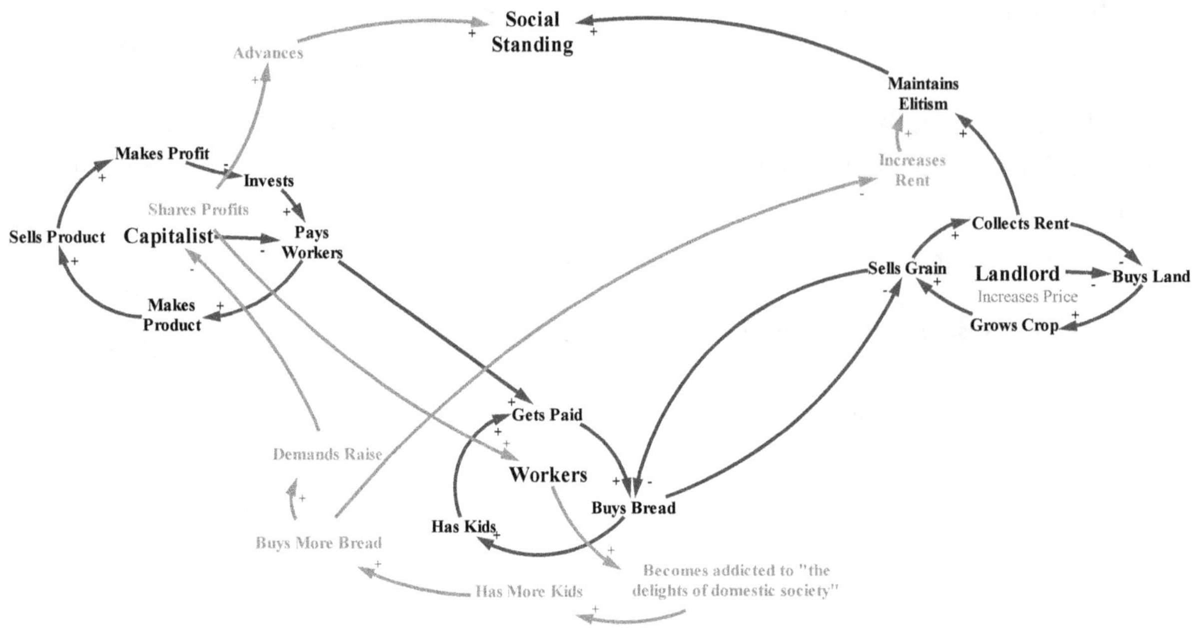

Figure 7-10d. Addition of new land owners.

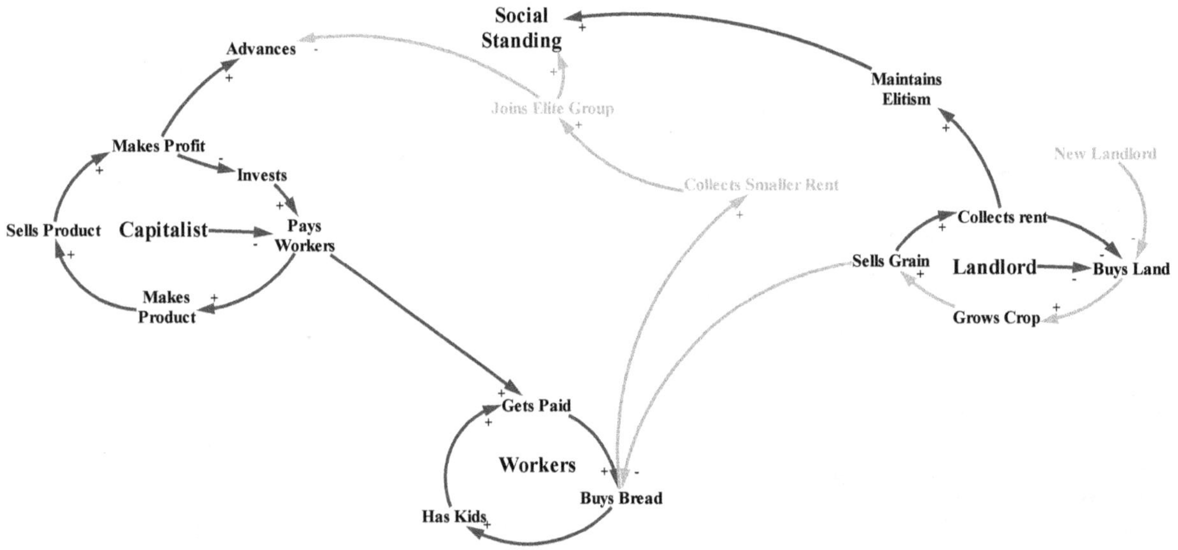

Figure 7-10a is the baseline situation when the landlord starts buying land and the capitalist pays wages – both are outflows of money. Value is accumulated at most events. The Sells Grain-Buys Grain couplet represents a decrease in stock and an outlay of money, so they are flagged with minus signs. Figure 7-10b captures what happens when the landlord raises the price for grain. This results in a direct impact to the capitalist, because the worker is forced to pay more for their bread, thus demanding a raise. Figure 7-10c details the negative feedback loop caused by profit sharing. This is a very old view that no longer applies, even though many managers seem to think so. The Millennial generation (the current one of child-bearing age)

consists of savvy decision makers with established priorities which sometimes include large families. Large families are a thing of the past as infant mortality rates are much lower now. While not totally efficient, the increase in expenditures by workers would be a one-time event and not result in the demand for a raise. The capitalist gains in Social Setting for altruistic efforts, but ends up in stress when confronted with a demand for a raise after being generous. Figure 7-10d introduces a new land owner into the mix, who then competes with the existing landowner by offering grain at a lower price. The workers benefit because of this competition, but the capitalist suffers in his Social Standing by the competition.

This suite of system dynamic models shows the tug and pull of various forces in an economic system. While deriving exact numbers would be difficult, these views show the direction of change to be expected by a certain action. It is imperative that these relationships be derived before more quantitative solutions are provided as they form a key step in the validation process.

Business models

Models have seen widespread use for business purposes since ancient times. Managers want to predict demand for the next time period so they can ensure they have stock on hand to make a sale. They also need to make a profit to invest for the future. Therefore, they want to optimize settings and have a prescription to ensure the best financial health. The classic models that have been proven apply across all sectors. While complaints are always raised that "our situation is unique" or "I don't trust it if it wasn't invented here," the reality is that the traditional mindsets of managers need to be changed and traditions broken so that standard methodologies can be adopted. Models that have been commonly deployed address:

- Cash Flow and Future Value Modeling,
- Supply Chain and Inventory Models,
- Optimization Routines that include Constraint Analysis,
- Genetic Algorithms and Rule Engines,
- Activity Based Costing,
- Zero Based Budget Building, and
- Black Scholes Equation and Hedging.

Davenport (2017) provides the following list of analytical techniques that are used in the marketing world:

- CHAID (Chi-squared automatic interaction detection),
- Conjoint Analysis,
- Econometric Modeling,
- Lifetime Value Analysis,
- Market Experiments,
- Multiple Regression Analysis,
- Price Optimization,
- Search Engine Optimization (SEO),
- Support Vector machines (SVM),
- Time Series Experiments, and
- Uplift Modeling.

This is quite a list. Each item has many software options that will aid in the use of the technique. Note that these techniques can be applied in many situations other than just marketing if the core concepts are similar. Political science and the election process is an example. Instead of expending currency on a product or service, a vote is exchanged for an appropriate candidate.

Barry Boehm's 1981 book on Software Engineering Economics introduced the Constructive Cost Model (CoCoMo). [48] This model represents a systemized approach for software development and uses heuristics to estimate the amount of labor required for various functions. It has proven to be a valid approach over the years. As software is now a big part of all that is done, vendors see model building as a profit center. They also like to pad estimates by pushing their proprietary development methodologies. Adoption of commercial-off-the-shelf solutions are typically more cost efficient than building from scratch, provided that a commitment is made to refactor data into the needed forms. Open source solutions exist as a cheap alternative, but lack rigor in performance and accountability.

A special class of models applies to network analysis. There are several existing solutions, such as petri networks and neural networks. The heart of these approaches is finding probabilistic outcomes and expected values rather than exact and prescriptive answers. This quantification of uncertainty is the heart of system dynamic models and is used to inform decision making.

There are many simulation packages that provide the user with a deep understanding of how their decisions affect the running of a company. These are various computer games such as *Zoo Tycoon, SimCity, Madden NFL* (in owner's mode rather than role-player mode), and *Wiz Khalifa's Weed Farm*. The code in each of these games is simply a model that processes the inputs selected by the user (player) in time-stepped manner.

---

[48] See https://en.wikipedia.org/wiki/COCOMO.

Appendix E is a case study of an analysis performed on the game *Final Stretch*, which is a simulation of being the owner of a racing horse. The intent of the case study was to determine the optimum settings to generate the most productivity. Instead of just changing one factor at a time, a Design of Experiments (DOE) approach was used so that the most information could be obtained from the least number of runs. While the results are fictitious, the methodology is a very powerful example of how to validate a model, to determine what the key process input variables are, and to define the expected range for results.

Model-based simulations are the experimental laboratory for system analysis. Empirical studies provide the foundation for creating, verifying, and validating the models. The goal of the experiments is to identify the key input variables, those that cause significant changes to the outputs. There is also the need to assess the costs and benefits in monetary, resource, and morale dimensions of achieving optimization. These efforts to achieve quantitative understanding of a system are the role of data scientists and a supporting team. Team size will reflect the complexity of the system and the rate of change in variables that inform the required decisions. Maturity will be recognized when discipline is shown to implement the technical solutions and suppress the popular gut decisions.

The following chapters discuss parameters that relate to human interactions and present predictions of future trends and developments. One sure thing is the next decade will see the use of quantitative analysis for system applications increase and become the key engine for decision making.

PART III

Looking to the Future

I have been impressed with the urgency of doing.
Knowledge is not enough; we must apply.
Being willing is not enough; we must do.

Leonardo da Vinci

CHAPTER **8**

Human Aspects

The first aspect for quantitative analysis is color blindness, then gender bias, then Ivy League bias, then not-invented-here bias, and so on. We all are people and quantitative analysis does not come naturally to any of us. We all must learn the skills and train our brains to have new pathways. We must also be on the defense not to be overly influenced by zeitgeist. It starts with flash cards and is reinforced if you do a Sudoku puzzle every morning at the breakfast table. It is like building arm or leg muscles – repetition and increasing the challenge leads to increased strength.

Back to color blindness…and going beyond skin color, one in nine males has a defect regarding red-green color discrimination. While most of us take traffic light colors for granted, some people are challenged. This condition must be accommodated when designing graphics. The default settings on most software packages already are accommodating. While some people are hesitant to ask a new boss if they have this challenge, I have found it is an ice breaker for show-casing my expertise.

In 2006, Thomas Davenport coined the term "analytics" in a Harvard Business Review article. He described four enabling factors: leadership, culture, people, and technology. When I teach college classes, I always assign this article in the first class and require the students to select a key factor and write a one paragraph defense of that selection. I invite you to do the same.

This chapter explores the soft side of quantitative analysis, or as recognized in Deming's Theory of Profound Knowledge, *Knowledge of People*. First is an exploration of the individual and how they maintain their skills. A discussion is provided regarding health and safety since sitting all day staring at screens is one of the deadliest professions in the modern world. Interactions with others and approaches to making decisions are presented in the last two sections.

Individuals

What makes a good data scientist? I have been asked this question many times and do not yet have a conclusive answer. The characteristics are intrinsic with being a scientist, and in having computer skills. I have been involved in multiple discussions about the Myers-Briggs personality test with the same conclusion always being reached – that the participants know about the same number of introverts as extroverts who could fill the role of data scientist. Motivation also is a factor and reflects leadership, culture, and fit. Extrinsic factors such as school, major, degree, work experience, and software mastery are not definitive in my experience.

One useful aid is the book *Now Discover Your Strengths* by Clifton and Buckingham and the related Strengths Finder assessment. Through a series of 180 question, the authors mapped responses to indicate how strong a person is in 34 categories that relate to the work place. While people might have athletic or artistic strengths, they are not assessed. The goal is not to limit or define a person, but rather to highlight his or her default approach to life. One of the strengths is Analytical – the desire to know how things work. It is a good skill to have to be a data scientist, but not definitive. Learner, not being afraid of the new, is another common strength that I have observed in successful data scientists.

These assessment results are a great tool to use as an ice breaker when a team forms, or to resolve conflict when a team storms. It is also a tool that breaks down age barriers. The commonality of shared strengths facilitates close working relationships regardless of age. Cross-generational training is important today as retiring baby boomers need to share their corporate memories with the twenty-somethings. I experienced a similar situation when I began working in the early 1980s and the World War II veterans were exiting the workforce.

A critical success factor is being a team player. Possessing statistical abilities is a specialized skill that's critical for the subject matter experts who solve problems or mature a process. It helps to be humble since it is easy to make a mistake and be required to reperform an analysis, or to receive a tranche of critical data at the 11th hour and start over. It helps to have separate leaders, scribes, presenters, database administrators, and subject matter experts as part of the team in support of the data scientist.

Computer skills are learned and are a function of time and money. How well we learn is related to imprinting. Imprinting is normally associated with goslings following their mother, or a human caretaker when they are raised in captivity. It also applies to how humans learn to use technology. Driving a car is a fitting example. Starting as toddlers, we learn to associate the red light with stopping and the green light with going. By the time we start driving, these responses are burned (hardwired) into the basal ganglia portion of our brains. So, if we are talking to someone in the car, or via a Bluetooth connection, our foot instinctively moves to the brake pedal when we see a red light. When it comes to using technology, people basically drop anchor on the first version they learn, and it is occasionally painful to learn a new system. Technology keeps moving forward requiring users to keep learning. Over time, you abandon mastering every aspect of a software package and accept a working knowledge to accomplish the task at hand.

How do I improve my skills involving data analytics?

If you have "connectedness" as a strength, there are some good books that provide a historical background:

- *The Lady Tasting Tea: How Statistics Revolutionized Science in the Twentieth Century*, by David Salsburg (2001, Henry Holt and Company)
- *The Statistical Sleuth: A Course in Methods of Data Analysis* 3rd edition, by Fred Ramsey and Daniel Schafer (2012, Brooks, Cole)
- *The Seven Pillars of Statistical Wisdom*, by Stephen M. Stigler. (2006, Harvard University Press).[49]

There are many learning opportunities for the person interested in expanding his or her skills. An internet search on "data analytics classes" yields over 19,800,000 hits. Some relate to classic courses at local brick-and-mortar institutions. Others relate to online schools, or traditional schools with online presences.

These schools are supplemented by vendor demonstrations, [50] training, and certification programs. Some allow 30-day trial versions to be downloaded[51] with some having embedded training modules or test data sets. Then there are the massively-open-online-courses (MOOC) available from vendors like Coursera, who will issue a completion certificate if you pay a fee but allows you to learn for free.

A popular data science specialization offering from Coursera (as evidenced by over 1.76 million registrations) is a sequence of nine classes offered by three professors of biostatistics

---

[49] They are Aggregation, Information, Likelihood, Intercomparison, Regression, Design, and Residual.

[50] https://tabsoft.co/2nomMnf.

[51] https://bit.ly/2npO1hv.

from the Johns Hopkins Bloomberg School of Public Health.[52] These classes use R and other open source packages to teach programming skills. Lectures are fast-paced and information-dense. Homework, quizzes, projects, and peer reviews all help to reinforce the concepts. The classes are described in Table 8-1.

Table 8-1. Data science course sequence available through Coursera.

Course	Observations
The Data Scientist's Toolbox	The accompanying book, *The Elements of Data Analytic Style*, provides a concise summary of all modules. However, they do not discuss the concept of "data ownership" and the security concerns from business and government about protecting and curating all data types.
R Programing	Hands-on demonstration and exercises with the software package. Available e-book is a useful reference.
Getting and Cleaning Data	The concepts presented are universal. While the examples are in R, the functions are seminal for any data project and can be performed in many other packages. It is a great module for understanding the workings of many modern data sources. However, verification and validation are not emphasized enough nor the "Mediation of the 10,000 Mouse Clicks" that is frequently used when data are only obtainable by cut-and-paste from a PDF table.
Exploratory Data Analysis	The first half covers the classic review tenants of assessing the data from practical, graphical, and analytical points of view. The second half is an advanced discussion of multivariate analysis that novices might find challenging.
Reproducible Research	This module focuses on the basics tenants of a scientific study and the commonalities of all computational endeavors. The case studies are quite interesting…even cringe-worthy at times.
Statistical Inference	The lectures are well-done and modern. This material must be mastered if you desire to be called a *data scientist*. The e-Book, *Statistical Inference for Data Science*, is informative and focused.
Regression Models	This module covers the work horse tools of linear regression and ANOVA. It is a detailed statistical treatment and software use is necessary to master the concepts.
Practical Machine Learning	This module covers cutting edge predictive models. The lectures after Week One are intended for the advanced learner. The class uses the caret package that runs in R. In use, this is the realm of professional grade software because security concerns trump cost of software. Major vendors are SAS, SPSS, and Salford Systems with some database companies (SAP, Oracle) having embedded capabilities.
Developing Data Products	This module presents various tools and techniques that build dynamic and interactive visualization products. This is one area where the suite of R products provides an easy and seamless solution.

---

[52] See https://bit.ly/2OmF6ZG.

Where can I continue to learn about analytics?

Newspapers like the *New York Times* and *The Economist* are leaders in using new visualization tools. Amazon adds new books on the subject every week. Many blogs exist (such as SimplyStatistics and StackOverflow) that have running commentary on the newest tools and techniques as well as problem solving help. Vendors provide numerous webinars regarding new releases and training materials specific to unique capabilities. The professional society INFORMS is a major information interchange and even hosts a certification credential program. Their free e-zine, *Analytics*, is quite informative. So, the main resource to learning is to be an inquisitive reader and try to implement prototypes in a progressively evolving manner. An up-and-coming approach is being delivered by 2U, a firm that has partnered with leading universities to provide distance education opportunities. They also have partnered with WeWork for continuing education certificates and the Flatiron School that offers a programming boot camp for those who need to master a new language quickly.

Table 8-2 is a list of supplemental readings that provide fundamental knowledge for being a data scientist. They focus on how to think. They are grouped into the themes from Deming's System of Profound Knowledge that was introduced in Chapter 1. For the philosophically inclined, these concepts are broadly contained in the branch known as *epistemology*. The reader can find a lifetime's worth of study material just by following the links in Wikipedia for this entry.

Table 8-2. Seminal readings for the data scientist.

Theme	Author	Title
Knowledge of Systems	Aristotle	*Categories*
	John N. Warfield	Twenty laws of complexity: science applicable in organizations. In *Systems Research and Behavioral Science* Vol. 16, pp. 3-40, 2000.
	John Godfrey Sacks	*The Elephant Poem*, 1872.
Knowledge of Variation	Stanley Kaplan and B. John Garrick	On the Quantitative Definition of Risk In Risk Analysis, Vol. 1, pp. 11-27, 1981.
	R.A. Fisher	*On the Mathematical Foundations of Theoretical Statistics.* 1921.
	G.E.P. Box	An Apology for Ecumenism in Statistics. In *Scientific Inference, Data Analysis, and Robustness*, pp. 51-84, 1983.
	Isaac Asimov	*Franchise*, 1955. Foundation series starring Harry Selden.

Theme	Author	Title
	Jared Diamond	*That Daily Shower Can Be a Killer.* New York Times, January 28, 2013.
Knowledge of Knowledge	Saunders Mac Lane	Mathematical Models: A Sketch for the Philosophy of Mathematics. In *The American Mathematical Monthly*, Vol. 88, pp. 462-472, 1981.
	Vannevar Bush	*As We May Think.* The Atlantic, 1945.
	Herbert A. Simon	Rational Choice and the Structure of the Environment. In *Psychological Review*, Vol. 63, pp. 129-138, 1956. [Satisficing]
	Herbert A. Simon	The Structure of Ill Structured Problems. In *Artificial Intelligence*, Vol. 4, pp. 181-201, 1973.
Knowledge of People	George A. Miller	The Magical Number Seven, Plus or Minus Two: Some Limits on Our Capacity for Processing Information. In *The Psychological Review*, Vol. 63, pp. 81-97, 1956.
	Herman Chernoff	The Use of Faces to Represent Points in k-Dimensional Space Graphically. In *Journal of the American Statistical Association*, Vol. 68, pp. 361-368.
	Daniel Keyes	*Flowers for Algernon, 1958.*
	Plato	*The Republic.*[53]

The table is a sampling of readings from the Great Conversation had by philosophers since Plato in 500 BC up to the 21st Century. It is neither an exhaustive treatment nor a step-by-step cookbook. They are just suggestions. Each reading was selected to bend (or, as my non-quantitative friends say, warp) the data scientist's mind so that massive amounts of data can be quickly processed into a strong signal that presents actionable knowledge to decision makers.

Health and safety

The person is the most important part of any quantitative analysis, but much more attention is paid to software selection and cyber-security issues than to the well-being of the data scientist.

---

[53] In the 21st Century, we don't burn campfires very often, nor do we live in caves. Instead, we sit on the couch and watch cable TV. Pick a channel at random and describe what perceptions of the world would result if "the Cave" was recreated and that screen replaced the shadows on the wall.

Ergonomics

The human-machine interface often is ignored. There are many websites that describe the correct posture and angles for working on a computer. Check them out. While convenient for working anywhere, laptops cause people to contort in many uncomfortable ways. Money must be spent to customize a set-up. The insoles I buy for my trainers cost more than the plastic laptop stand and wireless keyboard I use. All of the components fit in my roller bag so I can be mobile. Another aspect of ergonomics is stretching, especially of the wrist that holds the mouse.

Movement

Sitting is killing us! Humans are not made to sit and stare at a monitor all day. We must move. The following excerpt is from the book, *Ready Player One,*[54] and describes how Wade expanded his life from being a keyboard monkey before he could transform into Parzival every day. It is a lesson for all of us.

I usually got a little exercise while logged into the OASIS, by engaging in physical combat or running around the virtual landscape on my treadmill. But I spent the vast majority of my time sitting on my haptic chair, getting almost no exercise at all. I also had a habit of overeating when I was depressed or frustrated, which was most of the time. As a result, I'd gradually started to put on some extra pounds. I wasn't in the best shape to begin with so I quickly reached a point where I could no longer fit comfortably in my haptic chair or squeeze in to my XL haptic suit. Soon, I would need to buy a new rig, with components from the Husky line.

I knew that if I didn't get my weight under control, I would probably die of sloth before I found the egg. I couldn't let that happen. So I made a snap decision and enabled the voluntary OASIS fitness lockout software on my rig. I'd regretted it almost immediately.

From then on, my computer monitored my vital signs and kept track of exactly how many calories I burned during the course of each day. If I didn't meet my daily exercise requirements, the system prevented me from logging into my OASIS account. This meant that I couldn't go to work, continue my quest, or, in effect, live my life. Once the lockout was engaged, you couldn't disable it for two months. And the software was bound to my OASIS account, so I couldn't just buy a new computer or go rent a booth in some public OASIS café. If I wanted to log in, I had no choice but to exercise first. This proved to be the only motivation I needed.

---

[54] Permission requested.

While we are not quite that interconnected to our machines today, when you are a data scientist you need to mindful to get your daily steps in. A Fitbit is a useful tool.

Mental centering

Where do you work? Home, office, coffee shop, or down the hall in the computer lab are usual places. What cues help you be productive? Common ones are time of day, emotional state, sleep state, presence of other people, immediately preceding activity, spearmint chewing gum, and chemical intake.

How do you take mental breaks? Many days I dink around with a Sudoku puzzle from the morning paper. For two decades now, I have had a MyYahoo portal that shows headlines, email, stock prices, and sports scores and I glance at that when I stop and start an analysis session to help me change gears. When an analysis takes a month, these actions are required to maintain a healthy engagement. Just remember to design your work environment so that it creates a mindful atmosphere for you.

Elicitation

Unless you are a day trader, you will have to interact with others to acquire data. Collecting data is referred to as *elicitation*, which is defined as "to call forth or draw out (something, such as information or a response)."[3] The main output from the elicitation stage of an IT project is the list of requirements. Appreciative inquiry is a design principal that frames the question to influence the response. While not great for brainstorming, it is an aid to making interviews take less time. There is an art to uncovering all the tacit knowledge that a person has about nuances and exceptions regarding the processes they own. These special cases are the critical knowledge that must be incorporated into the design of "the Matrix" in order to assure success.

Managerial proverb: you cannot experience trust without first developing a relationship. So, how does one develop a relationship? By listening. This same skill applies to elicitation. The data provider is going to be defensive at the start, no matter how transparent you have been. In 2009, when my grandfather was 99, he still complained about the piece-work assignment he had in the factory he worked in during World War II, when the Time-Motion study guys came to re-standardize his process time in response to his having a good shift when he could beat the standard. This new standard time was then applied to everyone else in his department. My grandfather was a rather laid-back person, except for that grudge he harbored for 65 years when he was in a Catch-22 situation. It has been my career goal to never give anyone the reason to hate me for 65 years! I always try to preserve the person's dignity and assume that he or she is doing the best they possibly can. I also work up in my mind the answers to the following questions that are legitimate concerns of all process owners:

Why are you digging into…

- My data?
- My job?
- My business?
- My life?

I start with simple situations to showcase the tool so he or she will understand the methodology before I collect the data or build the model. That way I can sell them on what benefit we would gain, and that I have no preconceived ideas as to what the results will be. I then sit with them after the first successful run so that he or she can validate the results and ask follow-up questions that refine the analysis. I also encourage them to be the presenters to management, a true sign that they take ownership of the results. I still view myself as an offensive lineman, rather than a quarterback. I am happy seeing a job well done and the objective achieved, rather than being the glory hound who hogs all the credit. The real-world is always more important than the virtual one I create.

The key to success is always respecting human dignity. There are many ethical tenants of organized religions that can be applied. Christianity provides the Golden Rule – do unto others as you would have them do unto you. Jainism follows three jewels to achieve the goal of liberating the soul. The rules are:

- *Samyak Darshana* (right perception) – attempting to perceive the truth clearly without being swayed by superstition or prejudice,
- *Samyak Jnana* (right knowledge) – having accurate knowledge of the universe and scripture, and the mental attitude to use this knowledge, and
- *Samyak Charitra* (right conduct) – to live according to Jain ethics, and avoid doing any harm to other living creatures.

The goal of building any model that will be used to support decision making is to truthfully portray the situation. This can only be achieved by collecting high fidelity information from the process owners, who are busy, afraid that the model will replace them, and unmotivated to see you succeed. Therefore, building a rapport that leads to a trusting relationship is the key step of elicitation. The above examples are just the tip of the iceberg and you must develop your own approaches that fit your style and strengths.

Decision making under risk

Figure 8-1 presents a classic problem. What do we do first when we exit the highway for the evening? How is the decision made? What are the consequences and the corresponding probability that they will occur – the risk. The dictionary defines risk as "the chance of loss or the perils to the subject matter of an insurance contract; also: the degree of probability of such loss." The definition is refined by many disciplines with each having a preferred quantitative approach and scoring system. The commonality is that all risk assessment processes require data and utilize a computational methodology. Many times, models are run to give a range of probabilities using what is known as an *ensemble modeling* approach. Commonality of answers is good. Diversity of answers is good. The interpretation is a very nuanced art.

Figure 8-1. What do we do first?

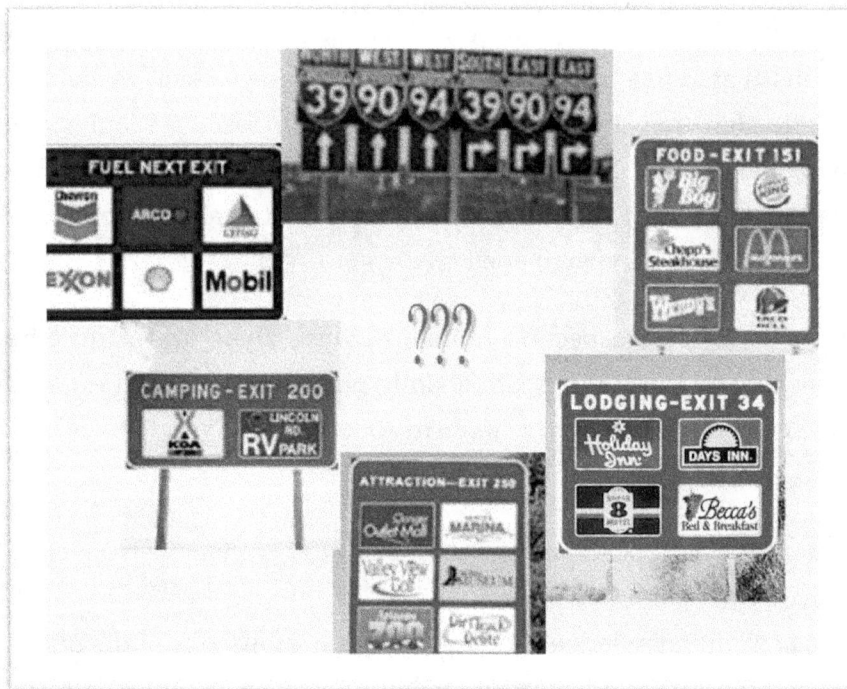

There is a dilemma in decision making, because pure and complete information is always assumed to be available. That is rarely the case. You must remember that a secret is knowable, while a mystery is unknowable and make that determination early in the process. Models are great tools to uncover secrets and can shed light on mysteries. Typically, qualitative and semi-quantitive methods are used when mysteries are encountered under a different approach – decision making with uncertainty.

The authors of the book, *Now Discover Your Strengths*, talked about the characteristics of an individual. A similar discussion can be made for an organization, using the term "culture." Ruth Benedict was a pioneer in this analysis with her 1938 book, *Patterns of Culture*. The focus of this discussion is the use of quantitative tools and decision making and how it is influenced by the culture(s) involved with the decision. This is nothing new. Caesar always consulted with the diviners before making a decision so that "the will of the gods" could be considered. The first categorization of an organizational culture regards how problems are solved: does management fully study the possible alternatives, or do they give in to the urge to reach a quick conclusion. Managers who consult modelers fall into the first camp.

Hofstede and others (2010) provide an international perspective and classify cultures across six dimensions:

- Power Distance (small to large),
- Collectivism versus Individualism,
- Femininity versus Masculinity,
- Uncertainty Avoidance (weak to strong),
- Orientation (long-term to short-term), and
- Indulgence versus Restraint.

As with so much in our world, technology changes are only adopted if they make sense to maintain the status quo of the culture of the people themselves. People do not want to be replaced by Artificial Intelligence. However, people will gladly accept a change that makes their life easier. I, for one, will never go back to using a landline telephone after tasting the freedom my cell phone has given me. Analytics will be absorbed into all cultures someday when the benefits become obvious to all stakeholders.

Formal social theories can be applied to the use of data-driven decision making. The processes deployed represent the culture of the group itself that result in order, chaos-reduction, normalizing roles, and social organization. Four main theories, and a bounding function, are introduced below. Just remember, technological advances can be hampered by cultural issues.

- **Secularization Theory**. Involves identification of the central functions of a group. Conflict is a normal part of the interactions between two groups: (1) small, geographically concentrated homogeneous tribal units; and (2) large, geographically-dispersed, heterogeneous national units that cause sharp differentiation of decision making approaches. See Martin (1978) for a deeper discussion.
- **Rational Choice Theory**. Is rooted in economics and free markets. In large heterogeneous environments, diversity in thinking is present and leads to competing

theories of selection and minimization of partiality in decision making. See Anand (1993) for a deeper discussion.

- **Practice Theory**. Assumes action is guided by experience, by what is familiar, and includes contingencies. A practice is developed by the activities social units experience over time that involves learned skills, habits, predispositions, and interpretations that reflect the unit's values. See Bourdieu (1990) for a deeper discussion.

- **Multilevel Theory**. Involves building models of behavior that address both micro and macro realms. Models consist of clusters of variables from each level and are designed to explain and describe complex interactions resulting from contingencies, constraints, and situations that cannot be defined by single variables, no matter how theoretically-rich or deductively-logical. Farming practices are one of the classic examples of where this theory can be applied. See Kozlowski and Klein (2000) for a deeper discussion.

- **Symbolic Boundaries**. Are categories that delineate patterns of social behavior. Boundaries can be reinforced, clarified, defined, or dramatized through narratives, collective rituals, and storytelling. These actions explain why boundaries exist, how people should behave, and why the categories are or are not legitimate. See Wimmer (2008) for a deeper discussion.

But human aspects of the decision maker can override culture. First, the amygdala is involved – the emotional center of the brain that can override the rational center. Second, biases exist such as those shown in the list of ~185 that can be found on Wikipedia.[55] Third, funding can be pulled by whoever controls the gold. All these paths eventually lead to a form of conflict. Decision making has never been for the weak.

Scoring is a straightforward way to make a decision and to avoid risk. It is commonly used for subjective categories and when the consequences of a mistake are low – like in reality TV. The results from an episode of the TV show *Iron Chef America* are shown in Figure 8-2 with the score resulting from three judges assessing three categories: one worth ten points and two worth five points. It is easy to distinguish a winner once the points are tallied. Alternatively, the scores could be averaged, and the final rating would be normalized back to the original point scale. Another option would be to weight the importance of each judge (either with an integer or a fraction) to generate a more nuanced outcome. It works as long as the judges are consistent and use the whole grading scale so that discrimination can be obtained.

Risk models build on the scoring approach because categories, hierarchies, weightings, and relative comparisons are combined along with more intensive mathematical treatments. Figure 4-3 represents a simple risk model built to answer a specific question: what are my chances of

---

[55] https://en.wikipedia.org/wiki/List_of_cognitive_biases.

making an "A" in this class? It is a weighted additive model that is useful and easy to explain. While it is fascinating to explore all the high end mathematical approaches possible, like multiplicative models, normalization, nonlinear functions, and feedback loops, they do not improve the precision associated with such a simple case. They also tend to obfuscate the message when it becomes hard to explain the complex behaviors that result when input terms are adjusted. Simple approaches are best.

Figure 8-2. A simple scoring grid example.

Case Study: Leadership selection data set

One of the toughest decision-making situations is selecting personnel to become the future leaders in an organization. It is complicated by the need to keep records showing there was no bias against, or partiality for, select groups. An example of the data challenges is presented. The data file can be found at https://technicspub.com/qasa and consists of 15,480 rows of data in normalized form for 24 people participating in 16 evaluations against five attributes using a 5-point scale as assessed by eight current managers. The file includes a pivot chart allowing drill down for each evaluation or for each person. The resulting scores are also an example of the realm of large numbers as no one evaluation has a large influence on the aggregate score. The measurements are on a 5-point scale and carried to the second decimal point when averages are calculated. Since each candidate theoretically can have up to 128 ratings for each of the five attributes, any one-point change in an evaluation only causes a 0.8% movement of the final value for that attribute, or 0.15% if all five attributes are combined. This approach minimizes the impact of extreme values.

The display of these results can also be a challenge. Figure 8-3 is a spider web diagram, or a radar plot, that allows comparison of the final attribute score for the 24 candidates. Extreme strengths and weaknesses are visible, as well as balance across attributes. In the end, these data were used to inform the final decision that relied on future-looking judgement from the experienced managers.

Figure 8-3. Spider web view of evaluation results for 24 candidates across five attributes.

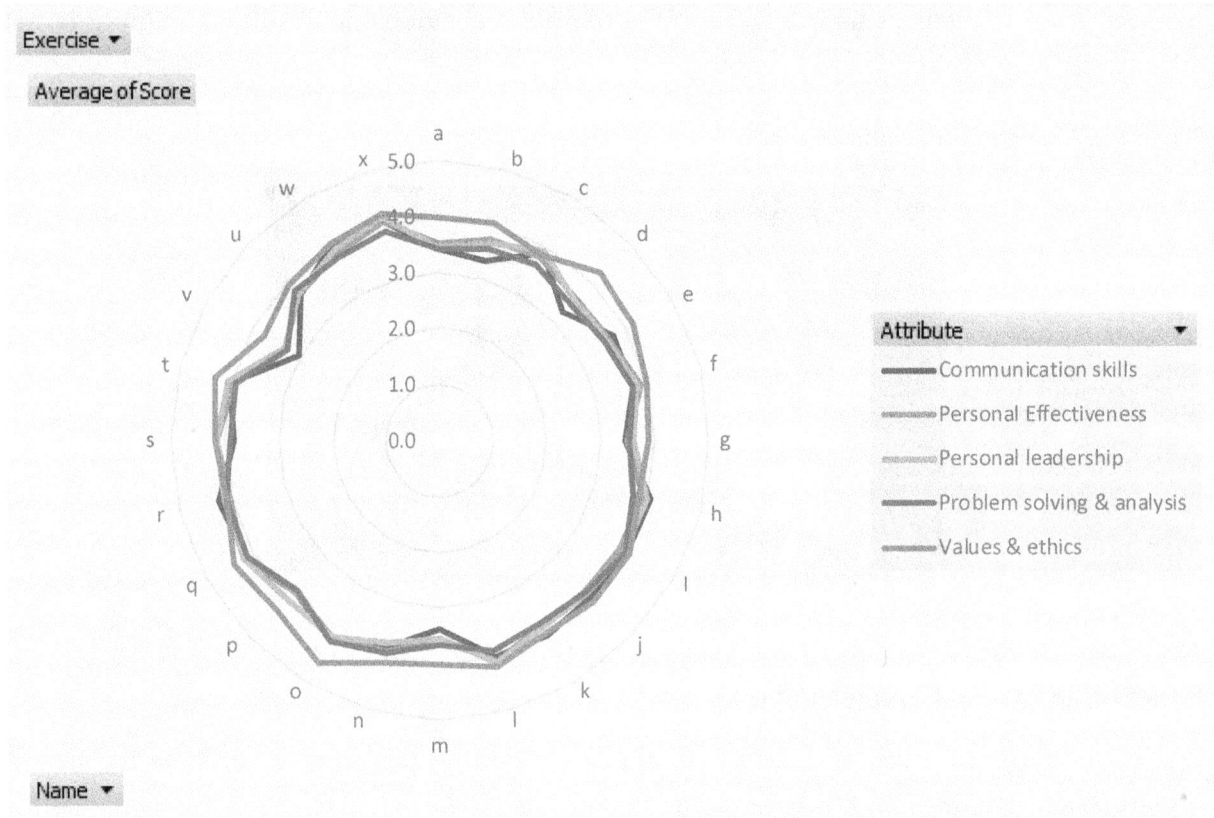

In all models, we make assumptions about the input terms. Sometimes these assumptions do not hold and we commonly say "we had a blowout." For example, we can assume that a value will always be less than 10. What happens if we get an 11? Typically, the model will run and we might or might not catch it, especially if it is a calculated value from other input data. It is better to automate a solution so that the model stalls and an analysis is needed to determine why, like "opening the hood." Here are three lines of logic code that captures this situation:

$$Blowout\_Count = 0$$
$$If\ x > 10,\ Blowout\_Count = 1$$
$$Add\ model\ term\ of:\ 1\ /\ (1 - Blowout\_Count)$$

These lines of code just simply set a flag and forcing a division by zero that will force the program to give an error message. This coding is an example of *poke yoke* – another Japanese term and means design for mistake proofing. Given all the moving parts involved with big data, while data screening and testing are necessary, they are not always sufficient to find all of the causes of failure. While more elegant programming approaches are available, this approach can be programmed into any syntax being used. This example also highlights the difference in how engineers think differently than marketers. Engineers focus on the micro view: what settings to a valve or a pump in a refinery could cause an explosion. Marketers focus on the macro view: they are happy to influence 30% of the 10,000,000 customers rather than 30.1% because those additional 10,000 customers are part of the measurement error of the effects of their tactics.

Another approach for managing risk involves probability trees. The data from Figure 6-5 is presented in a tree format in Figure 8-4. If there are two "yes" answers, the probability computes as 0.8 * 0.75 = 0.6. Note that the system concept of holism is preserved because the sum of the final probabilities equals unity. It is also valid to calculate ratios based on these final probabilities. This technique is applied in high-risk situations (like at nuclear power stations) using fault trees and stage gate diagrams. While computationally complicated because of the many variables, the complexity is similar to this probability tree. More complexity is encountered when non-linear relationships exist, or when feedback loops erode the independence of variables. In those cases, detailed procedures are followed so that believable numbers are generated.

Figure 8-4. Probability tree for two factors.

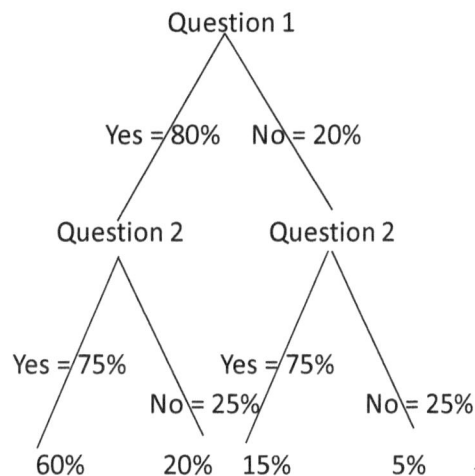

Remember, software has computational limits. When using Excel, you can only have 21 binomial terms before the computational precision limit of 15 significant figures is reached, shown in Table 8-3. Note that the likelihood values for factors 22 and 23 end in zero. This indicates that a buffer was exceeded and the 16th figure was truncated since it is a mathematical fact that a 5 multiplied by a 5 always results in another 5. This truncation of precision can be a problem in big data that uses probability models. Therefore, a calculation of one leg needs to be performed by hand just to validate that the software selected is adequate for the task at hand. A healthy skepticism is always required.

Table 8-3. Example probability tree calculations.

Factor	Chance	Likelihood
1	0.5	0.5
2	0.5	0.25
3	0.5	0.125
4	0.5	0.0625
5	0.5	0.03125
6	0.5	0.015625
7	0.5	0.0078125
8	0.5	0.00390625
9	0.5	0.001953125
10	0.5	0.0009765625
11	0.5	0.00048828125
12	0.5	0.000244140625
13	0.5	0.0001220703125
14	0.5	0.00006103515625
15	0.5	0.000030517578125
16	0.5	0.0000152587890625
17	0.5	0.00000762939453125
18	0.5	0.000003814697265625
19	0.5	0.0000019073486328125
20	0.5	0.00000095367431640625
21	0.5	0.000000476837158203125
22	0.5	0.00000002384185791015620
23	0.5	0.0000001192092895507810

There is also a handy approach for working with closed systems, or when the number of elements is so large, that holism is not required to make a profit. This approach is called the *Naïve Bayes Algorithm* with the calculations shown in Figure 8-5. [56] The technique was developed by Reverend Bayes to help him adjust odds while playing cards and was published in 1763. How you set it up is the key because the wording of the questions is the critical step. All statistical textbooks cover this concept in depth.

As the technological world becomes ever more complicated, it is necessary for humans to use technology as an aid for analysis. Machines, while capable of making decisions for complicated conditions, such as parallel parking between stationary objects, will never make complex

---

[56] Emmanuelle Rieuf. 6 Easy Steps to Learn Naïve Bayes Algorithm (with code in Python). Article posted to Data Science Central group of LinkedIn on September 3, 2017. https://bit.ly/2KCZp2E.

decisions for open systems. The main reason is machines will not be able to balance the inherent confusion between the priorities set by different cultures and that consensus decisions are, by definition, suboptimal.

Figure 8-5. Naïve Bayes example.

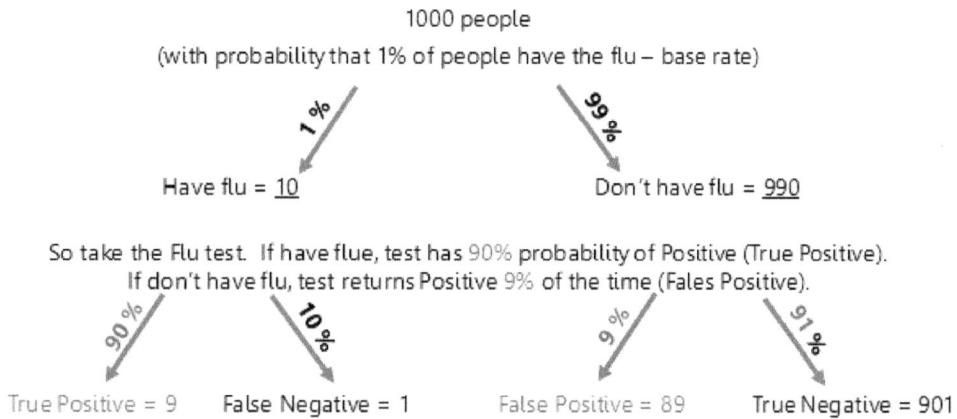

1000 people
(with probability that 1% of people have the flu – base rate)

1%   99%

Have flu = <u>10</u>   Don't have flu = <u>990</u>

So take the Flu test. If have flue, test has 90% probability of Positive (True Positive).
If don't have flu, test returns Positive 9% of the time (Fales Positive).

90%   10%   9%   91%

True Positive = 9   False Negative = 1   False Positive = 89   True Negative = 901

The probability that you have the flu given you tested <u>Positive</u> =

$$\frac{\text{\# of people who have flu and tested True Positive (9)}}{\text{\# of people who have flu and tested True Positive (9) + \# of people who don't have flu and tested False Positive (89)}}$$

9 % probability you have flu

The focus of the next chapter is futuristic and addresses the latest developments and trends regarding analytics and further fusion with big data in "the Matrix."

CHAPTER 9

Newest Developments

A cynical view of the future would predict only doom and gloom when analytics becomes more widely deployed. There are disturbing reports, like Target determining a teenage girl was pregnant before she had told her dad.[57] It got me thinking about what to call this next phase. The Gartner Hype Cycle[58] has its "valley of despair" for newly deployed technology, to which I gave some consideration. When I was flipping channels with the remote one evening, I stopped on a Star Wars movie and that is when it came to me: *Big Data Strikes Back*. Then I thought about the phase that will come after that and thought it could be called *Return of Big Data*. And then my brain went full nerd when I concluded that Darth Vader is the perfect poster child for big data. For both Darth and big data, it can be said, that they can be turned to the dark side, but deep down, there is still good present. This theme is explored below: how big data can be elevated to solve problems at the next level.

This chapter explores four futuristic themes: the internet of everything, machine learning as a segue to Artificial Intelligence, organizational challenges, and how organizations should be structured to leverage their strengths. The first step for designing the future is to have a set of rules to follow as shown in Table 9-1.

Table 9-1. Rules to govern design and deployment of analytics initiatives.

Rule #	Description
1	Order is Heaven's First Law!
2	Everything changes all the time!
4	Change is the father of twins: Progress and Problems!
5	What does the contract say?
6	Simpler is better, simplest is best!
7	Trust the network, but save frequently because you don't trust the network!
8	Always preview before printing!
9	Always inspect every data field after each and every change or download!

---

[57] Charles Duhigg, "How Companies Learn Your Secrets." *New York Times*. February 16, 2012.

[58] https://gtnr.it/2mo1fgT.

Rule #	Description
10	If a file does not exist in two places, it does not exist! (Simply email a working file to yourself at the end of the day to make replicates on multiple systems).
11	Avoid the common deceptions of data displays such as scale, aspect ratio, ungrounded ratios, and placement!
12	It takes more than one sitting to process big data.
13	Trust, but verify!
14	The elegant solution does not just answer the question, it dissolves the question, so you wonder why it was ever asked!
15	In any hierarchy of classification, the most important division is the first one!
16	Test everything. Keep all that is good!

Following these rules will not always guarantee success but ignoring them will almost certainly lead to failure.

What's next in technology?

There are always new software tools being released by vendors. A sampling of the cutting-edge products is shown in Table 9-2. It is essential to know what the packages do and how the capabilities are achieved. Sometimes it is helpful to know what they are replacing. This knowledge allows for better decisions regarding design parameters and technology refreshment triggers. If you have questions, there is an army of sales people ready to chat with you all day. The author of this table did not list JMP Pro from SAS that is becoming a major support package for data scientists.

Table 9-2. List of current cutting edge big data tools.[59]

Tool	Description
Avro	It was developed by Doug Cutting and used for data serialization for encoding the schema of Hadoop files.
Cassandra	Is a distributed and Open Source database. Designed to handle large amounts of distributed data across commodity servers while providing a highly available service. It is a NoSQL solution that was initially developed by Facebook. It is used by many organizations like Netflix, Cisco, and Twitter.

---

[59] "18 Big Data tools you need to know!" Posted by Sandeep Raut on May 27, 2017, to the Data Science Central group of LinkedIn.

Tool	Description
Drill	An open source distributed system for performing interactive analysis on large-scale data sets. It is like Google's Dremel, and is managed by Apache.
Elasticsearch	An open source search engine built on Apache Lucene. It is developed on Java, can power extremely fast searches that support your data discovery applications.
Flume	Is a framework for populating Hadoop with data from web servers, application servers, and mobile devices. It is the plumbing between sources and Hadoop.
HCatalog	Is a centralized metadata management and sharing service for Apache Hadoop. It allows for a unified view of all data in Hadoop clusters and allows diverse tools, including Pig and Hive, to process any data elements without needing to know physically where in the cluster the data are stored.
Impala	Provides fast, interactive SQL queries directly on your Apache Hadoop data stored in HDFS or HBase using the same metadata, SQL syntax (Hive SQL), ODBC driver and user interface (Hue Beeswax) as Apache Hive. This provides a familiar and unified platform for batch-oriented or real-time queries.
JSON	Many of today's NoSQL databases store data in the JSON (JavaScript Object Notation) format that's become popular with Web developers.
Kafka	Is a distributed publish-subscribe messaging system that offers a solution capable of handling all data flow activity and processing these data on a consumer website. This type of data (page views, searches, and other user actions) are a key ingredient in the current social web.
MongoDB	Is a NoSQL database oriented to documents, developed under the open source concept. This comes with full index support and the flexibility to index any attribute and scale horizontally without affecting functionality.
Neo4j	Is a graph database and boasts performance improvements of up to 1000x or more when in comparison with relational databases.
Oozie	Is a workflow processing system that lets users define a series of jobs written in multiple languages, such as Map Reduce, Pig, and Hive. It further intelligently links them to one another. Oozie allows users to specify dependencies.
Pig	Is a Hadoop-based language developed by Yahoo. It is relatively easy to learn and is adept at very deep, very long data pipelines.
Storm	Is a system of real-time distributed computing, open source and free. Storm makes it easy to reliably process unstructured data flows in the field of real-time processing. Storm is fault-tolerant and works with nearly all programming languages, though typically Java is used. Descending from the Apache family, Storm is now owned by Twitter.
Tableau	Is a data visualization tool with a primary focus on business intelligence. You can create maps, bar charts, scatter plots and more without the need for programming. They recently released a web connector that allows you to connect to a database or API thus giving you the ability to get live data in a visualization.
ZooKeeper	Is a service that provides centralized configuration and open code name registration for large distributed systems.

What will the future bring?

One of the biggest areas of change will result from studies of how the brain operates. Arlindo Oliveira[60] provides insightful glimpses into the future of progress that is being made, and that the tipping point has been passed in the realm of machine learning and artificial intelligence. Many 20-year-olds in today's workforce do not remember a time without computers. Computers have been around for his or her entire life. They think, and process information differently than I do. And then there are the toddlers I see in the shopping baskets at the supermarket playing on mama's smart phone. (My niece's 18-month old downloaded two e-books for her at the supermarket the other day, much to her surprise.) I cannot imagine what their work day will be like when they reach retirement age.

Vendors have much to say about the future. Some firms are stripping features to lower their costs because they have conceded that it is hopeless to duplicate all the features of Excel. They simply allow a back-door dump of a .csv file so that quantitative analysis can be performed on bulk data. Others are gearing up and including all features of Excel for either standalone or Analytics as a Service business models. Sometimes, they even supply the personnel to run the software. Think of it like taking your car to the transmission shop when you need a specialized mechanic and his unique tool box.

Shakespeare's *The Tempest* offers an apt metaphor, in which the spirit Ariel sings of a sea change – a dramatic and profound transformation. This idiom, while much like the 20th century successor, paradigm shift, succinctly describes the massive upheaval and ever-shifting landscape of customer needs and demographics, innovation processes, and regulatory environments. These changes apply to healthcare, TV and movie watching, shopping, listening to music, setting the thermostat at home, and most everything else with which we interact daily. The sea change will be even more profound in the work place when all inputs become plugged into "the Matrix." It will also bring security challenges, like when you buy a smart house that may or may not still be connected to the seller's phone.

Internet of Everything

GE's definition of the Internet of Things is too narrow. If I was buying a used aircraft engine, I would want to know more about its operating conditions. Things like how many West Texas

---

[60] *The Digital Mind: How Science is Redefining Humanity.* 2017 MIT Press.

sandstorms did it fly through, or how much volcanic ash was encountered that resulted from volcanic eruptions that littered the stratosphere with ash. I would want to know everything.

There is a critical need to have the external data to support model building, and maybe even a blockchain solution for maintenance and production lots. The same concern applies to oilfield equipment and whether it was used in Midland, Texas, or off-shore in the Gulf of Mexico, or on the North Slope of Alaska, to fairly assign value before being purchased for use on a Utica Shale well. You must know what the operating conditions were as a way to limit noise in any model building effort. This situational awareness will also be applied to people in the forms of surveys or, as in the case of Stich Fix (the clothes shopping service), your rating of the clothes you like and the type of clothes that you return because you dislike them, or they don't fit. Feedback is essential to validate that the right things are being accomplished.

It is estimated by the early 2020s, there will be over 40 billion devices talking to the internet – that will be five for each of the eight billion people on earth – almost all with a smart phone. Even infants will have their own number for their baby monitor app that can be accessed by parents at anytime from anywhere. Some of the machines will only talk to other machines. Some will be attached to us like Fitbits but instrumented to track dozens of biometric parameters. There might even be a chip embedded in my hand that will unlock my car and let me pay for purchases I make. And, thinking like a CIO again, how will all those transactions and consequent data be secured when they are stored and who will have access to mine it? And how about connectivity at Times Square on New Year's Eve with hundreds of thousands of people in the concrete canyons of Manhattan? The engineering challenges are daunting, and some companies will strike it rich if they can design the solutions, but the data itself will be overwhelming for analysis. How much will be done by bots? How much will be screened and only exceptions reported (think control charts)? Is there a way to play the stock market so that I hit it big during this disruptive revolution, and never have to work again, by using Machine Learning to unravel all the mysteries of market dynamics?

Case Study: The dairy farm system

I remember when I was in the First Grade and we took a field trip to Old McDonald's Farm. We got to go collect eggs in the chicken coop. Chase and pet the goats. Ride the horse. Milk the cow. Yes, it was done by hand way back in the 1960s and involved a wooden stool and a galvanized bucket and keeping your shoes dry. "The Cloud" still meant "rain," and servers were not part of the farm.

Today, things are quite different. The milk system is big business and production never ends since milking time comes around twice each day. Cows are fitted with RFID chips that allow all their movements to be tracked each time they go through a gate fitted with a reader. They have high-end Fitbits around an ankle that not only count steps, but records temperature and other biometrics. These data download to "the Matrix" during every trip to the milking parlor. Their destination after milking is controlled by the computer and a series of automated sorting gates that route them to where they are supposed to be. (So much for the cowboy sitting tall in his saddle and herding his little doggies.[61]) The guys at IMPRINJ make a lot of money from livestock. These sensors also bring up a whole new job function for someone to be the cows' tech support person.

The most quantification involves measuring the production and quality in the milking parlor. These data are processed real-time with a flag for medical attention set based on date or changes in output and quality. Automatic gates confine the cow. Modern veterinarians look like astronauts on a spacewalk, wearing coveralls with lots of full pockets, a mask, a backpack computer with an articulated arm holding the monitor, and an RFID reader gun on their hip. Everything they do is recorded and uploaded to "the Matrix." The most important flag concerns a change in milk production, since this can be an early warning indicator of pregnancy, or lack thereof, so that the appropriate actions can be taken.

The dairy herder himself is a user of "the Matrix." The details of the feed mix used each day is recorded, as well as the pedigree for all feeds and milk production tracked via blockchain, to be used to facilitate a recall. All of this information ends up on detailed dashboards customized to each user's role, including equipment maintenance, waste disposal, and water quality monitoring. The owner can achieve cost optimization via predictive analytics tools. He or she can consider price hedging with the futures market when linked up to COTS software such as AFIFARM-Dairy, Dairy One, and Milk Manager. It is mind boggling to explore the software capabilities and the totality of information management available for the milk production system. It is truly a glimpse of the future and how we will be able to feed the soon-to-be ten billion people here on Earth.

Ei-ei-ei-ei-ooo.

---

[61] However, for those ranches that are covered by cell phone towers, cattle can be fitted with a transmitter, like the one used in cars, and cattle movements can be tracked 24/7.

Where does blockchain fit into all of this?

Given all the hype about the meteoric rise of Bitcoin,[62] and the marketing push which states that blockchain technologies are the future for everything, it begs the question: How does this technology affect analytics? Not much so. Blockchain is intended for micro use and as a tracking tool. There is an algorithm that governs each and every Bitcoin that has been "minted." The hidden algorithm, which is tamper proof, tracks the date of creation and every time the coin changes hands or is split into pieces. Blockchain also serves to track production lots of crops from planting to harvest and then to delivery. Think of it as a pedigree for the bag of spinach you are buying. From a big data standpoint, the underlying transactions will always be tracked in databases and processed as we do today, except that there is a more rapid and powerful drill down capability so that the field that produced the *E. coli* contaminated lettuce can be found. It also can be used to track the provenance for a piece of art in an attempt to stop counterfeits. Blockchain is computationally intensive and NVIDIA chip sets are becoming the main computational engine. There is much competition to make faster chip sets, which will eventually benefit "big data analytics" and other high-end mathematical processing.

Models as Artificial Intelligence

Did you ever wonder how NASA makes spacewalks look so easy? It is simple. They practice the required activities at least a hundred times in a simulator. The design engineers try to include every failure mode that they can imagine so that the solution to the problem is already known before the actual spacewalk and decisions become an automatic response. Stated another way, they are creating intelligence. Another way to state this is they have created a model for decision making.

The definition of a simulator is "a device that enables the operator to reproduce or represent under test conditions phenomena likely to occur in actual performance."[3] This can apply to physical spacewalks being performed in a swimming pool in Houston, Texas, or to back testing a stock selection strategy using a model and supporting data that resides in "the Cloud." The definition of intelligence is:

---

[62] The creation of the Aloha Coin by the nation of Hawaii will be an interesting development and represents a dilemma in the open versus closed system of categorization. The goal is to replace the US dollar and open system concept, with a closed system of currency that is open to use and investment by the entire world. This is another system example for the Big Island as described in Chapter 1.

1) the ability to learn or understand or to deal with new or trying situations: reason; also: the skilled use of reason
2) the ability to apply knowledge to manipulate one's environment or to think abstractly as measured by objective criteria (such as tests).[3]

Simulators and models lead to the gain of intelligence. So, why do we need the word "artificial" to describe intelligence? Maybe it applies on Star Trek regarding the artificial life form, Data, and the intelligence he gains, but there was an episode which questioned if his life form was truly artificial. One could also make the case that all intelligence is artificial when compared to the concept of common sense in that additional information and more reasoning are required to reach a nuanced conclusion.

Which leads to the term *machine learning*. I do not think my car has learned to shift into 8th gear to save me money once I set cruise control on the highway on my way home every day. The sensors were designed to accept the signal that the speed is over 50 and that the acceleration over time is a constant and to respond by selecting the overdrive gear that results in a drop in RPMs and less consumption of fuel. In a word, functionality. As was shown in Figure 1-4, machines are within the realm of engineers, since severe consequences can occur if signals are not processed with a holistic view. This is validated every time there is a news story about another crash involving a self-driving car. Yes, users do learn, but they also can extrapolate for new situations which machines cannot do.

Once again, Tom Davenport comes to the rescue. He has named a fourth type of analytics. "Autonomous Analytics employs artificial intelligence or cognitive technologies (such as machine learning) to create and improve models and learn from data – all without human hypotheses and with substantially less involvement by human analysts" (2017, page 26). I like this wording because it refers to a tool to aid decision making rather than a creation from a science fiction novel.

I can see applying autonomous analytics in a virtual reality environment, such as Agent Smith in the movie *The Matrix*. One of the trends in education is gamification as a teaching tool, *sans* the graphics and violence. My kids learned about decision making by playing *Oregon Trail* when they were young. This game used a probability-based model to cause changes requiring decisions to be made. The newer tools used for teaching math track the categories that have frequent wrong responses and continue to present questions of that type until mastery is achieved. That might be considered artificial intelligence, but only as a means of augmentation for developing real intelligence. At a higher level, the Natural Resource Governance Institute has developed a game, *Petronia*, that allows you to be a minister of a country that just discovers oil.[63] Bechtel Corporation has a game that simulates your role as a project manager on a billion dollar build project.[64] Then there is a whole argument chain about ethics regarding big data and decision making involving human intelligence that must be resolved before the ethics conversation can even tackle artificial intelligence.

Is there a master tool that solves all of the problems for which analytics is heralded as the champion? Is there one master algorithm to rule them all? According to Domingos,[65] Table 9-3 provides descriptions from five tribes of algorithms that are called by a master algorithm, much as the equations of geometry are called for different shapes. They explain that these five tribes, plus machine learning, form the universe of artificial intelligence. The common machine learning tools are:

- pattern recognition,
- statistical modeling,
- data modeling,
- knowledge discovery,
- predictive analytics,
- adaptive systems, and
- self-organizing systems.

It figures, that if there is artificial intelligence, there must also be artificial *stupidity*. Does it ever seem to you that there is malicious coding of some websites? I heard that after Azkaban was closed, the Ministry of Magic enacted a re-training program for the Dementors that was focused on learning to code. I have found some of their work on some of the job applications sites I have accessed which sucked part of my soul into the ether. So much of the man-machine interface is cold and reductionist (just a check in the box or an avatar in VR), but the reality is modeling and AI is always about people (unless it is a truly natural system like a Hawaiian volcano).

The modern NVIDIA chips can crunch anything thrown at them, but the need for excellent quality data to begin with, as well as a detailed understanding of the process and the system, are critical for the benefits of the investment to be realized. As with any technological system, a rigorous cost-to-benefit analysis is required to support the go/no go decision. Even though

---

[63] https://bit.ly/2K6CT2N.

[64] https://bit.ly/2npmzjK.

[65] Pedro Domingos, 2015. *The Master Algorithm: How the Quest for the Ultimate Learning Machine Will Remake Our World*. Basic Books.

boards of directors may demand the newest and greatest innovations, they are still grounded by the need to be profitable.

Table 9-3. Components of the Master Algorithm Black Box.

	Tribe	Viewpoint	Roots	General-Purpose Learner Tool	Specific Tools	Examples
Master Algorithm	Symbolists	Learning as the inverse of deduction	Philosophy & Psychology & Logic	Inverse Deduction	Decision Tree (classifiers) Cognitive concepts	Credit approval Next chess move No free lunch theorem
	Connectionists	Reverse engineer the brain	Neuroscience & Physics	Backpropagation	Perceptrons and Neural Networks Hopfiled Machine Boltzman Machine Auto-encoder Logistic-Sigmoid-S Curve	Prospect Detector Credit Assignment Problem
	Evolutionaries	Simulate evolution	Genetics & Evolutionary Biology	Genetic Programming	Fitness Function Genetic Algorithms	Spam filters
	Bayesians	Learning is a form of probabilistic inference	Statistics	Bayesian Inference & Probabilistic Inference	Naive Bayes (Bayes Networks)	Medical diagnosis
	Analogizers	Learn by extrapolating from similarity judgements	Psychology & Mathematical Optimization	Support Vector Machine	Nearest Neighbor	Handwriting recognition Recommend books & movies

Organizational challenges

While marketers and TV commercials paint utopian pictures of what computers can do, the realities of implementation are different. The utopian view focuses on large operations. Progressive managers have already applied analytics to big data to keep their Boards happy. This book has focused on the needs of small and mid-sized operations, but also can be applied to the poorly-led large private companies and government agencies. Four themes are explored below, followed by a suggestion of a path forward.

Costs

The most important question all operations ask is: How does implementing analytics affect the bottom line? The stock market asks that question of IBM every quarter and Watson is still trying to earn his rent. A rigorous cost-to-benefit analysis is needed because the amount of overhead that can be expended will bankrupt many operations that do not have the luxury afforded by the Law of Large Numbers. There is a major difference in your approach if you have ten or ten million customers.

Hero-based

Is the managing director the wizard? This is true of start-ups and is a time-honored challenge when an operation attempts to scale up. As was discussed in Chapter 1, and elsewhere herein, processing big data requires a big team. The care and maintenance of "the Matrix" will become a significant expense.

Consolidation

A few years ago, I discovered a massive inefficiency at GM. I was looking for a job, and saw they had a Six Sigma opening, so I applied. I then started investigating for other openings around the country in their job posting database. When I went to apply for other openings, I discovered the HR systems were not connected and I had to make new accounts at every location that had an opening for which I was qualified. For grins, I applied at ten separate locations. I never had a response and have never followed up to see if they integrated their databases.

A successful example of consolidation involves water treatment services companies that support municipal water systems. This example of outsourcing has been in place for decades with the latest being to leverage the deployment of standardized computer systems that link billing and maintenance with production. This will be a growth area over the next few decades and will result in the creation of a few sanctioned monopolies that truly benefit the many by minimizing costs. Two consequences of this trend will be less use of consultants for short-term projects and the need for much more training. As next generation online educators, such as 2U, gain market share, modules for standardized corporate learning will be developed so that entire management teams become knowledgeable and experienced with the tools described herein.

Nationalization

Which industries do not gain value through competition and which ones truly waste money? Some industries would benefit society if they were nationalized – placed under the control of the Federal Government. Pharmaceutical companies are one of the poster examples because they spend a huge amount of money on advertising for drugs that benefit only a few, yet the costs are passed on to the many. While nationalization has been a non-starting proposition in the US in the past, the pendulum has started to reverse considering decreased workforce due to retirements, and centralization of all records to one super computer system and supporting analytic capabilities.

In the past, the Federal Government utilized government-owned contractor-operated facilities to maximize employment of the baby boomer generation rather than to maximize efficiency. This business model claimed it was based on competition, because non-performing contractors could be replaced every award cycle. Each round of contract renewal resulted in fewer participants because the bidders promised to achieve efficiency goals by consolidation across locations. When population pressures wane, so will this business model. Huge benefits will be gained from fewer executive salaries, elimination of duplicative services, elimination of large award fees for simply doing the contractually required job, and gains in the economies of scale for programming efforts when "the Matrix" is created to hold all data. It is this consolidation and control of data that will ultimately drive massive change.

Costs accumulate quickly on government projects from bid documentation, procurement, staffing, elicitation of requirements, audits, and progress status meetings. Once deployed, training costs can be substantial as well as required hardware and software upgrades. Change also occurs too frequently since every vendor lobbies his home Congressional delegation that they should be given a shot. The result is perpetual churning. Each new project is a risky undertaking with respect to schedule and budget because of excessive unknown conditions. Therefore, the government project manager negotiates for a promotion (ideally to the Senior

Executive Service) and subsequently moves on after the milestone is achieved. During all this churning, the CIO is totally ignored and all the rational decisions regarding interoperability and cyber security must be reexamined. The results for the CIO are goals not being meet, promotions missed, and excessive turnover. Once Congress realizes the true costs that have been wasted on computer applications, then a focused effort will be made to make "the Matrix." The corporate world is not that much better!

In 2017, Davenport and Kirby provided evidence of the common resistance to change regarding analytics.

> *Forty-three percent mentioned "lack of organizational alignment" as an impediment to their big data initiatives, forty-one percent pointed specifically to middle management as the culprit, the same percentage faulted "business resistance or lack of understanding." Eight-six percent say their companies have tried to create a data-driven culture, but only thirty-seven percent say they've been successful at it. (page 14)*

Now we return to Thomas Davenport, the father of analytics, for his take on the future. His 2016 book, with Julia Kirby, is entitled *Only Humans Need Apply: Winners and Losers in the Age of Smart Machines*. The authors assert that the number of architects and power users in analytics will be limited to a few select people. However, smart machines will be used to augment the workforce by granting superpowers and deploying leverage. The two superpowers they describe involve information retrieval and essential decision making. Both have been around for years, such as how we use IMDB to research an actor or all the smart systems in our car including cruise control, anti-lock brakes, and parallel parking mode. They describe two types of knowledge leverage. The first allows mundane tasks to be performed (like asking Alexa to order more toothpaste). The second focuses on the individual and our health (like the high-end Fitbits). They warn that human liberties may be compromised to gain these powers. They also present five paths of new work to support these developments.

The Five Options for Augmentation

For those fixated on the threat of automation, there is essentially one move available (and only to an increasingly small set of people): a step up to cognitively higher ground. Continued employability depends on being able to occupy those rarefied realms of rational decision-making not yet conquered by computers. Reframing the challenge as augmentation opens up a broader range of strategies for individual job holders and seekers. In place of that one possible step, now multiple steps reveal themselves as viable:

Stepping Up: Moving up above automated systems to develop more big-picture insights and decisions that are too unstructured and sweeping for computers or robots to be able to make.

Stepping Aside: Moving to a type of non-decision-oriented work that computers aren't good at, such as selling, motivating people, or describing in straightforward terms the decisions that computers have made.

Stepping In: Engaging with the computer system's automated decisions to understand, monitor, and improve them. This is the option at the heart of what we are calling augmentation, although each of these five steps can be described as augmenting.

Stepping Narrowly: Finding a specialty area within your profession that is so narrow that no one is attempting to automate it – and it might never be economical to do so.

Stepping Forward: Developing the new systems and technology that support intelligent decisions and actions in a particular domain.

Davenport and Kirby, 2016, pp. 76-77

Data Analytics Support Office (DASO)

How will companies organize to implement analytics, quantitative analysis, and even systems analysis? I doubt they will gather talent straight from a bachelor's program. They might recruit them from a master's program that included formal internships. Work experience is almost as necessary as quantitative skills. Some operations will re-train employees who have the capacity to shine. Other operations will pick-up retirees who have the skills but may not know the latest software tools. The workers all will have to be brave, mature, and willing to tell management that, even though a small fortune was spent on software, the quality of the data causes the width of the error bars to mask the true signal being sought. Data collection and storage improvements will always be needed to have high fidelity data for use. Sycophants need not apply when the survivability of an operation is at risk.

Quantitative analysis needs to be ingrained into the culture of an operation. Leadership needs to demand detailed analyses and to reward high performers. The entire management chain needs to be aware of the importance, the goal, and the process being followed. Figure 6-11 was an example of who needs to be involved in what steps. I am sorry for those managers who skipped through the early part of life without building mathematical muscles. The digital revolution runs on data, and managers must trust the decisions made by algorithms, rather than freelance a "cowboy solution" because it feels good. Many managers are only marginally capable with analytics, but fully willing to adopt this new normal. They require a support network for success to be ensured. One approach is to establish a Data Analytics Support Office (DASO).

There are four main functions that must be assigned to a DASO. First, a standardization of tools and techniques must be pushed out to the entire operation. The features to be emphasized are a consistent presentation template, a common language, software selection and implementation, and requirements applicable for training and certification purposes. The benefits of standardization include efficient time use by management when reviewing outputs, focus on implementation, minimization of distraction involving stylistics approaches, robust documentation, and trainability. If speed and budget allow, an off-the-shelf implementation from a consulting firm could be adopted.

Second, the DASO is charged with knowledge retention. There are many formal approaches regarding knowledge management. They all require discipline, sharing, and working for the common good. While the formal argumentation approach taken by Saint Thomas Aquinas in *Summa Theologica* may be overkill for some operations, it is a good template for most. The CRISP-DM methodology presented in Chapter 2 follows this approach, using questions, plans, feedback, and refinement. Given the pace of personnel turnover, a repository of all attempts to achieve a quantitative solution must be recorded to facilitate progress. A benefit is that wheels are only invented once.

Third is a function that makes competency determinations for both personnel and supporting software. Prior to assessing competency, a formal requirements analysis is needed to design a measurement rubric and the training function. The determination of competency is a two-pronged role because different approaches are needed for the initial demonstration of competency, and for on-going maintenance. The degree of formality is governed by marketing needs and the expectations of the governing body. INFORMS has a certification program for individuals[66] as well as a maturity assessment tool for organizations.[67] A robust knowledge management system is a prerequisite to demonstrating competency.

Lastly, the DASO will provide a home for a cadre of experts who provide support, deliver training, curate "the Matrix", and judge competency of individuals and technologies. A hidden agenda item for this cadre is to find the shining stars who have analytical skills and talents that need to be taken to the next level. The senior members also will have a role in mentoring management as to how to read and interpret analytical outputs, as well as how to make decisions. Once bodies are assigned to the DASO, the cost commitment becomes real since this is an overhead function. A large cadre will be needed at the start, but a plan for downsizing must be rapidly deployed. The goal is to provide support, not to develop a hidden factory.

---

[66] https://www.certifiedanalytics.org/.

[67] https://analyticsmaturity.informs.org/.

In keeping with the hidden factory theme, some pundits are calling for a new C-Suite executive. Some would like to see a Chief Data Officer. Others would like a Chief Analytics Officer. These positions would be wastes of money. The head of the DASO needs to be an executive with experience pushing cultural change and a great relationship with the top tiers of the management chain. They do not need super-star skills, because they have a cadre of experts on their team. If an operation truly wants to thrive by making data-driven decisions, then everyone in the C-suite needs to have knowledge and skills so that they can utilize the DASO. The goal is to embed a data-centric approach, not to acquire trophies to show the board.

Remember, the overall goal of a DASO is to achieve sustainability of the transformation. Some operations require slogans and glitz for reinforcement, while others require an individualized touch. A pragmatic view that is unique to each organization and operation is required. Hence the requirement for an experienced leader to head the DASO. History must be accommodated, while the future is shaped. Someday, it might be ten or twenty or more years from now, but the computer revolution will reach its full potential and the concept of work will be redefined around autonomous systems that blur the difference between real systems and digital systems. That will be the time that data scientists shine with their quantitative analysis skills.

This book has provided a comprehensive review of systems concepts, how they are measured and modeled, and how decisions are made using this information. Most of the tools and techniques for quantitative analysis have their roots in scientific investigations. The rest are from human systems such as marketing and economics. The limiting factor of any analysis will be related to the quality of the data itself, with Big Data providing opportunities for spectacular successes and disappointments. There are also several mathematical principles that govern how quantitative analysis can be performed and how the results can be interpreted. Remember that Ishikawa's seven tools were developed for small data but need to be applied to Big Data before the advanced tools can be used. Armed with this knowledge, you are now ready to call yourself a data scientist and start experiencing the trials and joys of quantitatively analyzing systems. Good luck!

You don't need a weatherman
To know which way the wind blows.

Bob Dylan, Subterranean Homesick Blues

Detailed Description of the 10 T's for Data Mining

The 10 T's is a standardized approach for data mining to guide the data scientist in all activities. These are listed in Table A-1 and fully explored in Table A-2. Rather than a detailed process, they represent mental attitudes that will help ensure success. There is always a strong possibility that second and third passes of the mining activities will be needed for complex situations.

Table A-1. Summary of the 10 T's for data mining.

The T's	Objective
Time (Attention)	Long, focused sessions.
Time (Duration)	Make multiple passes.
Transparency	Be open – no hidden agendas.
Tenacity	Data must be integrated from multiple, fragmented sources.
Testing	Need for cynical and skeptical views while striving for the truth in a non-judgmental manner.
Type	Determine the true nature of the data holdings.
Tools	Evaluate and select what is needed from the available options.
Treatment	Practical, Graphical, and Analytical reviews.
Transport	Presentation of the results.
To Learn More	Keep up-to-date with evolving technology.

Table A-2. Detailed description of the Ten T categories.

1. Time (Attention)	
Objective	Long, focused sessions.
Purpose	To strategically evaluate and plan how the analysis will be performed. Mining data is not like changing a tire. A long reflection period is required to determine how to proceed.
Inputs	Database schema. Entity Relationship Diagram. Sample versus population for a "First Look" analysis.

1. Time (Attention)	
Tools and Techniques	Classification. Affinity Grouping. Clustering. Description. Estimation. Prediction.
Outputs	Project plan. Decision on granularity, chunking, and distributions.
Checkpoint	Will this approach answer the question being asked? If available hours become limited, can the plan be trimmed to just a "proof of concept" or "pilot study" scope?

2. Time (Duration)	
Objective	Make multiple passes.
Purpose	To prepare to spend sufficient time to rework approaches after false starts are made. It takes long focused sessions. MRI research has shown it takes the brain 15 minutes to refocus once interrupted by a phone call or checking email. So, the old computer lab with a ban on phones still makes sense.
Inputs	All data! This may take a while to gather and to clean. Data exhaust – the trail of clicks (metadata) that internet users leave behind from which value can be extracted is an example of turning dross into gold.
Tools and Techniques	Classic Extraction Transformation Loading (ETL) function. Five Passes: 1st – To judge quality – usually errors are found that need fixed in the data warehouse itself. A "First Look" analysis. 2nd – To assess goodness of measurement – formal gauge R&R study and other similar approaches. 3rd – To identify prospects – Interim results are shared with the Subject Matter Expert (SME) to confirm validity of the approach. Wrong approaches are documented so they are not reperformed. This is the stage when the analysts says, "I wonder", "what if", and "if I hypothesize I will get x, do I?" These answers cause new projects to be started or prototypes of results to be shared with the customer. 4th – To acquire all holdings – First Pass analysis is performed and shared with SME. 5th – To verify 4th pass and generate presentable information – apply shine and polish so that future queries can be automated and to generate ratio, cumulative, and time series views. Meditation of 10,000 Mouse Clicks – The Zen-like condition reached during deep dives into massive data holdings.

2. Time (Duration)	
Outputs	Data of known quality with respect to: completeness, validity, consistency, timeliness, and accuracy.
	List of approaches that did not yield results.
Checkpoint	Are all the idiosyncrasies documented?

3. Transparency	
Objective	Be open – no hidden agendas.
Purpose	To assure all stakeholders of the scope and purpose of the data mining activities.
Inputs	List of all stakeholders and participants.
Tools and Techniques	Team building.
	Quality Circle.
	Blog posts.
Outputs	In Brief Slide Deck.
	Out Brief Slide Deck. (A fast track step is to spend the last third of the Kickoff Meeting framing the deck and reviewing and updating at all subsequent team meetings.)
Checkpoint	Who has a legitimate reason to complain?
	If details were leaked to the TV show *60 Minutes*, would I still have a job?

4. Tenacity	
Objective	Data must be integrated from multiple, fragmented sources.
Purpose	To alert the data scientist that not all the processes were built to handle a rigorous review and that the interim results will guide the method and duration.
	If the answer was already known, you wouldn't be doing this. The easy stuff is already done!
Inputs	Integration strategy.
Tools and Techniques	Mashups.
	Prototypes.
	Patience!
Outputs	Changes to project plan.
	Changes to In Brief Slide Deck.
	Changes to Out Brief Slide Deck.
Checkpoint	Am I open in my view, or do I have preconceived ideas and biases?

5.	Testing
Objective	Need for cynical and skeptical views while striving for the truth in a non-judgmental manner.
Purpose	To look at the results both cynically and skeptically. Sometimes, a mess is found because of incompetence, loss of control, willful manipulation, or the desire to deceive. These instances are not fun. People typically lose their jobs. Data scientists need to be a fair and truthful and assertive enough to report findings up the management chain.
Inputs	Test plan. Quality Assurance Plan.
Tools and Techniques	Verification and Validation strategies. Data understanding process step from CRISP-DM model.
Outputs	Confidence statements. Documents, documents, documents to CYA.
Checkpoint	Are these data suitable to provide answers to the questions being asked? Am I willing to bet my job on the validity of these holdings?

6.	Type
Objective	Determine the true nature of the data holdings.
Purpose	To categorize and classify the data contained in each field. This is a true IT function so that data types and widths can be assigned for transformation and loading functions.
Inputs	All data.
Tools and Techniques	Auto-filters. Pivot tables. Traditional mathematical and logical approaches.
Outputs	Scorecard describing characteristics of all fields. Also called a *data dictionary*.
Checkpoint	Are these data suitable to provide answers to the questions being asked?

7. Tools	
Objective	Evaluate and select what is needed from the available options.
Purpose	To assess holdings and determine what is need for ETL and the First Pass analysis (Excel, SAS/JMP, R, Specialty Packages).
Inputs	Schema. Entity Relationship Diagram. System Configuration details. List of Approved Software.
Tools and Techniques	COTS. GOTS. Open Source. Proprietary.
Outputs	Plan of attack.
Checkpoint	Is there anything new that would work better? Do I have funds to buy something else, or to upgrade, or to refresh? Which team member can apply these tools?

8. Treatment	
Objective	Practical, Graphical, and Analytical reviews.
Purpose	To critically and systematically digest everything that is obtained. This processing is done in conjunction with each pass and is the true realm of the data scientist. Remember practice makes perfect! It also gives serendipity a chance to find the giant golden nugget.
Inputs	All data. Stakeholder requirements and preferences.
Tools and Techniques	Practical assessment (ho-ho test). Graphical displays. (Include every graph for every field and document its suitability. This is the chance to be complete.) Analytical drill downs using all relevant approaches. (Document all approaches that fail in case the manager is from Missouri and demands "Show Me." Formatting is not important.)
Outputs	First Look write-ups. (Minitab's Report Writer is a convenient place where graphs can be posted and your observations, questions, and subsequent answers can be journaled. The write-up can be included in Final Reports as appendices.)
Checkpoint	What did I get? Does this make sense? Will this work?

9. Transport	
Objective	Presentation of the results.
Purpose	To explain to stakeholders what was obtained and how it can be applied.
Inputs	Stakeholder requirements and preferences.
Tools and Techniques	Memos.
	Reports.
	Slide Decks.
	Briefing packages.
	Classic communications model.
Outputs	Final version of Out Brief Slide Deck.
Checkpoint	Who is the audience?
	Did I answer the question that was asked?

10. To Learn More	
Objective	Keep up-to-date with evolving technology.
Purpose	To evaluate the state-of-the-art to maintain your currency.
Inputs	Professional societies.
	Blogs.
	Job postings by the industry leaders (What new skills are they looking for and should I also be looking at them?)
	Vendor marketing materials.
Tools and Techniques	Read-read-read.
	Listen-listen-listen.
Outputs	Recommendation to "change horses mid-stream" if something better is found.
	Learning Plan.
Checkpoint	Is there a better or faster or cheaper way to re-do this work?
	Where else can I use what I now know?
	Do I need to go back to school?

Million Row Data Audit (MRDA) Process

This procedure describes the steps required to perform an audit of a million rows of data from a big data source. The process is conducted in five distinct phases: 1) preparation, 2) review, 3) drill down analysis, 4) visualization, and 5) reporting. It was designed with the assumption that analysis would be performed in Excel 2010 on a multi-chipped machine with a large RAM capacity and fast processors running on a 64-bit operating platform.

Preparatory phase

1. Save file with working name to allow recovery back to original data if needed. Cycle working file with sequence number prior to high risk activity in order to allow recovery.

2. Remove all header and footer entries to a new worksheet labeled "Notes" and name first worksheet "Data." Document all special events or discoveries relevant to the structure of the data on the notes page when encountered. This includes questions to ask of the data owner regarding inconsistencies and exceptions. Once the audit is complete, an extract can be obtained for coordination and disposition. Save file![68]

3. Ensure Column A is labeled "Index" and a sequence number listed in case data needs to be sorted back to an "as received" state. Advance to bottom of page and delete all rows beyond 1,000,001 to be true to the MRDA name. Advance across Row one and inspect for identical names, blank headers, and blank columns. Adjust to remove blanks and uniquely identify headers. Save file!

---

[68] Yes, million row spreadsheets save slowly, especially after pivot tables are created. This pause (20-40 seconds) is a convenient time when the muscles can be stretched.

4. Insert split below Row 1 to allow viewing of headers while scrolling. Enable auto-filters for Row 1. If super-headers used, adjust row number accordingly. Save file!

5. Create new worksheet labeled "Schema." Copy column titles from "Data" worksheet and Paste Special/Transpose at Cell A3 on "Schema." Proceed to generate column headings as shown in Table B-1. Draw outline around each cell for the entire column-row matrix. Spellcheck. Save file!

Table B-1. Example schema worksheet.

Field	Number of Records	% Returned	Type	Notes	-	-	Number Unique	Average	Median	Maximum
							Pivot Table Review			
Index	1,000,000	100%	Number	Record Counter	-	-	-	-	-	-
Name	886,575	89%	Alpha	First_Name Middle_Name Last_Name	-	-	244,770	4	6	2,467
B	21,584	2%	Alphanumeric	447 categories. See Table xxx	-	-	-	-	-	-
C	1,000,000	100%	Date	Text as 20121204	-	-	-	-	-	-
D	1,000,000	100%	DTTM	Machine generated date and time stamp	-	-	-	-	-	-
Gender	576,839	58%	1 letter code	56% male, 44% female	-	-	-	-	-	-
E	631,175	63%	Alpha	5 categories as	Count	% Found	-	-	-	-
				aaa	299,769	47%				
				bbb	209,682	33%				
				ccc	121,712	19%				
				ddd	12	0.002%				

6. Select Print and add Headers and Footers to include Project Name, Data Set Name, Date, Page x of xx, and any endorsements of sensitivity/classification, like Personally Identifying Information, Proprietary, and For Official Use. Preview the printed page to ensure readability and adjust as needed. Save file! Print. This hard copy becomes a useful operator aid for maintaining awareness of the schema.

Review phase

7. This phase involves a process that is applied to each column of data on the "Data" worksheet. It is described below with the intent that the sub-steps will be repeated for each column. The functionality of Excel is utilized.

   a. Click on the B in column B. Note the count in the box at the right side of the lower tool bar. Mentally subtract one (for the header name) and record as "Number of Records" on the Schema worksheet.

   b. Click on the drop-down box in cell B1. Inspect the entries in their entirety. Record "Type" and add a description of anything unusual under "Notes." This step also will identify the purity of the data (consider Senior Administrative

Clerk, Senior Admin. Clerk, Senior Admin Clerk, and Sr Adm Clerk), and what cleaning steps are needed prior to identifying categories. In extreme cases, the solution might be a fix in the database itself and then generation of a second extraction. If the data are not useable, note on the Schema to format the cells red.

c. The final review step is to look at the categories themselves. For simple binomial categories like Yes/No or Male/Female, it is best to determine the number of one category and calculate percentages while assessing that field. This result is notated under results along with antidotal descriptions of exceptions. When 3-10 categories are present, select each category and observe the number of records. A judgment call is needed if the categories provide information germane to the question that initiated the analysis and should be decomposed or if the inherent complexity must be processed. It is never a good idea to remove any data. It is simpler to add a field (or column in Excel) and refactor the complexity into simpler categories. An example of decomposition can be found in Table 1 for Field E. The entry with 12 records needs further review to determine if they are errant entries, a legitimate black swan category, or the proverbial "needles in the haystack." Flag the fields with more than ten categories for drill down analysis in the next phase.

d. If coded data are encountered, three questions must be asked. Is there a key to the codes? Were the codes used consistently? Should the codes be replaced with a more logical name using the VLOOKUP function? Note the answers. Contact the data owner for a legend if needed. Schedule a time for additional processing if needed.

e. Repeat for Columns C until the end.

f. Enter details annotated on printed schema. Spellcheck. Save file!

Drill down analysis phase

8. A pivot table needs to be built to allow a detailed review of fields that have many categories. This is a standard feature within Excel. The easiest way is to select cell A1 on the Data worksheet, after ensuring there are no blank rows or columns in the table and all headers have a unique name, and then use the Insert Pivot button. Once established,

change the name on the worksheet tab to "Pivot." Ensure a field that has a complete set of entries ("Index #") is selected for the 'Values' portion of the table. Save file!

9. Occasionally, an adjustment to a hierarchical category is needed either by reduction or expansion. Reduction is when various subcategories are lumped into a parent category. Affinity diagrams are a useful tool for this activity. Expansion is when a category is split into subgroups. A typical example of both operations is cities and states in the appropriate context. Changing a date into a DTTM field is an example of an expansion and allows embedded features in Excel to present the results in a hierarchical manner to the requisite granularity. When adjusting:

    a. Add a row on the Schema worksheet with the new field name and paint yellow. Reprocess the new field in accordance with Step 7 in the Review Phase above. Save file!

    b. Add Comment in header cell describing what changes were made. Include any formulas used by executing the Paste function. Replace the formula using the Paste Values function to suppress refreshing when other activities are performed. Save file!

    c. It is helpful to keep a Word file open while the data are being reviewed to capture processing steps, discoveries, interesting facts, and questions for further investigation. This documentation becomes more critical as the length of analysis proceeds from days to weeks to months, or if analysis is conducted in a disjointed manner.

10. Other situations require the creation of metadata. This is common when analyzing network nodes with airports being a common example. Each airport can represent either an arrival or a departure location. If flight leg is of interest, a new field can be created by using the Concatenate function in Excel. Concatenate can also be used to make a stronger entity resolution construct such as between First_Name/Last_Name/Date_of_Birth. Other times the age of the record is needed. This is easily calculated in a new field (column). The formula is =today()-"Record_Date". Note that the answer is returned in number of days, with the default format being Date. The format needs to be changed to Number for ease of use. See sub-steps under Step 9.

11. The most common activity is the inclusion of additional data from other sources. This is easily done within the VLOOKUP function after source data has been added as a new worksheet as long as a valid common key can be constructed. Again, replacement of the function using Paste Values is imperative to suppress refreshing after machine

completion which is then immediately followed by a Save. If running a multi-processor machine, change the number of processors from "Use all processors on this computer" to manual and set at n-1 at the radio buttons under "Enable multi-thread calculation" in the Formulas section of the Advanced page on the Excel Options screen in order to minimize non-responsive faults when meshing two Million row worksheets in a multi-hour run. See sub-steps under Step 9.

Visualization phase

12. Graphical views are the most effective way to highlight data contained in a tabular format. A judgment call is needed as whether all fields need to be developed into a graphical view, or if only the key variables should be displayed and the entirety shown in tables. Sub-tables, nested tables, and tree diagrams are effective tools in some situations. A few common types are shown below.

    a. Pie chart

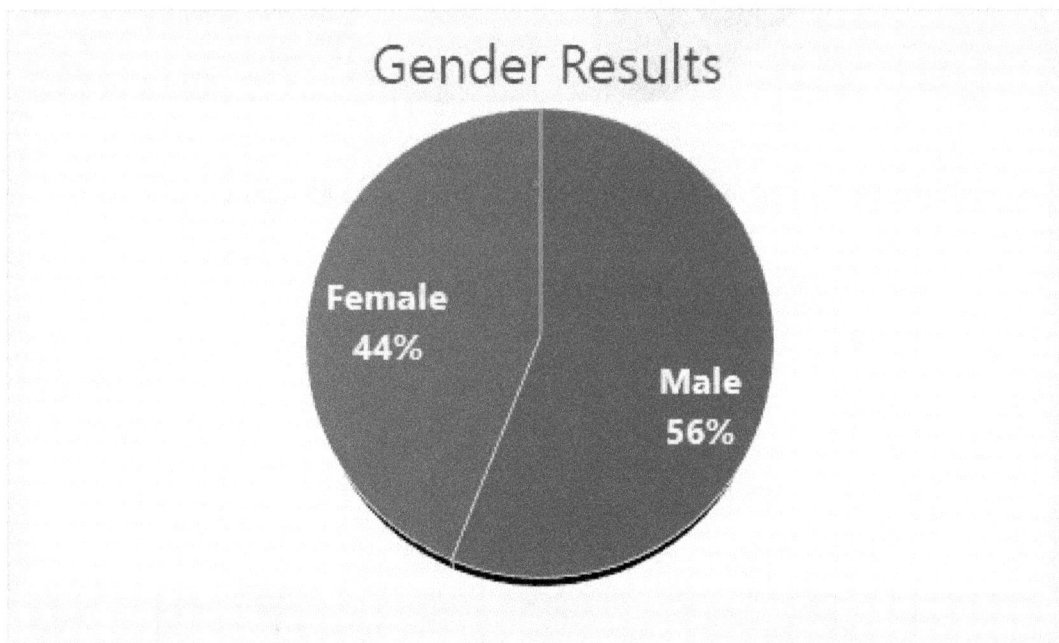

Gender Results

Female 44%

Male 56%

b. Sunburst chart

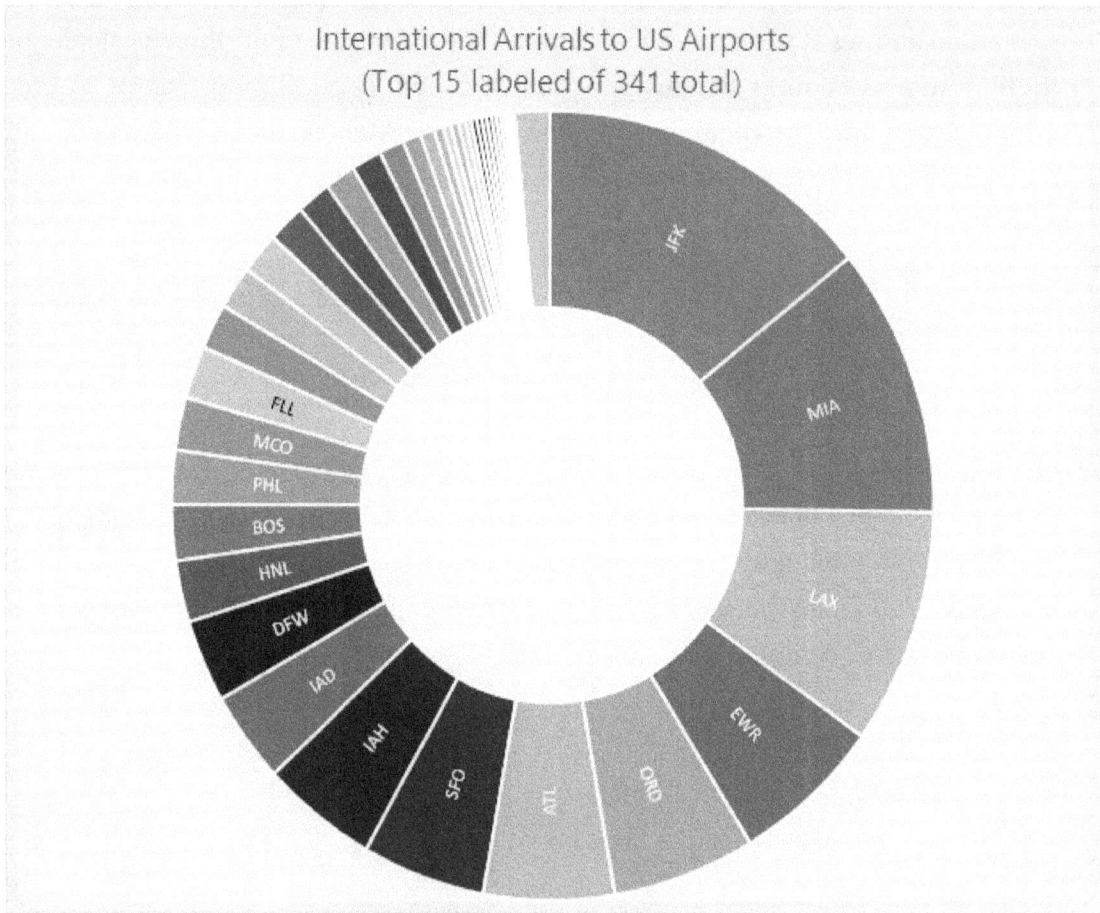

International Arrivals to US Airports
(Top 15 labeled of 341 total)

c. Boxplot (Generated from an ANOVA analysis in Minitab)

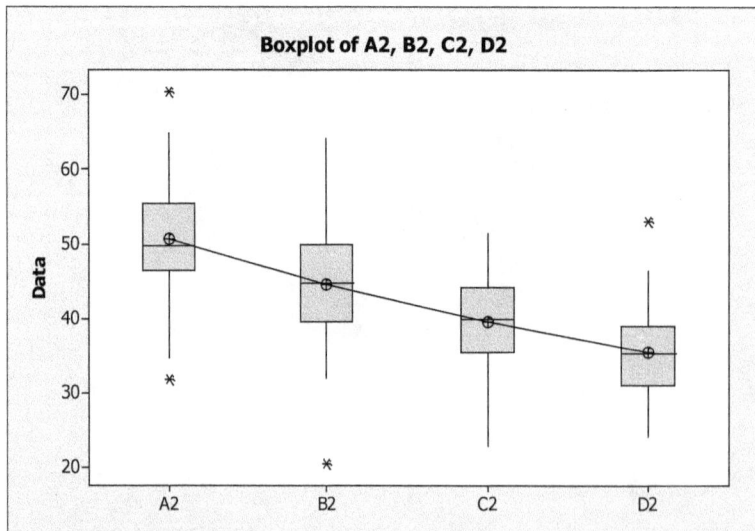

Boxplot of A2, B2, C2, D2

d. Distribution (bimodal)

The first figure below shows a complex distribution for a variable. The second figure highlights the data by group with a definite separation visible. The third figure provides descriptive statistics regarding each pattern. Because the two groups are distinct, the data needs to be separated and analyzed independently.

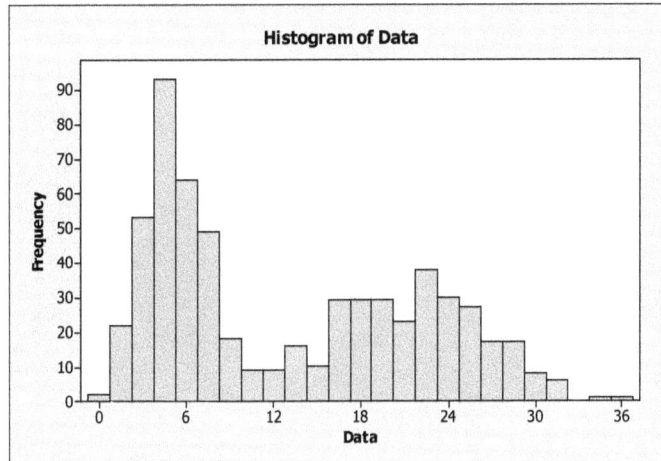

Histogram of Data

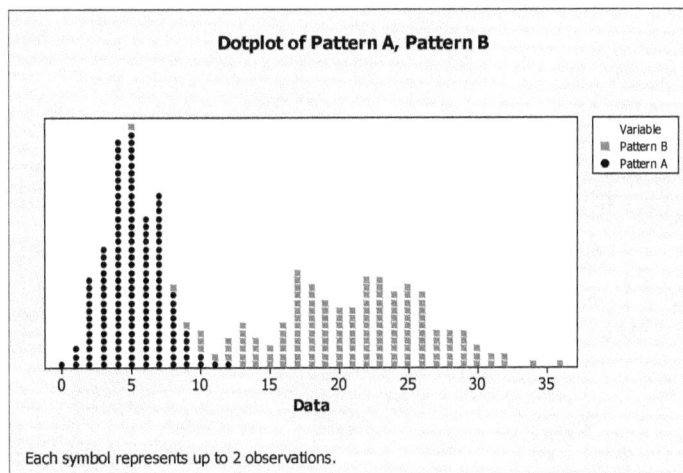

Dotplot of Pattern A, Pattern B

Each symbol represents up to 2 observations.

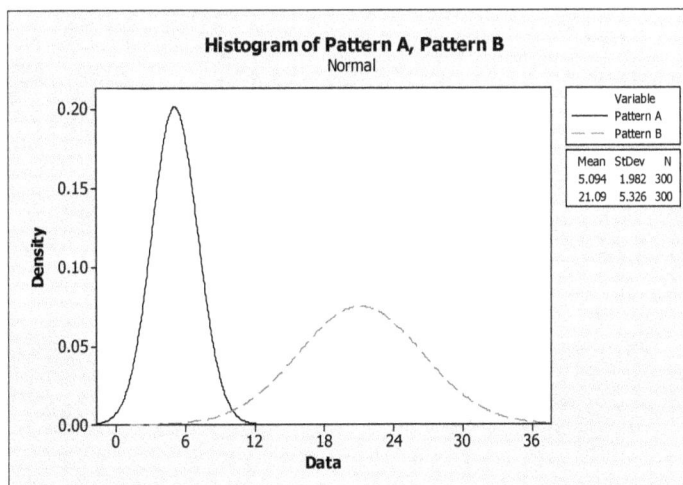

Histogram of Pattern A, Pattern B
Normal

13. Pareto Analysis, also known as the 80-20 rule, is a powerful screening tool used to identify the critical few categories of a variable from the trivial many. Simply stated, the Pareto rule describes the situation where 80% of the records are captured by 20% of the categories. The standard visual example is:

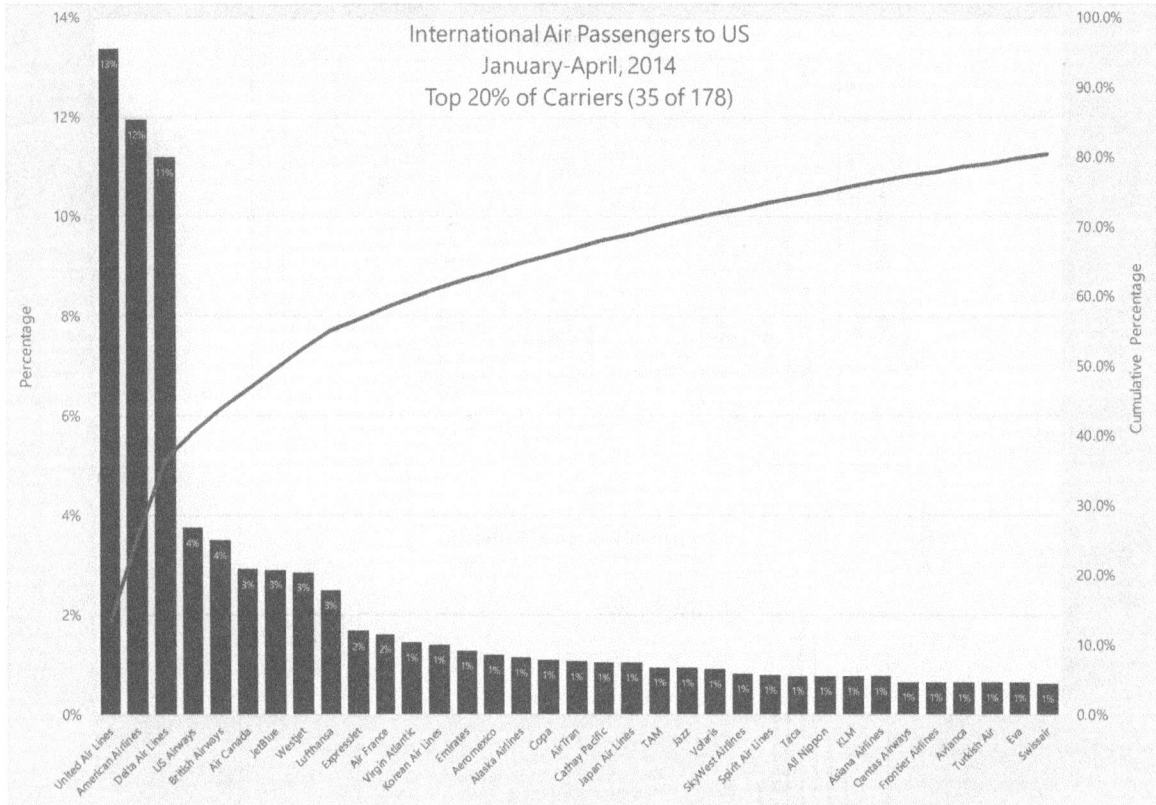

International Air Passengers to US
January-April, 2014
Top 20% of Carriers (35 of 178)

Reporting phase

14. A golden rule activity – whoever has the funds to pay for the report will determine what is necessary and sufficient.

Analysis of Rainfall Data for Amarillo, Texas (1880-2017)

There are many instances when rainfall data are needed for modeling purposes. Typical applications include agriculture, civil engineering, and geosciences. Many manufacturing processes are indirectly affected by rainfall. Depending upon the need, two levels of granularity are appropriate – annual and monthly. Prior to the internet, the Amarillo Globe-News would publish annual rainfall data from the National Weather Service (NWS) one Sunday every January with monthly data going back to 1880. I have seen this table in numerous barns and garages belonging to people whose livelihood is governed by the weather.

I had originally entered these data into Lotus in 1990 and performed various statistical analyses. There were some gaps, but I was able to find values either through internet pages or searching historic documents at a US Government document depository. At some point, I migrated the data to Excel. Over the years, I have updated the table with the data for the last 138 years available for download.

Currently, the NWS only uses a 30-year base for historical comparison even though there is 138 years available. The NWS also avoids combining data from the first 13 years as the monitoring station was moved a few miles in location. In order to use this oldest data, I performed a two-sample t-test[69] for each month and found there were no significant differences between the two locations (p values were all much greater than 0.05).

Overall, accuracy between annual total and sum of monthly data are 99.92% for the source data. The table I use presents the sum of monthly values for the annual value. Given the large spatial variability of precipitation amounts associated with any one event, this level of accuracy is more than sufficient to develop a reasonable predictive distribution.

---

[69] Two-sample t-test, two tailed, heteroskedastic (unequal variance).

Observations for annual data

Figure C-1 is a time series view of annual data displayed on a control chart generated by Minitab with mean and approximately ±3 standard deviation reference lines displayed. The control limits are based on a normal distribution. As no data points transcend the lower control limit and 4 of 138 exceed the upper control limit, the assumption of normality must be challenged. Figure C-2 is a default descriptive statistical summary from Minitab. The bell curve is for reference only. Normality is rejected because the Anderson-Darling p-value is less than 0.1. Figure C-3 demonstrates the value of normality in both tails, and to a lesser extent for the center because most of the data points are above the calculated probability line.

Further analysis in Minitab was performed. Autocorrelation was run with a lag of up to 32 calculated. The observed cyclicity may be correlatable to the solar flare cycle, and subsequent harmonics. Various data smoothing routines were explored to determine the best way to show the autocorrelation. When n=6, a distinct cyclicity is observable as shown in Figure C-4. When n=13, cyclicity is further demonstrated as shown in Figure C-5. The dampening of variation since the mid-1960s is an interesting phenomenon.

Observations for monthly data

Figures C-6 through 17 present descriptive statistics by month. Practical observation suggests that stratification for summer and winter months may exist to accommodate thunderstorm and snow events, respectively. Note that the data are only one-sided since values cannot drop below zero. These distributions are best portrayed with exponential, Weibull, or logistic distributions with an example for March shown in Figure C-18. Data were analyzed with Minitab's ANOVA routine with results shown in Table C-1 with full knowledge that the assumption of normality was not correct.

Significant differences and stratification exist as evidenced by the dashed distributions being non-inclusive of each other. Monthly data were compared in a pair-wise manner using a Mann-Whitney[70] test to determine when differences were significant. It was found that winter (November-December-January-February) and summer (May-June-July-August) seasons could be grouped together with the resultant combined data shown in Figures C-19 and C-20,

---

[70] The Mann-Whitney test is a nonparametric test that allows a distribution independent analysis similar to a Students t-test via calculations that depend upon the median. An alpha of 0.90 was used.

respectively. Figure C-21 is an interesting find. It represents a time series portrayal of 1,656 months of data. Graphs of this size become hard to read, consume a large share of computer resources, are difficult to print, and are common when working with big data. Sometimes graphics will have to be segmented so that clarity can be obtained. The increased noise and related higher center portrayed for the middle time period is an interesting phenomenon. Further research using either krieging or autoregressive conditional heteroscedastic time series models would be appropriate.

Figure C-1

I Chart of TOTAL

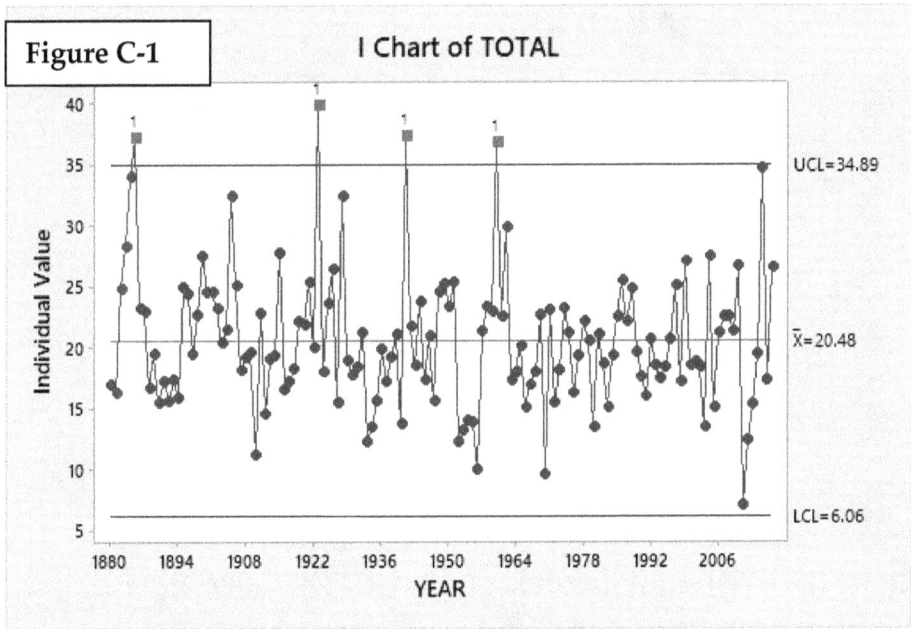

Figure C-2

Summary Report for TOTAL

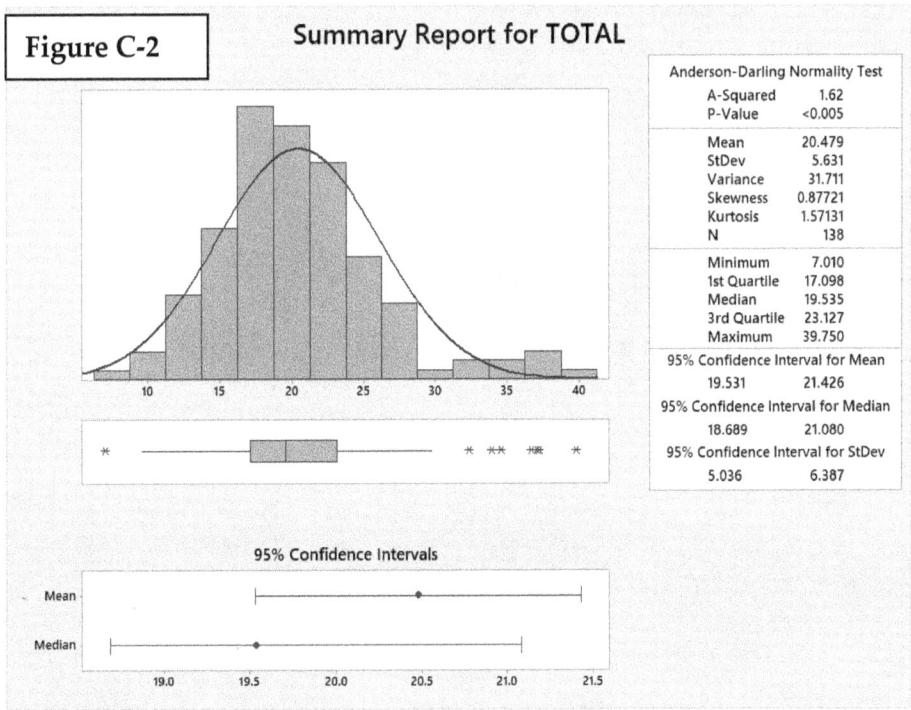

Figure C-3

Probability Plot of TOTAL
Normal - 95% CI

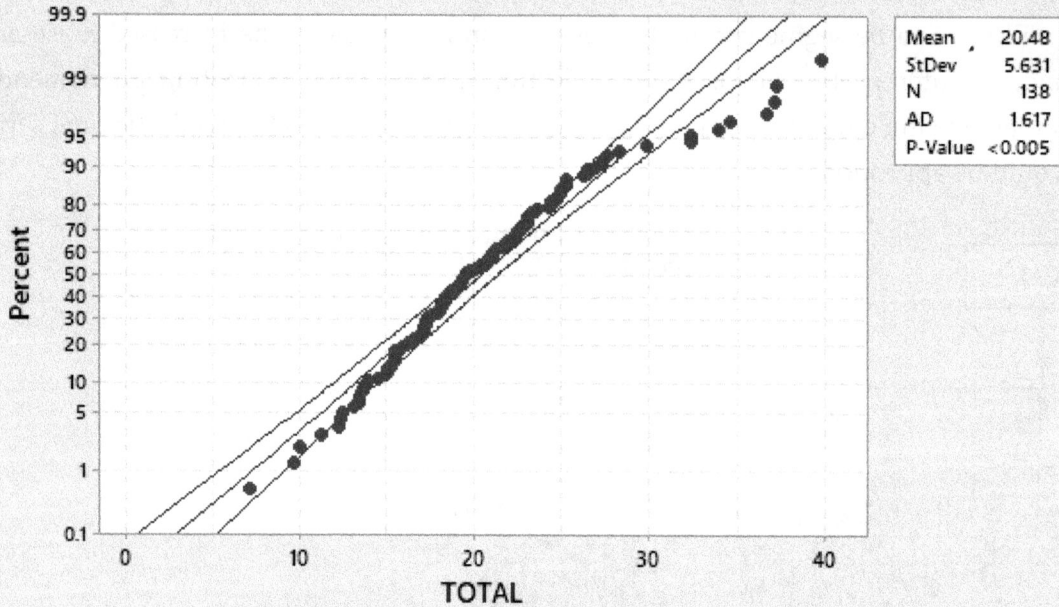

Mean	20.48
StDev	5.631
N	138
AD	1.617
P-Value	<0.005

Figure C-4

Moving Average Chart (n=6) Annual Rainfall - Amarillo, TX - 1880-2017

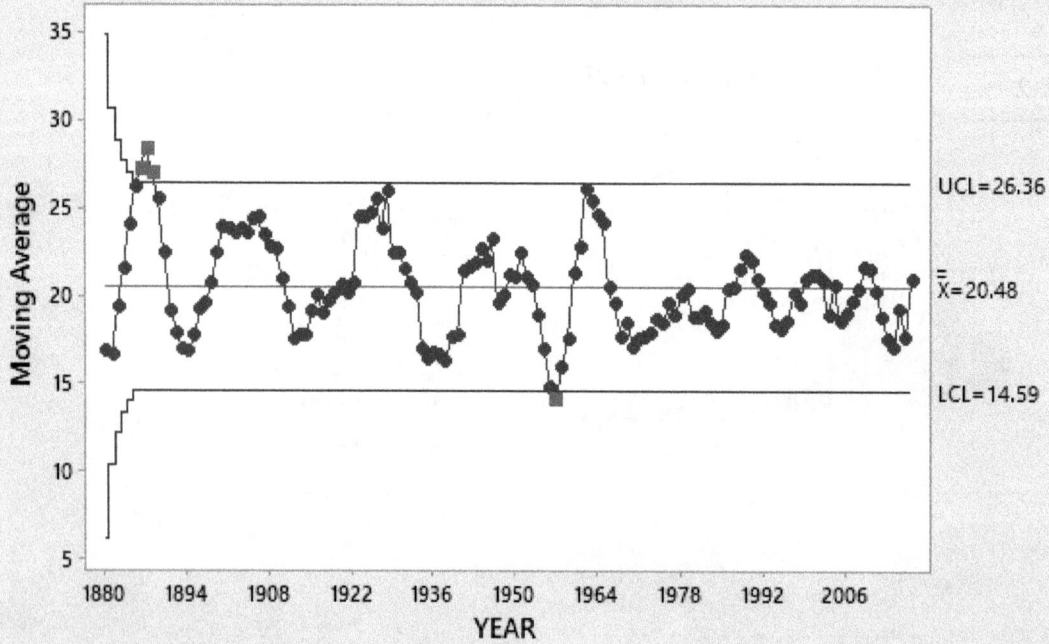

UCL=26.36
X̿=20.48
LCL=14.59

Figure C-5

Moving Average Chart (n=13)

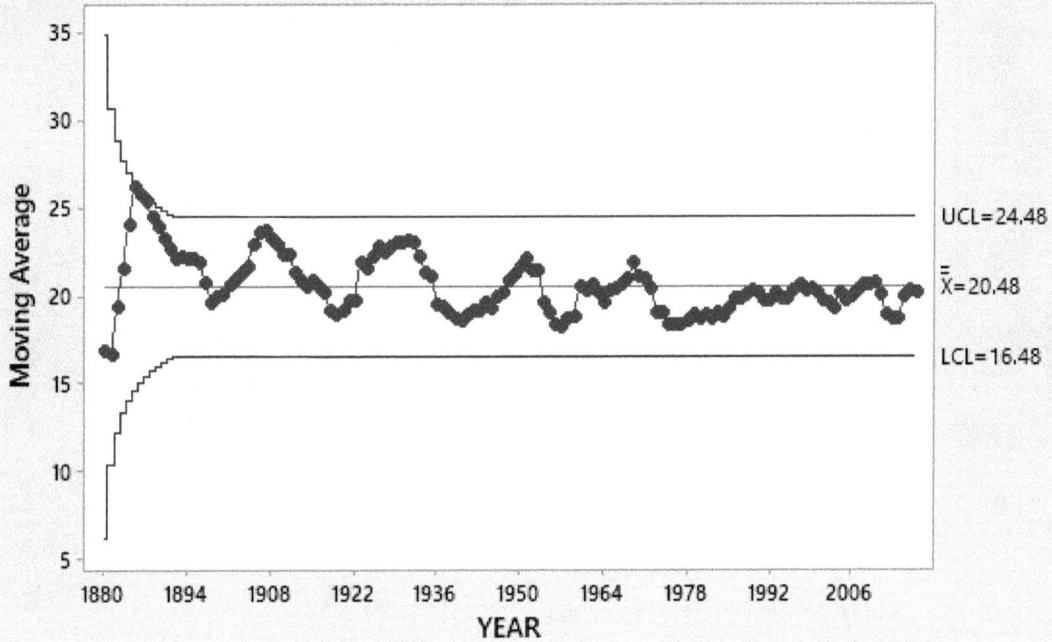

UCL=24.48
$\bar{\bar{X}}$=20.48
LCL=16.48

Figure C-6

Summary Report for JAN

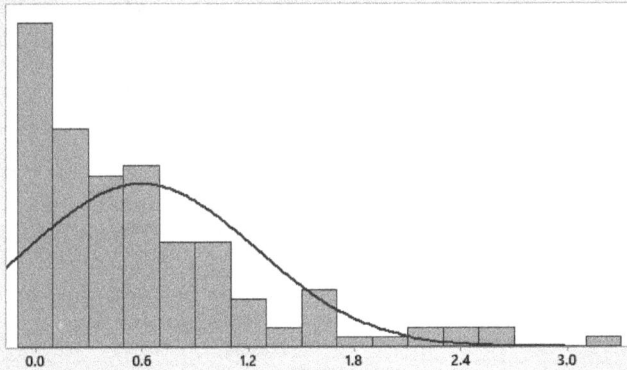

Anderson-Darling Normality Test	
A-Squared	6.74
P-Value	<0.005
Mean	0.59739
StDev	0.64272
Variance	0.41309
Skewness	1.61158
Kurtosis	2.61138
N	138
Minimum	0.00000
1st Quartile	0.09750
Median	0.41500
3rd Quartile	0.86000
Maximum	3.17000

95% Confidence Interval for Mean
0.48920   0.70558
95% Confidence Interval for Median
0.30980   0.55010
95% Confidence Interval for StDev
0.57479   0.72900

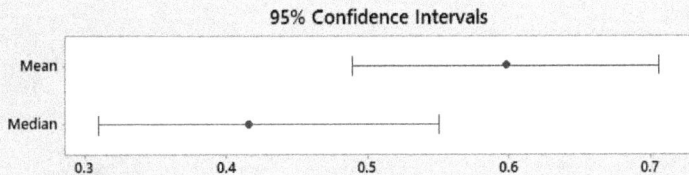

95% Confidence Intervals

Mean

Median

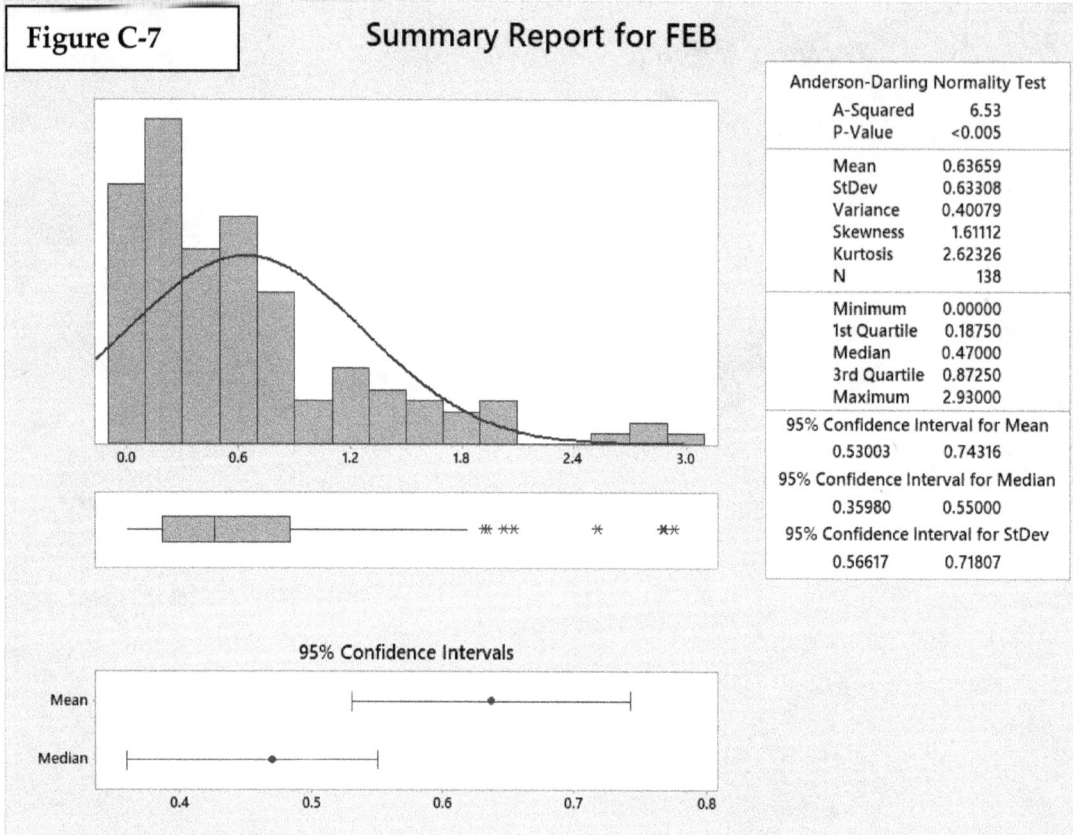

Figure C-7

Summary Report for FEB

Anderson-Darling Normality Test	
A-Squared	6.53
P-Value	<0.005
Mean	0.63659
StDev	0.63308
Variance	0.40079
Skewness	1.61112
Kurtosis	2.62326
N	138
Minimum	0.00000
1st Quartile	0.18750
Median	0.47000
3rd Quartile	0.87250
Maximum	2.93000

95% Confidence Interval for Mean
0.53003   0.74316

95% Confidence Interval for Median
0.35980   0.55000

95% Confidence Interval for StDev
0.56617   0.71807

95% Confidence Intervals

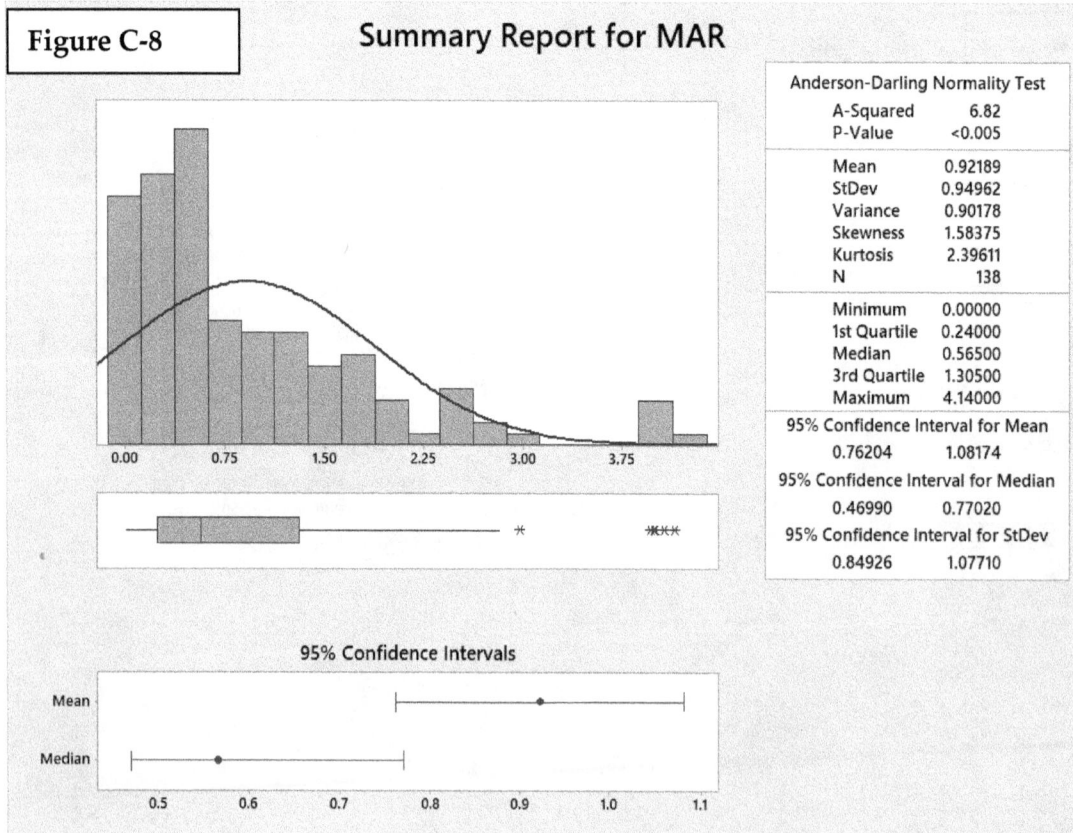

Figure C-8

Summary Report for MAR

Anderson-Darling Normality Test	
A-Squared	6.82
P-Value	<0.005
Mean	0.92189
StDev	0.94962
Variance	0.90178
Skewness	1.58375
Kurtosis	2.39611
N	138
Minimum	0.00000
1st Quartile	0.24000
Median	0.56500
3rd Quartile	1.30500
Maximum	4.14000

95% Confidence Interval for Mean
0.76204   1.08174

95% Confidence Interval for Median
0.46990   0.77020

95% Confidence Interval for StDev
0.84926   1.07710

95% Confidence Intervals

Figure C-9

Summary Report for APR

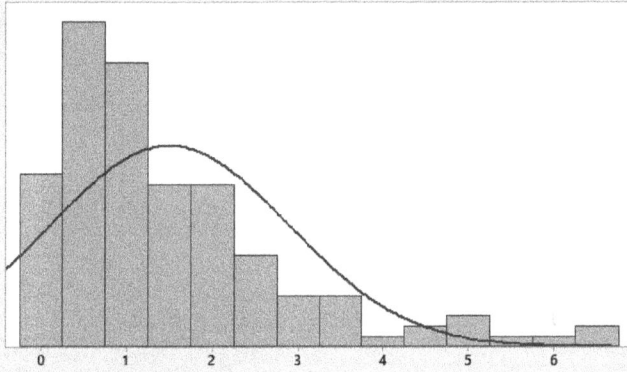

Anderson-Darling Normality Test
A-Squared 6.25
P-Value <0.005

Mean 1.5013
StDev 1.3889
Variance 1.9291
Skewness 1.57549
Kurtosis 2.41124
N 138

Minimum 0.0000
1st Quartile 0.5500
Median 1.0650
3rd Quartile 2.0100
Maximum 6.4500

95% Confidence Interval for Mean
1.2675 1.7351
95% Confidence Interval for Median
0.8300 1.2902
95% Confidence Interval for StDev
1.2421 1.5754

95% Confidence Intervals

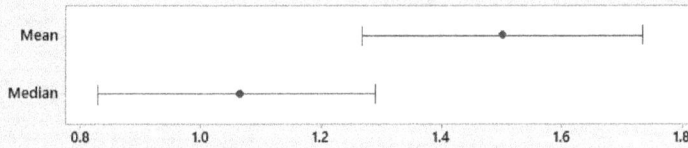

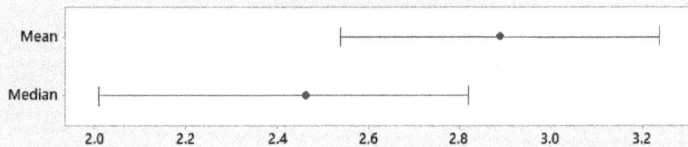

Figure C-10

Summary Report for MAY

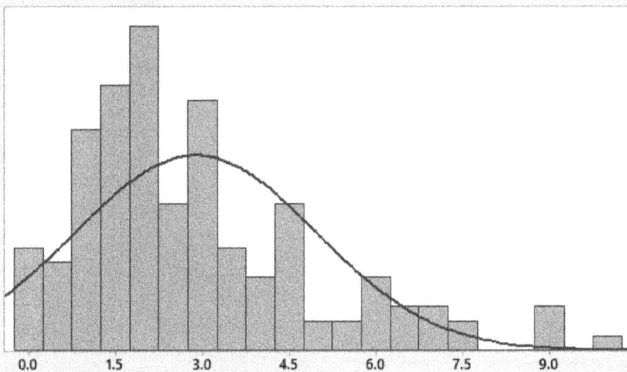

Anderson-Darling Normality Test
A-Squared 3.66
P-Value <0.005

Mean 2.8875
StDev 2.0757
Variance 4.3086
Skewness 1.19262
Kurtosis 1.26885
N 138

Minimum 0.0400
1st Quartile 1.4250
Median 2.4600
3rd Quartile 3.9875
Maximum 9.8100

95% Confidence Interval for Mean
2.5381 3.2369
95% Confidence Interval for Median
2.0098 2.8202
95% Confidence Interval for StDev
1.8563 2.3544

95% Confidence Intervals

Figure C-11

Summary Report for JUNE

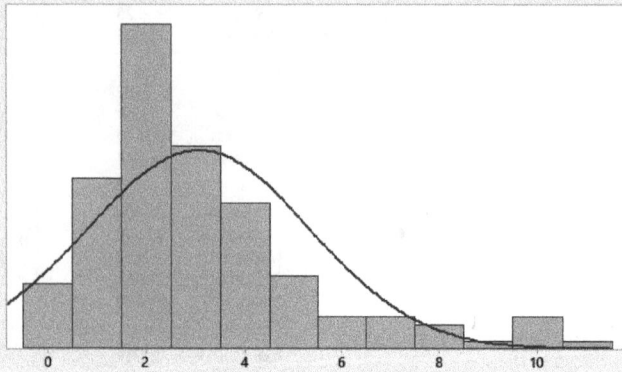

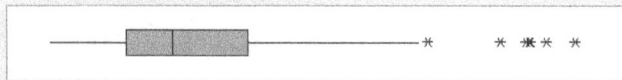

Anderson-Darling Normality Test	
A-Squared	4.43
P-Value	<0.005
Mean	3.0731
StDev	2.2514
Variance	5.0690
Skewness	1.39966
Kurtosis	2.01286
N	138
Minimum	0.0100
1st Quartile	1.5925
Median	2.5200
3rd Quartile	4.0450
Maximum	10.7300

95% Confidence Interval for Mean
2.6941   3.4521
95% Confidence Interval for Median
2.0699   2.9300
95% Confidence Interval for StDev
2.0135   2.5537

95% Confidence Intervals

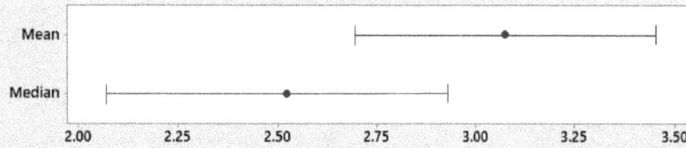

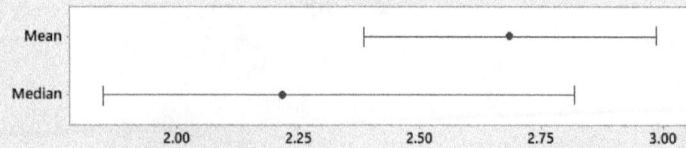

Figure C-12

Summary Report for JULY

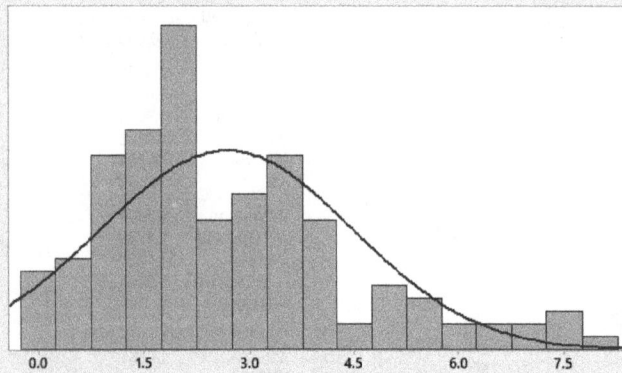

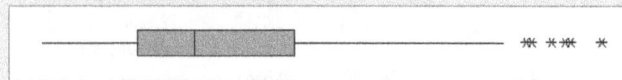

Anderson-Darling Normality Test	
A-Squared	2.79
P-Value	<0.005
Mean	2.6843
StDev	1.7920
Variance	3.2114
Skewness	0.971601
Kurtosis	0.576470
N	138
Minimum	0.0400
1st Quartile	1.4000
Median	2.2150
3rd Quartile	3.6225
Maximum	8.0200

95% Confidence Interval for Mean
2.3826   2.9859
95% Confidence Interval for Median
1.8500   2.8203
95% Confidence Interval for StDev
1.6026   2.0326

95% Confidence Intervals

Figure C-13

Summary Report for AUG

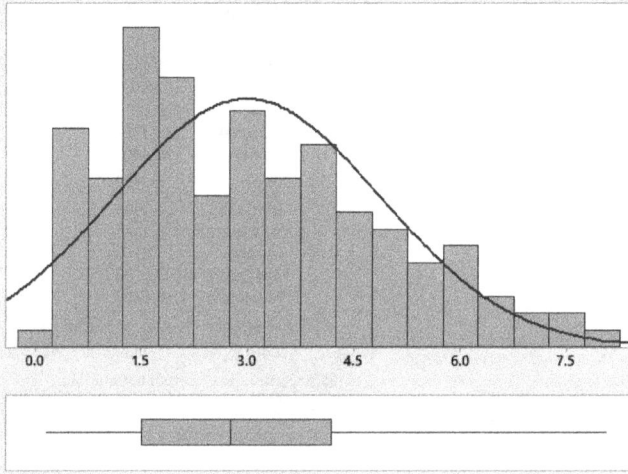

Anderson-Darling Normality Test
A-Squared	1.76
P-Value	<0.005
Mean	2.9935
StDev	1.8705
Variance	3.4986
Skewness	0.584617
Kurtosis	-0.443690
N	138
Minimum	0.1500
1st Quartile	1.5175
Median	2.7700
3rd Quartile	4.1825
Maximum	8.0700

95% Confidence Interval for Mean
2.6786   3.3083
95% Confidence Interval for Median
2.1900   3.2101
95% Confidence Interval for StDev
1.6728   2.1215

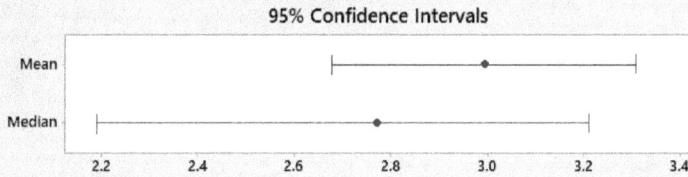

95% Confidence Intervals

Figure C-14

Summary Report for SEPT

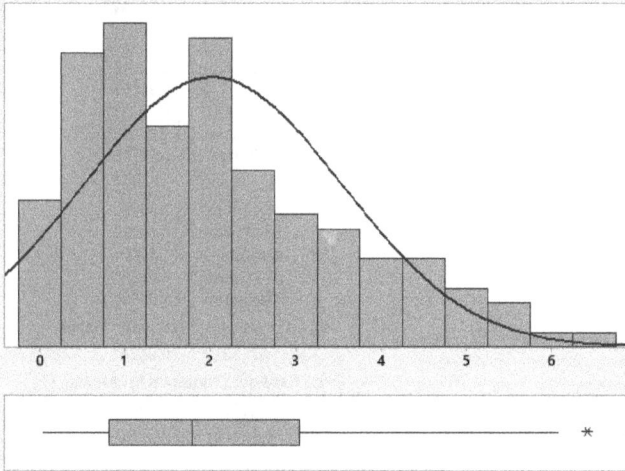

Anderson-Darling Normality Test
A-Squared	2.90
P-Value	<0.005
Mean	2.0291
StDev	1.5038
Variance	2.2614
Skewness	0.821049
Kurtosis	-0.058948
N	138
Minimum	0.0300
1st Quartile	0.8125
Median	1.7950
3rd Quartile	3.0500
Maximum	6.4200

95% Confidence Interval for Mean
1.7759   2.2822
95% Confidence Interval for Median
1.4096   2.0001
95% Confidence Interval for StDev
1.3449   1.7057

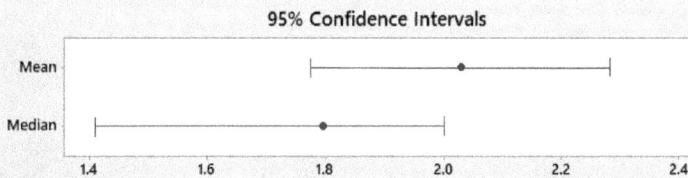

95% Confidence Intervals

Figure C-15

Summary Report for OCT

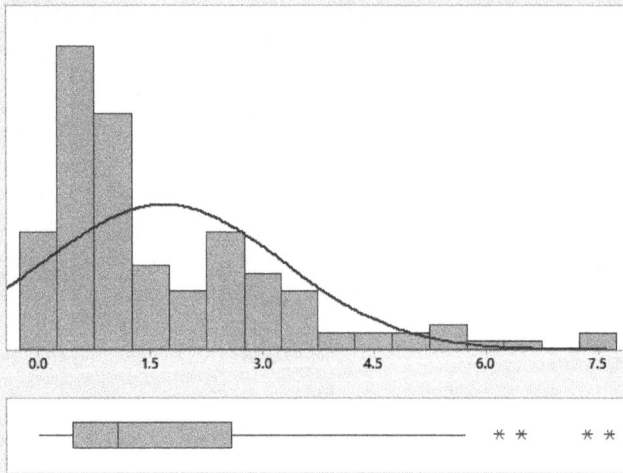

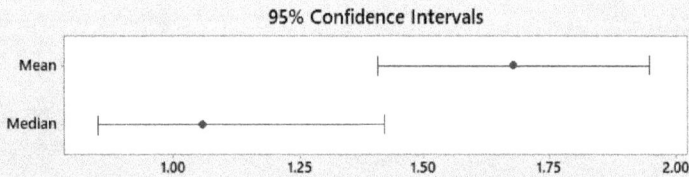

Anderson-Darling Normality Test	
A-Squared	6.40
P-Value	<0.005
Mean	1.6774
StDev	1.6010
Variance	2.5633
Skewness	1.51267
Kurtosis	2.24901
N	138
Minimum	0.0000
1st Quartile	0.4575
Median	1.0550
3rd Quartile	2.5700
Maximum	7.6400

95% Confidence Interval for Mean
1.4079   1.9469
95% Confidence Interval for Median
0.8500   1.4212
95% Confidence Interval for StDev
1.4318   1.8159

Figure C-16

Summary Report for NOV

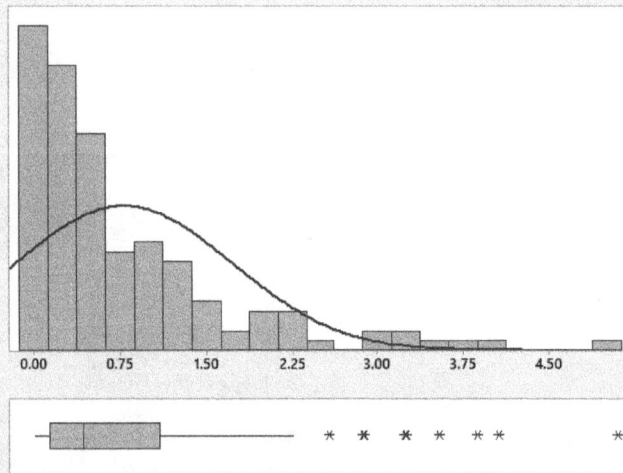

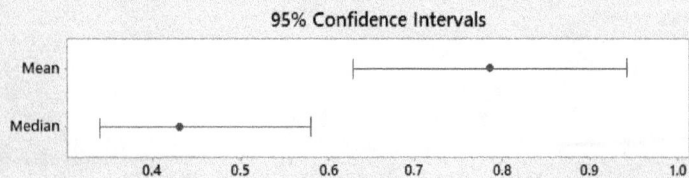

Anderson-Darling Normality Test	
A-Squared	9.41
P-Value	<0.005
Mean	0.78515
StDev	0.93657
Variance	0.87717
Skewness	2.02310
Kurtosis	4.58728
N	138
Minimum	0.00000
1st Quartile	0.14000
Median	0.43000
3rd Quartile	1.09250
Maximum	5.09000

95% Confidence Interval for Mean
0.62750   0.94281
95% Confidence Interval for Median
0.34000   0.58010
95% Confidence Interval for StDev
0.83759   1.06230

Figure C-17

Summary Report for DEC

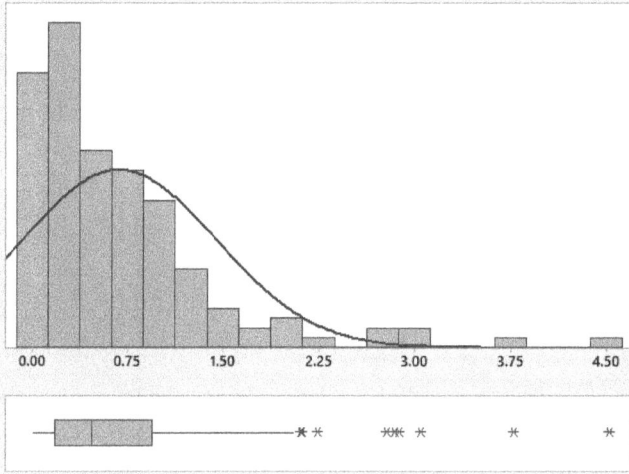

Anderson-Darling Normality Test	
A-Squared	8.07
P-Value	<0.005
Mean	0.68414
StDev	0.76144
Variance	0.57979
Skewness	2.29034
Kurtosis	6.72063
N	138
Minimum	0.00000
1st Quartile	0.17750
Median	0.46500
3rd Quartile	0.95000
Maximum	4.52000

95% Confidence Interval for Mean
0.55596   0.81231
95% Confidence Interval for Median
0.32000   0.59041
95% Confidence Interval for StDev
0.68097   0.86366

95% Confidence Intervals

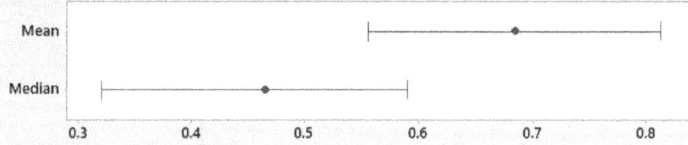

Figure C-18 - March

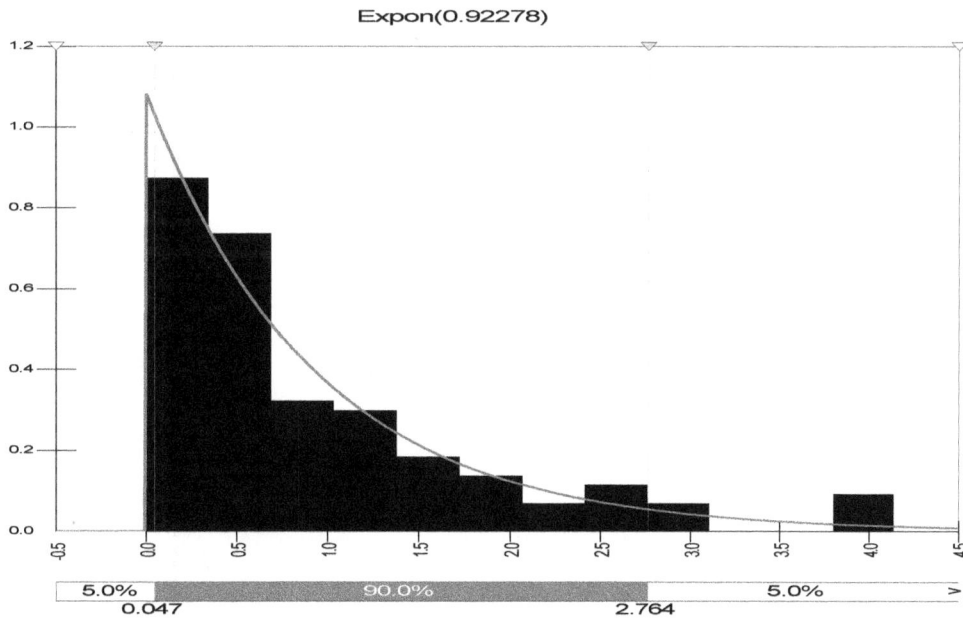

Expon(0.92278)

5.0%   90.0%   5.0%
0.047   2.764

Table C-1 ANOVA

One-way ANOVA: JAN, FEB, ... , NOV, DEC

Method

Null hypothesis	All means are equal
Alternative hypothesis	Not all means are equal
Significance level	α = 0.05

Equal variances were assumed for the analysis.

Factor Information

Factor	Levels	Values
Factor	12	JAN, FEB, MAR, APR, MAY, JUNE, JULY, AUG, SEPT, OCT, NOV, DEC

Analysis of Variance

Source	DF	Adj SS	Adj MS	F-Value	P-Value
Factor	11	1505	136.825	63.12	0.000
Error	1644	3564	2.168		
Total	1655	5069			

Model Summary

S	R-sq	R-sq(adj)	R-sq(pred)
1.47235	29.69%	29.22%	28.66%

Means

Factor	N	Mean	StDev	95% CI
JAN	138	0.5974	0.6427	(0.3516, 0.8432)
FEB	138	0.6366	0.6331	(0.3908, 0.8824)
MAR	138	0.9219	0.9496	(0.6761, 1.1677)
APR	138	1.501	1.389	(1.255, 1.747)
MAY	138	2.888	2.076	(2.642, 3.133)
JUNE	138	3.073	2.251	(2.827, 3.319)
JULY	138	2.684	1.792	(2.438, 2.930)
AUG	138	2.993	1.870	(2.748, 3.239)
SEPT	138	2.029	1.504	(1.783, 2.275)
OCT	138	1.677	1.601	(1.432, 1.923)
NOV	138	0.7852	0.9366	(0.5393, 1.0310)
DEC	138	0.6841	0.7614	(0.4383, 0.9300)

Pooled StDev = 1.47235

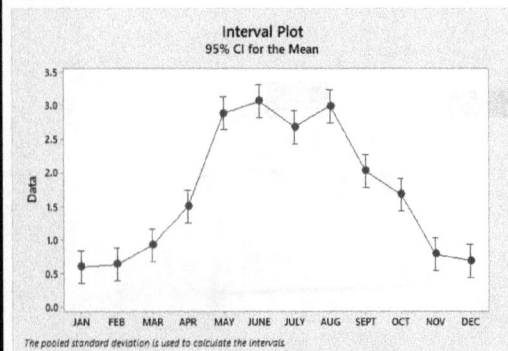

Interval Plot
95% CI for the Mean

The pooled standard deviation is used to calculate the intervals.

Figure C-19

Summary Report for JFND

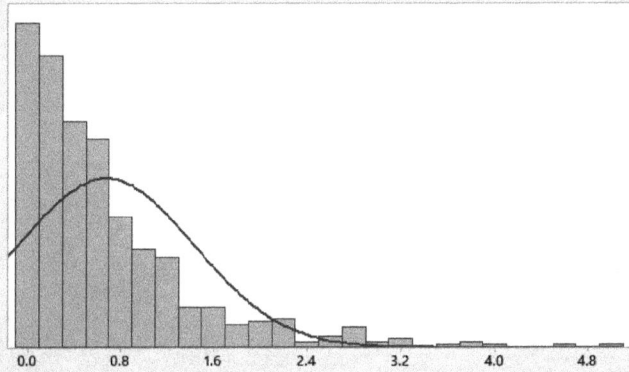

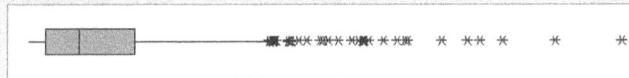

Anderson-Darling Normality Test
A-Squared 31.06
P-Value <0.005

Mean 0.67582
StDev 0.75469
Variance 0.56956
Skewness 2.10019
Kurtosis 5.68113
N 552

Minimum 0.00000
1st Quartile 0.16000
Median 0.45000
3rd Quartile 0.92750
Maximum 5.09000

95% Confidence Interval for Mean
0.61272 0.73891
95% Confidence Interval for Median
0.39000 0.51000
95% Confidence Interval for StDev
0.71264 0.80205

95% Confidence Intervals

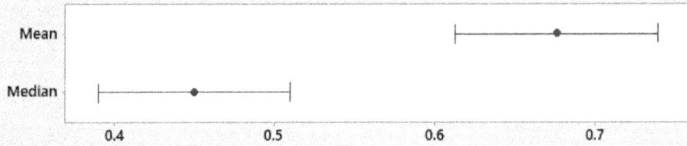

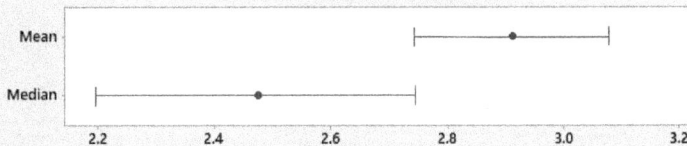

Figure C-20

Summary Report for MJJA

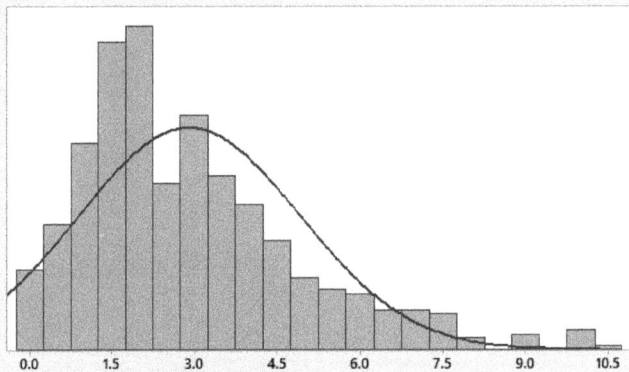

Anderson-Darling Normality Test
A-Squared 11.28
P-Value <0.005

Mean 2.9096
StDev 2.0053
Variance 4.0213
Skewness 1.12761
Kurtosis 1.27817
N 552

Minimum 0.0100
1st Quartile 1.5025
Median 2.4750
3rd Quartile 3.9225
Maximum 10.7300

95% Confidence Interval for Mean
2.7419 3.0773
95% Confidence Interval for Median
2.1950 2.7450
95% Confidence Interval for StDev
1.8936 2.1312

95% Confidence Intervals

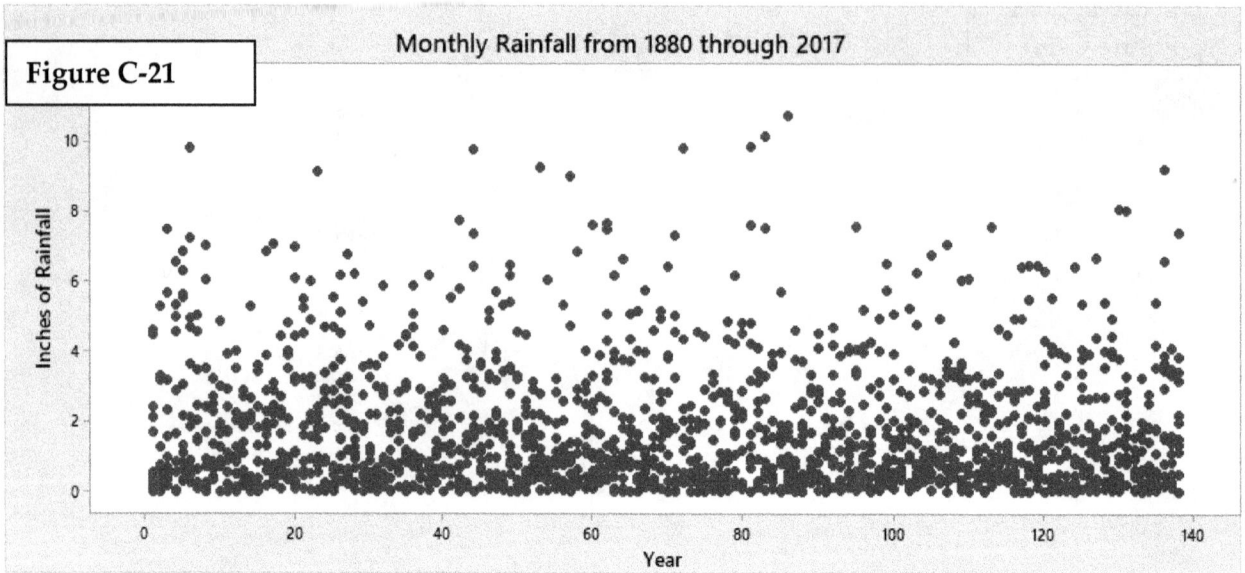

Figure C-21

Monthly Rainfall from 1880 through 2017

Use of Simulation Models for Evaluating Traffic Flow Options at Special Events Using Arena

The situation:

- Annual 4th of July Fireworks Show,
- Minor league baseball team stadium located on the fringe of a small town,
- Only one road leads to the stadium grounds with traffic entering and exiting in both directions as shown in Figure D-1,
- 5 parking lots that each hold 300 cars and feed into one exit road,
- 6,000 people arriving in 1,500 cars, and
- Average of four people per car.

These situations are quite common for emergency managers to evaluate. A big concern has always been an extreme thunderstorm over a giant sporting stadium where 100,000 people can be at risk. This example was based on a real-life occurrence with me and my family. Everyone else fell asleep and once I quit fuming and sputtering, I started thinking about what would make it better. I wrote down notes on some napkins I found under the seat and built and ran the model during my lunch break the next day on my work computer where I had Arena software loaded.

For the baseline model shown in Figure D-2, I had to estimate how long the walk times were for each segment as shown in Figure D-3. These estimates were in the form of a triangular distribution because there were sprinters and laggards in each cohort. Driving times were more uniform and varied based on the distance to the common merging point. The exit road became the bottle neck and a giant queue formed when five lanes all merged into one. At the county road, about ⅓ of the cars made a right turn that did not require them to stop. The left-hand turn took about 30 seconds longer, on average, when occasional cars came up the road. A police officer was directing traffic.

Figure D-4 shows the results from the Arena baseline run for total time, waiting time, and driving time. My total time of 120 minutes was on the high end. At the time, I mentally validated that the model was a good approximation of reality. Ideally, it would be nice to have video footage of the exiting traffic so that exact numbers could be compared.

The option that I devised is shown in Figure D-5. It involves the construction of two gravel roads and some signage. The signs would label that cars parking in the upper overflow lot would only be allowed to exit to the left and that cars in the soccer fields lot would only exit to the right. Cars in the other three lots would then flow sequentially through the permanent exit road. The schematic for the Option 1 model is shown in Figure D-6, and the results are presented in Figure D-7. The new resulting times are substantially better, with average total and waiting times reduced around 70%. Driving times increased slightly for some unknown reason. This increase may not be significantly different.

In conclusion, adding two lanes greatly reduced waiting times. Even more of a gain could have been realized if the lower lot traffic could be diverted to the gravel roads. When multiple scenarios are run, and the results compared, the name "ensemble modeling" is applied to the results. An indirect impact would result to people traveling the county road, because they would encounter three lanes of turning traffic instead of one, resulting in more wait time for them. Even though I guessed at the driving times, the percentage change is a realistic number.

As I said, it took about an hour to build and run this model in Arena. The actual model run time was ~3 seconds. Building this model is a quick and straightforward way to evaluate potential changes and could be run for many small special events in a cost-effective manner if trained personnel were assigned to the project.

Figure D-1. The layout.

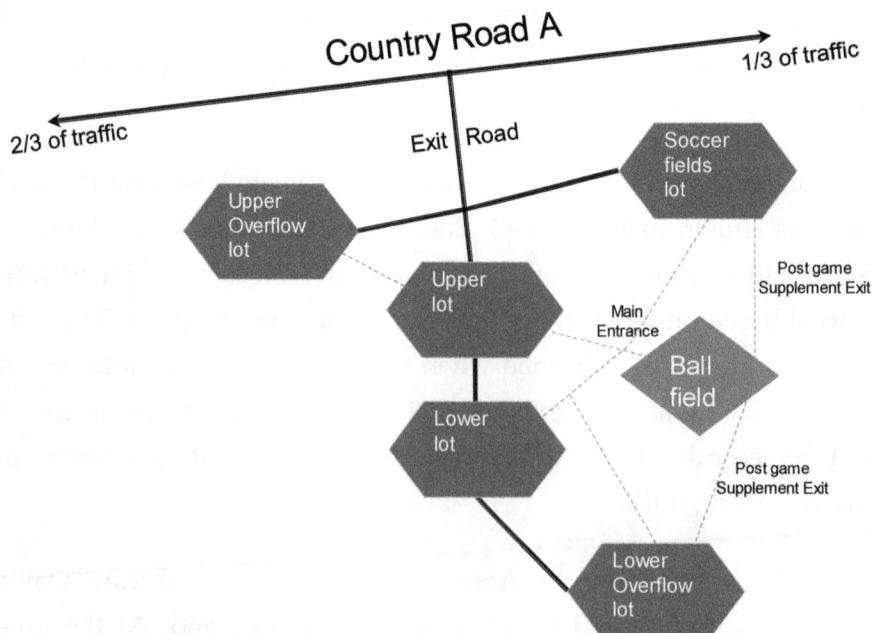

Figure D-2. Baseline model.

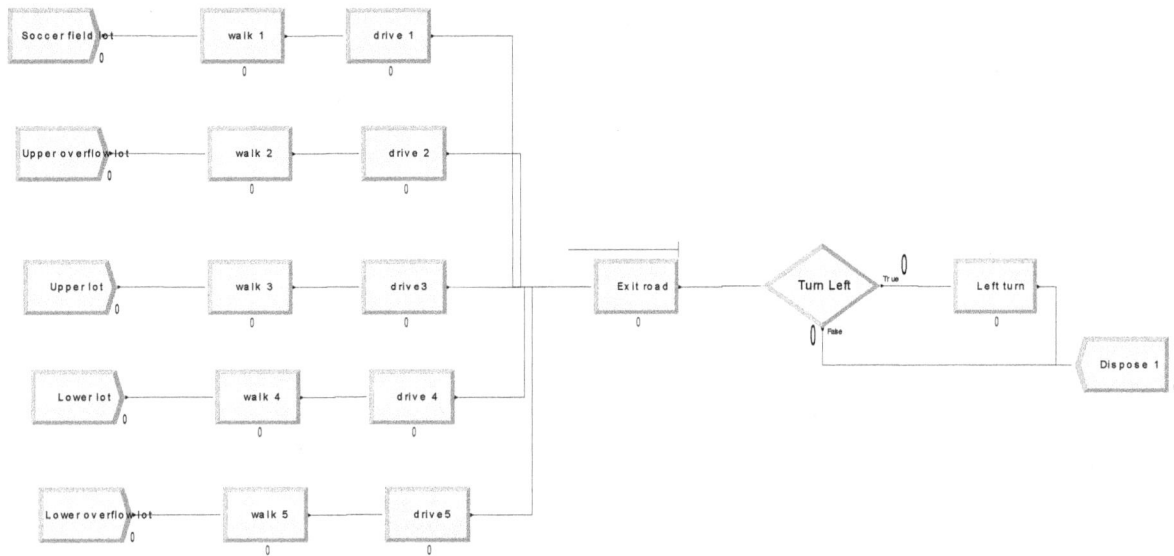

Figure D-3. Model input numbers (in minutes).

Location	Details	Walk Time	Drive time
Exit Road	One lane road	NA	Unif(0.5, 1.5)
Soccer field lot	300 cars	Tri(3, 8, 15)	Tri(2, 3, 4)
Upper overflow lot	300 cars	Tri(10, 15, 20)	Tri(2, 3, 4)
Upper lot	300 cars	Tri(5, 7, 15)	Tri(3, 4, 5)
Lower lot	300 cars	Tri(6, 9, 15)	Tri(4, 5, 6)
Lower overflow lot	300 cars	Tri(3, 8, 15)	Tri(5, 6, 7)
Leaving stands	6,0000 fans	Tri(0.5, 2, 5)	NA

Left turn adds 30 seconds driving time

Figure D-4. Baseline results.

	Baseline
Total Time	
Average	77
Minimum	10
Maximum	143
Waiting Time	
Average	62
Minimum	0
Maximum	120
Driving Time	
Average	5
Minimum	3
Maximum	9

Figure D-5. Option 1 layout.

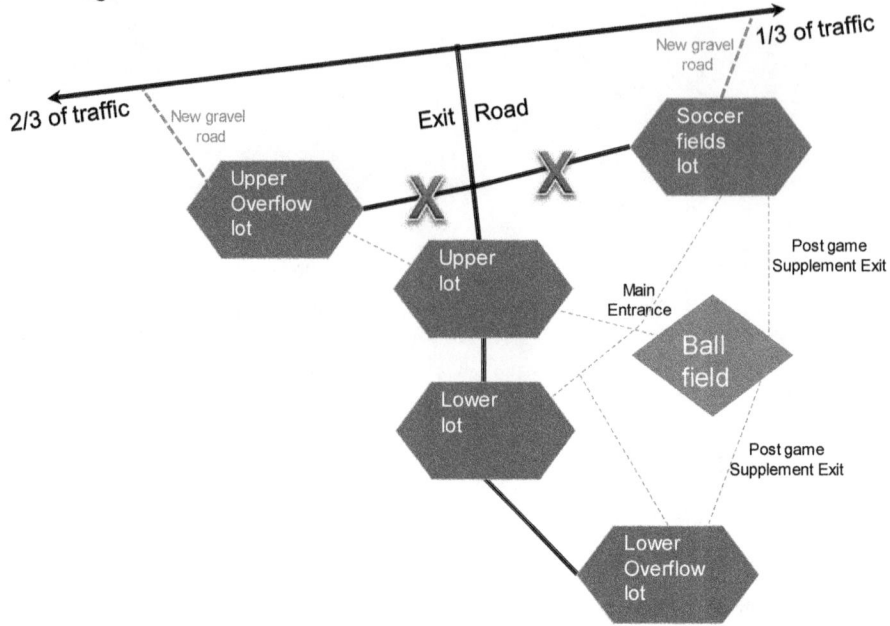

Figure D-6. Option 1 model.

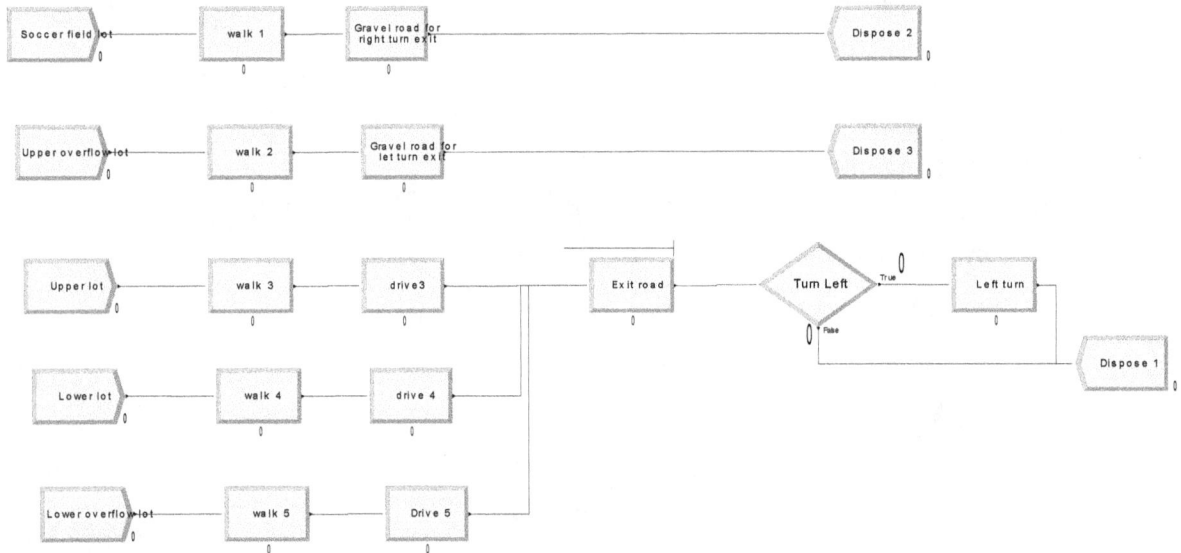

Figure D-7. Option 1 results.

	Baseline	Option 1	% Change
Total Time			
Average	77	28	64%
Minimum	10	10	0%
Maximum	143	54	62%
Waiting Time			
Average	62	11	83%
Minimum	0	0	0%
Maximum	120	35	71%
Driving Time			
Average	5	7	-33%
Minimum	3	4	-40%
Maximum	9	10	-7%

Walking time unchanged.

Application of DOE to Simulation Model *Final Stretch*

Appendix E is a case study of an analysis performed on the game *Final Stretch*,[71] which is a simulation of being the owner of a racing horse. The intent of the case study was to determine the optimum settings to generate the most fun (as in wins), money, and productivity (as defined by 11 measures[72]). Instead of just changing one factor at a time, a Design of Experiments (DOE) approach was used so that the most information could be obtained from the least number of runs. While the results are fictitious, the methodology is a very powerful example of how to validate a model, to determine what the key process input variables are, and to define the expected range for results.

Preliminary simulation runs were made to understand which settings caused the most changes and which settings to set as constants. In order to gain a high level of statistical power, six terms were selected that could be assigned three settings. In order to perform an efficient analysis, a formal Design for Experiments approach was taken. Rather than a full factorial, or even a partial factorial analysis, an advanced orthogonal approach[73] (named after Professor Taguichi) was used. A Taguichi L27 design was selected to establish a robust testing scenario that would cause as many different perturbations as possible. By using 27 runs, some factors were duplicated and the variation due to the model itself could be identified. Interactions could also be assessed with this amount of data. However, given the amount of time the first run took, the design was scaled back because I did not want to spend all of my free time running the game. The revised L9 design only used four factors, still with three settings, and was truly a screening design because direction and scale of change to the outputs could be identified. In hindsight, the L27 design would have been an example of over processing, given

---

[71] https://amzn.to/2vyIUQp.

[72] Based on the concepts in David Sumanth's book, *Total Productivity Management*.

[73] https://en.wikipedia.org/wiki/Taguchi_methods.

the coarseness of the magnitudes of change and the noise inherent in the modeled results themselves.

Next is a storyboard that describes the settings selected, the process followed, and presents the results. Also included are the Minitab output screens from the analysis of the experimental design results for main effects. Given the small number of degrees of freedom from only nine runs, the effects of interactions were not analyzed. All files, including the model run results, can be found on the website https://technicspub.com/qasa. After the experimental design was analyzed, a confirmation run was made using the median settings (the simulation model would not accept finer detail for any of the factors). That model presented several previously unseen challenges that impacted the results. More work is needed to fully understand the model.

The application of the Design of Experiments approach for assessing model performance is a cost-effective means to maximize knowledge of the functionality of a system. The essential features for a successful experimental study are an extensive knowledge of all the processes in the entire system including all interfaces and interactions, a complete collection of system-related data, a deep understanding of the facts and knowledge, and a crew of people who value data collection, technical analysis, and rational decision making. These essential elements match those that Deming identified in his Theory of Profound Knowledge.

There are three distinct advantages to this disciplined approach:

- Mature methodology with broad software support,
- Track record of supporting high-risk decisions in pharmaceutical and food supply applications, and
- Diverse cohort of trained data scientists.

Successful deployment will result in substantial benefits including:

- Focus on optimizing controllable variables that cause response in output while still accommodating uncontrollable variables and noise,
- Cost savings realized since unnecessary experiments will not be run,
- Explores the interactions on all variables when any one is changed and the ultimate effect on the response to the output variable, and
- Determine settings for input variables that both center the output in its target range and minimize the variability of the output.

PowerPoint storyboard

El Rancho de Oro
Horse Boarding/Racing Feasibility Study

Prepared by

DAM Consultants, Ltd

1

Mission

- Our client has decided to pursue horse racing as his retirement hobby with the goal being to breakeven (before taxes) and to have as much fun as possible.

- Fun is defined as winning races that bring in the most money. A win is any finish in the money.

- Innovation was the hallmark of our clients career. His long-term goal is to breed and train a Triple Crown winner. Therefore, breeding potential of his herd is a critical success factor. Actual long-term breeding is beyond the scope of this study.

2

Approach

- Prepare simulations of various scenarios using the software Final Stretch to
  - (1) screen the 3-levels of the clients 4 preferred variables (Taguichi L9 designed experiment) via 6 month runs starting after construction complete
  - (2) allow the simulation to run for 6 months to validate the preferred option.
- Baseline is to invest the $500,000 capital fund at 3% annual interest.

3

Baseline Scenario

- $500,000 invested at 3% annual interest
- Clock starts at 3/1

4

Settings
Customers 4 Variables

- Number of Horses Owned (2 4 6)*
- Race Frequency (2/month 1/month 0.5/month)
- Jockey Salary (<$100 $100-1,000 >$1,000)
- Bets ($1,000 $5,000 $10,000)

*Maximum is eight in any scenario with horses from other owners boarded in all empty stalls as an additional revenue stream.

5

El Rancho de Oro Location
(Summary of Preliminary Study Results)

- Ranch purchased in USA
  - Runner-up countries: Germany, England, Australia, Canada, United Arab Emirates, France, Italy, Japan, Sweden
- Ranch located outside of Salt Lake City has 6 stalls at level $ and purchased for $95,300
  - Runner-up cities: Miami, Dallas, Memphis, Phoenix, Denver, Cincinnati, Las Vegas, Nashville, Austin
  - Key decision criteria was "the greatest snow on earth" to accommodate the clients winter skiing passion.

6

Constants (1 of 3)

- Breeding excluded
- Facilities will be built to base quality level using as much of the clients time as available, $500,000 initial capital (easy setting in Final Stretch), and a $100,000 loan will be obtained after construction and before horses are purchased. Loan is to cover the initial purchase price. Working capital will earn 3% annualized interest.
- Owner will perform as much maintenance and other activities as his free time allows so that staff morale will be maximized.

7

Constants (2 of 3)

- Trainer, stable hand, and maintenance guy will be constant in all runs, selected based on initial morale, and hired week construction finished.
- Computer will run each race as a simulation
- Horse will be flat sprinters (no steeplechasers or trotters).
- Management attitude set to: participate
  - as opposed to: delegate, directorial, and authoritarian.
- Horses trained for 2 months before first race and default training options will be selected.
- All horse preferences purchased (e.g. blinders, ear plugs)

8

Constants (3 of 3)

- A side bet will be made for all horses raced for win-place-show.
- Feed consultants will be used for all owned horses, all vet bills will be paid, and shoes kept in good condition.
- Races to be run at preferred tracks if possible. Schedule has precedence. Race selected based on maximum purse.
- If owner has more than 3 staff, assistant must be hired, so staff set at 3.
- Default farrier, bone doctor, and vet used.

9

Noise Factors

- Horses
  - performance
  - injuries
  - sickness
- Timing of when horses sold and needing boarding
- Track conditions

10

Outsourcing

- Training facilities
- Lads
- Travel

11

Race Tracks Available in Final Stretch

- Aqueduct (New York) *sand*
- Belmont (New York) *sand*
- Churchill Downs (Kentucky) *grass*
- Delaware Park *sand*
- Del Mar (California)* *grass*
- Fair Grounds (New Orleans) *grass*
- Fairplex Park (California)* *grass*
- Gulfstream (Florida) *grass*
- Hawthorne (Chicago) *grass*
- Saratoga (New York) *sand*   *Preferred tracks

12

Jockey selection criteria

- Jockey behavior = inflexible, authoritarian, rough, relaxed, understanding, considerate, benevolent, thoughtful, tolerant, gentle, coarse, calm, hard, quiet, firm, laid-back, bossy, flexible, excessive, accommodating
- Control = 31-98
- Fame = 6 - 99
- Winnings = $567,002 - $16,448,676
- Experience = 3-99
- Salary = $70 - $9,370

- Since salary is a variable, selection will be made to maximize "control" in that salary range.

13

Horse Selection Criteria

- Gallop vs Trotters
- Rating = Listed, Level I, Level II, LIII, International
- Age = >3 year old, so race eligible
- Gains of the Year
- Career Gains
- Speed
- Acceleration
- Endurance
- Price
- Total Races
- Number Firsts, Seconds, Thirds

- Since time is a major factor in this study, all first available gallop racers will be purchased as long as cost is below $40,000.

14

Training Options

- Can train to improve strengths, decrease weaknesses, or both.
- Training options:
  - endurance
  - speed
  - behavior
  - technique
  - acceleration
  - preferences (always first after purchase)
- Training will be auto-generated by the software.

15

Construction Phase

- The following facilities must be built before horses can be boarded:
  - 2 boxes (to bring total to eight)
  - staff quarters
  - paddock
  - horsemanship facility
  - club house
- Facilities will not be upgraded to level 2 or level 3 plushness. Sensitivity analysis showed upgrades would cost an additional $150k and $100k, respectively, with only a 50% increase in boarding rate. Pay back period exceeded 7.5 years.
- Facilities not built: hanger, mechanization, saddlery, paddock upgrade, riding alley, straight line tack, loop tack, circle of tether, mechanical walker, grooming room, solarium, pool.

16

Starting Point

	$
Cash on Hand	$443,264
Building Value	$152,950
Line of Credit	$45,885
Fame	9
Construction End	5/22
First claiming race	5/29
6 month date	11/20

17

Staff Hired

	Stable hand	Trainer	Maint. Man
Name	Javier Pascual	Jacky Plaza	Ted Mancebo
Organizational Ability	2	2	198/510
Weekly Salary	$500	$510	$422
Initial Morale	56	56	61

18

Financials Screen*

- General
  - Cash
  - Net Income
  - Budgeted Cash

- Income of the Week
  - Boarding-Training
  - Race Gains
  - Betting Gains
  - Horse Sales
  - Loan (already established)

- Loan
  - Borrowed
  - Duration left (weeks)
  - Rate (%)
  - Weekly Reimbursement
  - Total cost

- Expenses
  - Horse's food
  - Salaries
  - Building maintenance
  - Loan payment
  - Vet bills
  - Training
  - Purchases (horse-building)
  - Races (declarations - bets)
  - Outsoucing

*Tax considerations not addressed.

19

Output Measures

- Model Generated
  - Wins, Financials, Fame Score, Morale Score
- Productivity
  - Total Productivity
  - Partial Productivities (10)

20

Total Productivity Measure

- Productivity = Income / Expenses
  - Income = Boarding fees + Winnings (race+bet) + Interest on Working Capital
  - Expenses = all expenses
    - Human = Payroll for trainer, stable hand, maintenance guy, jockey
    - Horse = feed, training, vet bills
    - Capital = monthly payment on loan
    - Material = horse purchases, building maintenance.
    - Energy = *
    - Overhead = entry fees, bets
    - Outsourcing = training facility, lads, travel
- Software provides all detail. Only aggregate number collected

*Client is retiring from an electrical utility and ranch usage is paid using a different account.

21

Partial Productivity Measures

- Efficiency #1 = Number Wins / Race
- Efficiency #2 = Number Wins / Rest Ratio
- Effectiveness = Number Wins / Average Weekly Expense
- Quality #1 = Fun = Number Wins
- Quality #2 = Average Fame
- Quality of Worklife #1 = Horse morale = Average Form
  - times injured, injuries, endurance not captured
- Quality of Worklife #2 = Average Staff morale
- Innovation = Breeding potential (not captured)
- Profitability #1 = Final Cash Amount / Baseline Value at end
- Profitability #2 = Average Weekly Income / Average Horses Boarded

22

Process

- Time step to be weekly
  - fame, financials, race results, morale, and horse status collected weekly
  - owners free time minimized at each time step
  - After each horse bought
    - training and race schedule set
    - farrier, vet, and feed consultant scheduled
    - sickness over rides training and race schedules
- First horse purchased at first race after construction, then from ads. All horse less than $40,000. Category = listed. If ad posted before scheduled race, horse will be purchased.
- News not read.

23

Designed Experiment Matrix

Number Horses	Races/Month	Jockey Salary	Bets
2	0.5	<$100	$1,000
2	1	$101-999	$5,000
2	2	>$1000	$10,000
4	0.5	$101-999	$10,000
4	1	>$1000	$1,000
4	2	<$100	$5,000
6	0.5	>$1000	$5,000
6	1	<$100	$10,000
6	2	$101-999	$1,000

*Design set by Minitab.

24

Example Data Sheet

See spreadsheet for all results.

25

Example Productivity Calculations

See spreadsheet for all scenario results.

26

Scenario results

run	prod	effic 1	effic 2	effect	qual 1	qual 2	QWL 1	QWL 2	Prof 1	Prof 2
1	0.15	0	0	0	0	17	51	65	0.61	221
2	0.21	0.28	3.2	0.0003	2	19	82	65	0.59	25
3	0.1	0	0	0	0	21	31	65	0.33	199
4	0.3	0.4	4.7	0.000224	2	18	76	65	0.56	751
5	0.07	0.2	5	0.000165	2	19	51	65	0.28	238
6	0.08	0.16	4.4	0.00029	3	16	40	65	0.38	218
7	0.2	0.25	9.5	0.00017	2	17	45	65	0.38	1524
8	0.21	0.25	7.2	0.00017	4	21	48	65	0.07	2124
9	0.13	0.13	3.6	0.000222	3	17	33	65	0.26	832

27

Lessons Learned

- Staff morale was a constant
- Rest helped winning percentage
- Scenario 6 greatly affected by injuries
- Scenario 7 greatly affected by delayed acquisitions

28

DOE Results

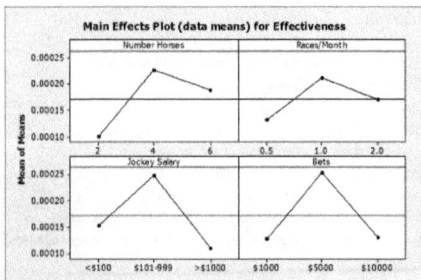

See Appendix E.2 for all results.

29

DOE Rank Results

(Factor with lowest number is most influential on the measurement listed)

	# Horses	Race Freq	Jockey Salary	Bets
Prod	4	1	2	3
Effic 1	1	2	3	4
Effic 2	1	3	4	2
Effect	2	4	1	3
Qual 1	1	2	3	4
Qual 2	3	2	4	1
QWL 1	3	1	2	4
QWL 2	2.5	2.5	2.5	2.5 #
Prof 1	1	2	3	4
Prof 2	1	3	4	2
mean	1.9	2.2	2.9	3.0
#1	5	2	1	1
#4	1	1	3	4
All 10	1	3	4	2
8 of 10*	4	1	2	3

*Excludes Effectiveness and Profitability #2 due to scalar inequalities.
Not carried forward

30

DOE Settings to Maximize

	# Horses	Race Freq	Jockey Salary	Bets
Prod	3	1	2	3
Effic 1	2	2	2	2
Effic 2	3	2	3	2
Effect	2	2	2	2
Qual 1	3	2	2	2
Qual 2	1	2	3	3
QWL 1	2	2	2	2
QWL 2	2.5	2.5	2.5	2.5
Prof 1	1	1	2	2
Prof 2	3	1	1	3
mean	2.2	1.7	2.1	2.3
All 10	3	1	1	3
8 of 10	2	2	2	2

31

Optimized Scenario

- Client will have the most fun and do financially best with the median* settings:
  - Number of Horses owned = 4
  - Race Frequency = 1/month
  - Jockey Salary = $100-1,000
  - Bets = $5,000
- Validation Run was less than optimal
  - all purchases made in first month, which hurt cash flow
  - boarders fees were below average
  - 4 bouts of sickness (most of any scenario)

*Simulation does not allow finer divisions of the settings.

32

Results

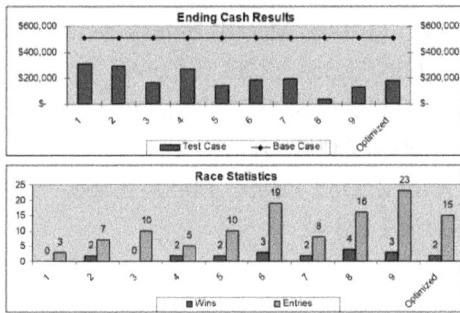

33

Recommendation

- Client will have difficulty making money with this arrangement leading to a tax loss.
- Building additional training facilities may help.
- Breeding may help long-term money concerns
- Would be a fun retirement hobby.

34

```
3 0.1800 0.1033 0.1233 0.2033
Delta 0.0300 0.1133 0.0900 0.0867
Rank 4 1 2 3
```

DOE output

Exhibit E.3-1

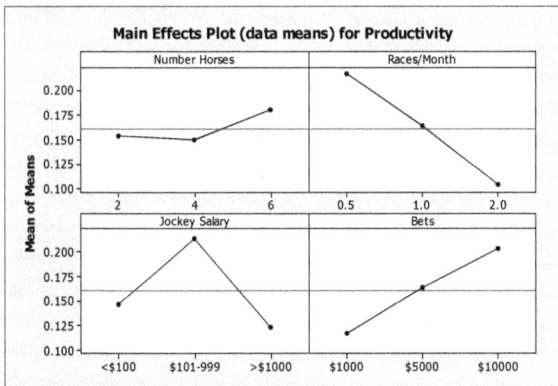

Response Table for Means

Level	Number Horses	Races/ Month	Jockey Salary	Bets
1	0.1533	0.2167	0.1467	0.1167
2	0.1500	0.1633	0.2133	0.1633

Exhibit E.3-2

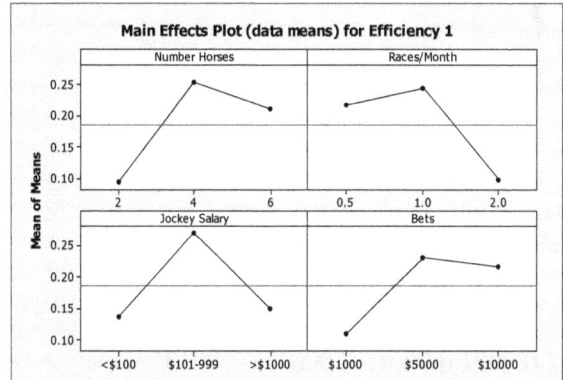

Response Table for Means

Level	Number Horses	Races/ Month	Jockey Salary	Bets
1	0.09333	0.21667	0.13667	0.1100
2	0.25333	0.24333	0.27000	0.2300
3	0.21000	0.09667	0.15000	0.216

Delta 0.16000 0.14667 0.13333 0.1200
Rank 1 2 3 4

Exhibit E.3-3

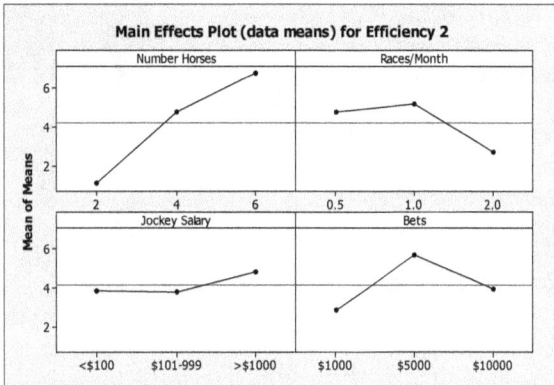

Response Table for Means
Number Races/ Jockey
Level Horses Month Salary Bets
1 1.067 4.733 3.867 2.880
2 4.713 5.147 3.833 5.700
3 6.767 2.667 4.847 3.967
Delta 5.700 2.480 1.013 2.820
Rank 1 3 4 2

Exhibit E.3-4

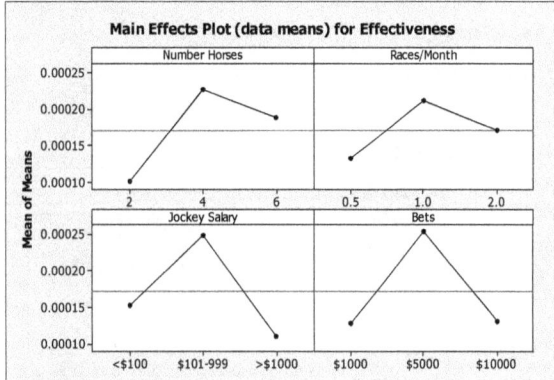

Response Table for Means
Number Races/ Jockey
Level Horses Month Salary Bets
1 0.0001 0.00013 0.00015 0.00013
2 0.0002 0.00021 0.00025 0.00025
3 0.0002 0.00017 0.00011 0.00013
Delta 0.0001 0.00008 0.00014 0.00012
Rank 2 4 1 3

Exhibit E.3-5

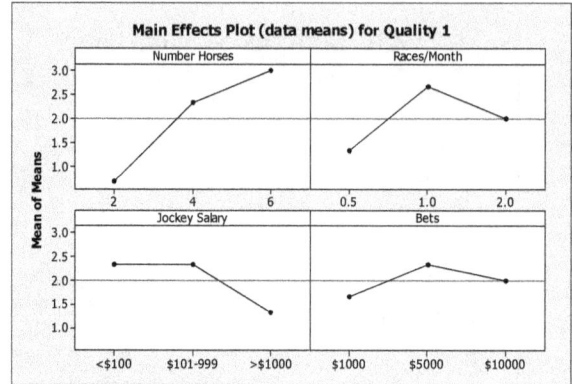

Response Table for Means
Number Races/ Jockey
Level Horses Month Salary Bets
1 0.6667 1.3333 2.3333 1.6667
2 2.3333 2.6667 2.3333 2.3333
3 3.0000 2.0000 1.3333 2.0000
Delta 2.3333 1.3333 1.0000 0.6667
Rank 1 2 3 4

Exhibit E.3-6

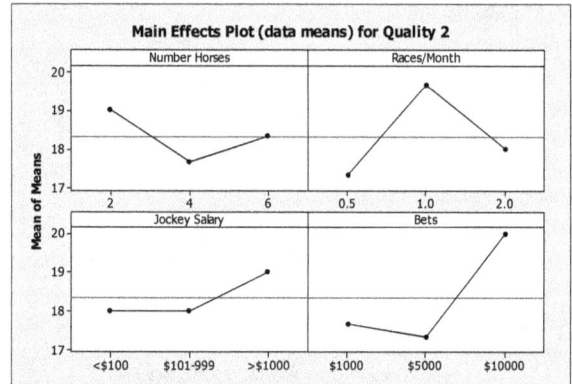

Response Table for Means
Number Races/ Jockey
Level Horses Month Salary Bets
1 19.00 17.33 18.00 17.67
2 17.67 19.67 18.00 17.33
3 18.33 18.00 19.00 20.00
Delta 1.33 2.33 1.00 2.67
Rank 3 2 4 1

Exhibit E.3-7

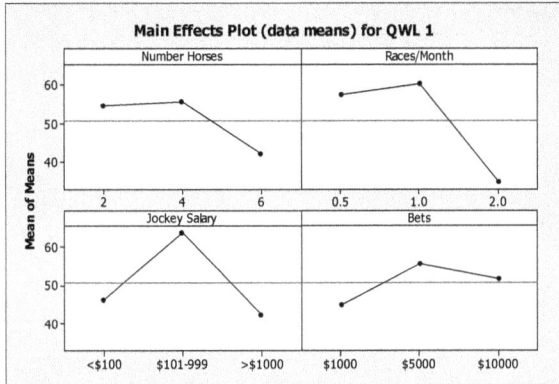

Main Effects Plot (data means) for QWL 1

Response Table for Means
Number Races/ Jockey
Level Horses Month Salary Bets
1 54.67 57.33 46.33 45.00
2 55.67 60.33 63.67 55.67
3 42.00 34.67 42.33 51.67
Delta 13.67 25.67 21.33 10.67
Rank 3 1 2 4

Exhibit E.3-8

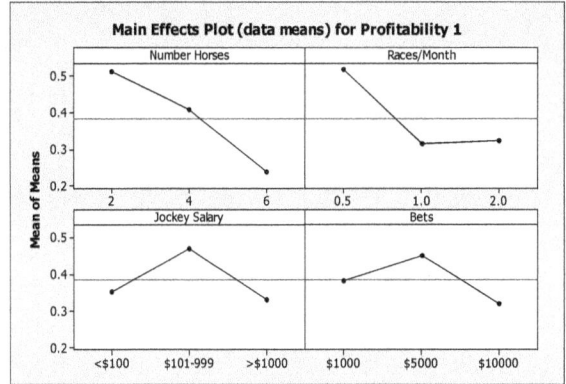

Main Effects Plot (data means) for QWL 2

Response Table for Means
Number Races/ Jockey
Level Horses Month Salary Bets
1 65.00 65.00 65.00 65.00
2 65.00 65.00 65.00 65.00
3 65.00 65.00 65.00 65.00
Delta 0.00 0.00 0.00 0.00
Rank 2.5 2.5 2.5 2.5

Exhibit E.3-9

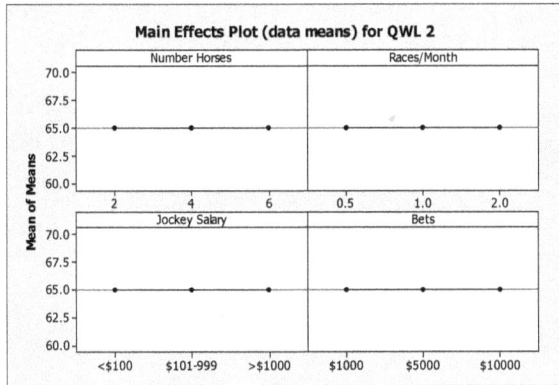

Main Effects Plot (data means) for Profitability 1

Response Table for Means
Number Races/ Jockey
Level Horses Month Salary Bets
1 0.5100 0.5167 0.3533 0.3833
2 0.4067 0.3133 0.4700 0.4500
3 0.2367 0.3233 0.3300 0.3200
Delta 0.2733 0.2033 0.1400 0.1300
Rank 1 2 3 4

Exhibit E.3-10

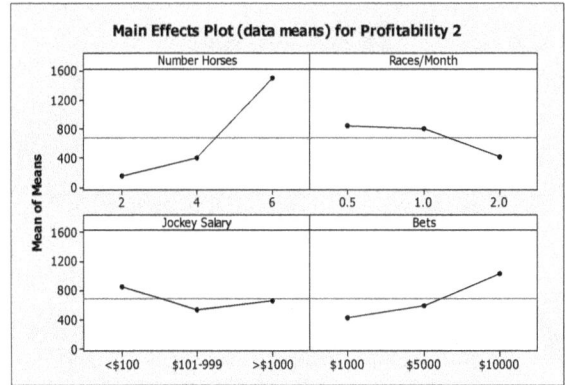

Main Effects Plot (data means) for Profitability 2

Response Table for Means
Number Races/ Jockey
Level Horses Month Salary Bets
1 148.3 832.0 854.3 430.3
2 402.3 795.6 536.0 589.0
3 1493.3 416.3 653.7 1024.7
Delta 1345.0 415.7 318.4 594.3
Rank 1 3 4 2

Exhibit E.3-11

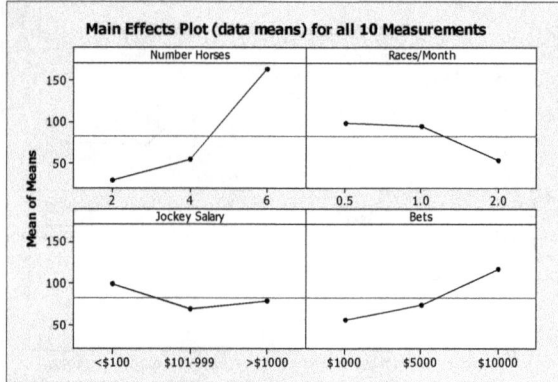

Main Effects Plot (data means) for all 10 Measurements

Response Table for Means

Number Races/ Jockey

Level Horses Month Salary Bets

1 28.95 97.87 99.05 56.32

2 54.85 94.92 68.98 73.58

3 162.91 53.92 78.68 116.80

Delta 133.96 43.95 30.07 60.49

Rank 1 3 4 2

Exhibit E.3-12

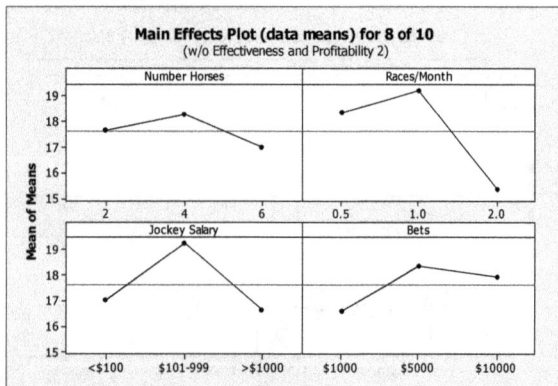

Main Effects Plot (data means) for 8 of 10
(w/o Effectiveness and Profitability 2)

Response Table for Means

Number Races/ Jockey

Level Horses Month Salary Bets

1 17.64 18.34 17.02 16.60

2 18.27 19.19 19.22 18.36

3 16.97 15.36 16.64 17.92

Delta 1.31 3.83 2.58 1.76

Rank 4 1 2 3

Acknowledgements

First, I want to thank the unknown librarian at the Carnegie Library who sent home a copy of *Cheaper by the Dozen*, the biography of motion-study expert Frank Gilbreth who had invented the QWERTY keyboard. I was in middle school and home for a week with walking pneumonia and my mother asked my dad to stop at the library to check out some books, so I would not watch game shows all the time. It was a book about efficiency and saving time and squeezing the most possible out of a process. It resonated with me and I have spent my life trying to reach the Nirvana he was striving for. My wife and kids might even say I obsess a bit on it. This desire for not wasting time has lead me down the path that landed in the garden of analytics that I have described herein. It has been a fun journey.

Second, I want to thank my Dad. He was a Quality Control manager at a steel mill who was obsessed with perfection. My desires do not hold a candle to his absolute focus. Some people thought the way he taught me to mow the grass when I was 11 was strange, but I accepted it as the normal and natural way to do things. He had a detailed, written procedure of exactly what to do, how to compensate for spring wetness, and inspection criteria. I never understood why my friends laughed at me over it because it made total logical sense to me. As a matter of fact, the results were so good that I got gigs from the neighbors to do their yards! Thanks Dad!

Third, I want to thank my geology classmate who told me this oilfield fable when we were sitting around a campfire during a field trip. Once upon a time, there was a small oil company that needed to hire a replacement for the CEO who had announced a retirement date. The stated goal was to promote from within and they would interview the head geologist who lead development, the head engineer who lead production, and the head geophysicist who lead exploration. The geologist showed up wearing a jacket and tie with field boots and jeans and smoking on a pipe. As stereotypical as you would expect. After chatting with the board for an hour, the chairman said, "we have one last question" and proceeded to ask, "what is 2 + 2"? The geologist harrumphed, walked to the window, relit the pipe, and with a lowered head eventually said, "it is an integer between 3 and 5." The next interview was with the engineer who arrived in a conservative suit with a bulge in the coat pocket where the ever-present calculator was carried. When asked what "2 + 2 was", the calculator was produced with a smile and a remark like "I always carry this since one never knows when it will be needed." After mashing buttons for a minute, the answer was stated as exactly 4.0000. Finally, it was the geophysicist's turn. In walked a disheveled mess of hair, in a T-shirt, woolly socks, and sandals, and carrying a crumbled map and some printouts. When asked what "2 + 2 was" the

271

eyes behind the thick glasses blinked and blinked and blinked, then both printouts were flipped open, and out came mumbles: "I was just working on a problem like that", and "the printout I need is on my desk – excuse me while I fetch it!" Back in a flash and with a flurry of flipping pages and highlighting some numbers and circling others, an answer was ready: "what do you want it to be?" The moral of the story is you must know when to hedge, when to be exact, and when to be flexible. In the end, they hired the new CEO out of marketing once again; someone who understood how to be adaptable. A goal my inner engineer is still trying to master.

Fourth, I need to thank my network. Over the years many co-workers, classmates, training-session attendees, clients, and students served as my beta-testers while I tried to find the words to explain quantitative analysis. Sometimes I had yet to master the technique. Other times the data were so convoluted it almost defied explanation. There were some people whose conversations helped me shape this book: Mario Beruvides, Mary Jane Davis, Gary Shipman, Evan Chiacchiaro, Charles Key, John Francis McGrath, and JD Booker. Thanks!

This book would not be possible without the professionals at Technics Publications. Steve Hoberman and his team were phenomenal in how fast and how precisely they transformed a clunky draft into a polished product. I appreciate the care that was shown.

Finally, I must thank my wife, Betty, and three daughters for putting up with me while working on this book. I journeyed down the proverbial rabbit hole multiple times while building the examples and when spooling off text that resulted in me being less than attentive. They have already stated that they don't care to ever hear about million row spreadsheets or European soccer teams ever again.

Bibliography

Anand, P. (1993). *Foundations of Rational Choice Under Risk*. Oxford University Press.

Antonopoulos, A.M. (2017). *Mastering Bitcoin: Programming the Open Blockchain*. 2nd Edition. O'Reilly Books.

Aristotle, *Categories*. Any edition.

Asimov, I. (1950). *I, Robot*. Spectra, reprint edition.

Auburn, D. (2001). *Proof*. Faber and Faber.

Ayers, I. (2007). *Super Crunchers: Why thinking-by-numbers is the new way to be smart*, Bantam Books.

Baker, S. (2008). *The Numerati*, Houghton Mifflin Company.

Baltagi, Y. and Kussener, F. (2013). *Advantages of Bootstrap Forest for Yield Analysis*. SAS White Paper.

Banks, J., Carson, J.S. II, Nelson, B.L., and Nicol, D.M. (2005). *Discrete-event System Simulation*, 4th *Ed*. Pearson Education, Inc.

Bardossy, G. and Fodor, J. (1997). Assessment of the Completeness of Mineral Exploration by the Application of Mineral Exploration by the Application of Fuzzy Arithmetic and Prior Information. *Acta Polytechnica Hungarica*, Vol.2, pp. 15-31.

Batini, C. and Scannapieco, M. (2006). *Data and Information Quality: concepts, methodologies and techniques*. Springer.

Bayes, T. (1763). An Essay Towards Solving a Problem in the Doctrine of Chances. *Philosophical Transactions of the Royal Society*, Vol. 53, pp. 370-418.

Benedict, Ruth (1934). *Patterns of Culture*. Houghton Mifflin.

Bertalanffy, Ludwig von (1968). *General Systems Theory: Foundations, Development, Applications*. George Braziller.

Beruvides, M.G. and Canto, A.M. (2004). *The Myth of Organism and Mechanism. Some Thoughts on Organizational Structure*. Paper presented to American Society for Engineering Management.

Berry, M.J.A. and Linoff, G.S. (1997). *Data Mining Techniques: for marketing, sales, and customer relationship management*. Wiley Computer Publications. John Wiley and Sons, NY.

Birkenshaw, J. (2010). *Reinventing Management: Smarter Choices for Getting Work Done*. Jossey-Bass.

Boehm, B.W. (1981). *Software Engineering Economics*. Prentice Hall.

Bookstaber, R. (2007). *A Demon of Our Own Design: Markets, hedge funds, and the perils of financial innovation*. John Wiley & Sons, Inc.

Bourdieu, Pierre (1990). *The Logic of Practice*. Stanford University of Press.

Brassard, M. (1988). *The Six Sigma Memory Jogger: A Pocket Guide of Tools for Continuous Improvement*. GOAL/QPC.

Brown, J.R., Ed. (1987). *Soil Testing: Sampling, Correlation, Calibration, and Interpretation*. Soil Science Society of America Special Publication Number 21.

Buckingham, M. and Clifton, D.O. (2001). *Know, Discover Your Strengths*. The Free Press.

Camm, J.D., M.J. Fry, and J. Shaffer. (2017). A Practitioner's Guide to Best Practices in Data Visualization. *Interfaces,* Vol. 47, Issue 6, pp. 473-488.

Chamberlin, T.C. (1890). The method of multiple working hypotheses: *Science* (old series) Vol. 15, pp. 92-96; reprinted 1965, Vol. 148, pp. 754-759.

Chapman, Pete, Julian Clinton, Randy Kerber, Thomas Khabaza, Thomas Reinartz, Colin Shearer, Rüdiger Wirth (2000). *CRISP-DM 1.0: Step-by-step data mining guide*. SPSS White Paper.

Conover, W.J. (1999). *Practical Nonparametric Statistics*. Third Edition, Joh Wiley and Sons.

Covey, Stephen R. (1989). *The Seven Habits of Highly Effective People: Powerful Lessons in Personal Change*. Simon and Schuster

Daimler-Chrysler Corp, Ford Motor Company, General Motors Corporation. (2002). *Measurement Systems Analysis: Reference Manual*. Daimler-Chrysler Corporation Ford Motor Company, General Motors Corp.

Darnall, R.W. (1994). *Achieving TQM on Projects: The Journey of Continuous Improvement.* Project Management Institute.

Davenport, T. H. (2005). *Thinking for a Living,* Harvard Business School Press.

Davenport, T. H. (2006). *Competing on Analytics,* Harvard Business Review. January.

Davenport, T.H. (2014). *Big Data @ Work: Dispelling the myths, uncovering the opportunities.* Harvard Business School Press.

Davenport, T. H. and Harris, J.G. (2007). *Competing with Analytics: The New science of winning.* Harvard Business School Press. Second edition 2017.

Davenport, T.H., Harris, J.G., and Morrison, R. (2010). *Analytics at Work: Smarter decisions, better results.* Harvard Business School Press.

Davenport, T.H. and Kim, J. (2013). *Keeping Up with the Quants: Your guide to understanding & using analytics.* Harvard Business School Press.

Davenport, T.H. and Kirby, J. (2016). Only Humans Need Apply: Winners and losers in the age of smart machines. Harper Collins Publishers.

Davenport, T.H. and Manville, B. (2012). *Judgement Calls: 12 Stories of big decisions and the teams that got them right.* Harvard Business School Press.

Davenport, T. and Redman, T.C. (2015). *Build Data Quality into the Internet of Things.* Wall Street Journal, August 31.

Davis, K. and Patterson, D. (2012). *Ethics of Big Data.* O'Reilly Media.

DeFusco, R.A. et al. (2007). *Quantitative Investment Analysis, 2nd Ed.* John Wiley & Sons Inc.

Deming, W.E. (1982). *Out of the Crisis,* MIT Press.

Deming, W.E. (1994). *The New Economics for industry, government, education,* MIT Press.

Dennis, A., Wixom, B.H., and Tegarden, D. (2015). *System Analysis and Design: An Object-Oriented Approach with UML.* (5th Edition). Wiley.

Domingos, P. (2015). *The Master Algorithm: How the quest for the ultimate learning machine will remake our world.* Basic Books.

Duhigg, C. (2012). *How Companies Learn Your Secrets.* New York Times, Feb. 16.

Ewing, M. (2017). *Once Upon an Algorithm: How Stories Explain Computing.* MIT Press.

Fenton, N.E., Whitty, R.W., and Iizuka, Y. (1995). *Software Quality Assurance and Measurement: A World Perspective.* International Thomson Computer Press.

Fox, C., Levitin, A., and Redman, T.C., (1994). The notion of data and its quality dimensions. Information *Processing and Management.* 30(1):9-19.

Galin, D. (2004). *Software Quality Assurance.* Pearson/Addison Wesley.

Geary, D.C. (1994). *Children's Mathematical Development.* American Psychological Association.

Gigenrenzer, G., Swijink, Z., Porter, T., Daston, L., Beatty, J., and Kreger, L. (1990). *The Empire of Chance: How probability changed science and everyday life,* Cambridge University Press.

Godbole, N.S. (2004). *Software Quality Assurance.* Alpha Science.

Goldratt, E.M. (1992). *The Goal: A process of ongoing improvement.* Second Revised Edition. North River Press.

Granville, V. (2014). *Developing Analytic Talent: Becoming a data scientist.* John Wiley & Sons, Inc.

Grayson, J., Gardner, S., and Stephens, M. (2015). *Building Better Models with JMP© Pro.* SAS Institute.

Han, J. and Kamber, M. (2006). *Data Mining: Concepts and techniques,* Elsevier.

Hansen, Rick (2009). *Buddha's Brain: The practical neuroscience of happiness, love & wisdom.* New Harbinger Publications, Inc.

Hastie, T. and Tibshirani, R. (2009). *Elements of Statistical Learning.* Springer 2nd Ed.

Heilbroner, R.L. (1953). *The Worldly Philosophers.* The Time Reading Program printing, Simon and Schuster.

Hillier, F.S. and Lieberman G.J. (1974). *Operations Research.* Holden-Day, Inc.

Hoberman, Steve. (2015). *Data Model Scorecard: Applying the Industry Standard on Data Model Quality.* Technics Publications.

Hofstede, G. Hofstede, G.J., and Minkov, M. (2010). *Cultures and Organizations: Software of the Mind.* McGraw Hill.

Hotelling, H. (1947). Multivariate Quality Control, Illustrated by the Air Testing of Sample Bombsights. Chapter 3 in Eisenhart, C., Hastay, M.W., and Wallis, W.A., Eds. *Selected Techniques of Statistical Analysis for Scientific and Industrial Research and Production and Management Engineering.* Statistical Research Group, Columbia University. New York, McGraw-Hill Book Company, Inc.

Hybertson, D. (2006). *Using Models and Abstraction to Extend and Unify Systems.* IEEE/SMC International Conference on System of Systems Engineering.

Johnson, E.G. and Tukey, J.W. (1987). *Graphical Exploratory Analysis of Variance Illustrated on a Splitting of the Johnson and Tsao Data.* In *Design, Data and Analysis,* C.L. Malloe, (Ed.), pp. 171-244.

Juran, J.M. (1974). *Quality Control Handbook, 3rd Ed.* McGraw-Hill Book Co.

International Atomic Energy Association (1988). *Manual on Quality Assurance for Computer Software Related to the Safety of Nuclear Power Plants,* Vienna, #282.

Ishikawa, K. (1986). *Guide to Quality Control.* Asian Productivity Organization, 2nd Ed.

Kaplan, R.D. (2018). *The Return of Marco Polo's World: War, Strategy, and American Interests in the Twenty-first Century.* Random House.

Klösgen, W. and Zytko, J.M., Eds. (2002). *Handbook of Data Mining and Knowledge Discovery.* Oxford University Press.

Knaflic, Cole Nussbaumer (2015). *Storytelling with Data: A Data Visualization Guide for Business Professionals.* Wiley.

Kotz, S. and van Dorp, J.R. (2004) *Beyond Beta: Other Continuous Families of Distributions with Bounded Support and Applications.* World Scientific.

Kozlowski, S.W.J. and Klein, K.J., Eds. (2000). *Multilevel Theory, Research, and Methods in Organizations.* Jossey-Bass.

Krugman, P. (2013). *The Excel Depression.* New York Times, April 18.

Kuhn, T.S. (1962). *The Structure of Scientific Revolutions,* University of Chicago Press.

Lapointe, F. and Legendre, P. (1984). A Classification of Pure Malt Scotch Whiskies. In *Applied Statistics,* Vol. 43, pp. 237 – 257.

Law, A.M. and Kelton, W.D. (2000). *Simulation Modeling and Analysis, 3rd Ed.* McGraw-Hill, Inc.

Lee, Y.M., L.L. Pepino, R.Y. Wang, and J.D. Funk, (2006). *Journey to Data Quality.* MIT Press.

Lewis, C.I. (1924). *Mind and the World Order.* Dover Publications.

Line, Ernest (2011). *Ready Player One.* Crown Publishers.

Lohr, S. (2012). *The Age of Big Data.* New York Times, February 11, 2012.

Loveman, G. (May, 2003). *Diamonds in the Data Mine.* Harvard Business Review, Harvard Business School Publishing Corporation.

Martin, David (1978). *A General Theory of Secularization.* Harper Row.

May, T. (2009). *The New Know: Innovation Powered by Analytics*, Wiley.

Mazalov, V. (2014). *Mathematical Game Theory and Applications.* John Wiley & Sons, Ltd.

McGilvery, D. (2008). *Executing Data Quality Projects: Ten Steps to Quality Data and Trusted Information.* Morgan Kaufmann.

McGrath, D.A. (2009). *On the Application of Non-Zero Sloping Analysis to Complex Systems: A Case Study Involving Mutual Funds.* Ph.D. Dissertation, Texas Tech University.

Montgomery, D.C. (2013). *Introduction to Statistical Quality Control*, Seventh Edition, Wiley.

Nagle, T., Redman, T.C., and Sammon, D. (2017). Only 3% of Companies' Data Meets Basic Quality Standards. Harvard Business Review, September 11.

Nisbet, R., Elder, J., and Miner, G. (2009). *Handbook of Statistical Analysis and Data Mining Applications.* Elsevier, Inc.

Oliveira, A. (2017). *The Digital Mind: How Science is Redefining Humanity.* MIT Press.

Oreskes, N., Shrader-Frechette, K., and Belitz, K. (1994). Verification, Validation, and Confirmation of Numerical Models in the Earth Sciences. *Science*, Vol. 263. Pp. 641-646.

Pasquale, F. (2015). *The Black Box Society: The secret algorithms that control money and information.* Harvard University Press.

Patterson, Scott (2008). *The Quants: How a New Breed of Math Whizzes Conquered Wall Street and Nearly Destroyed It.* Crown Business.

Perket, C.L., Ed. (1986). *Quality Control in Remedial Site Investigation: Hazardous and Industrial Solid Waste Testing.* ASTM STP 925, Fifth Volume.

Pfluger, S.L. and McGowan, C.L. (1990). Software metrics in a process maturity framework. *Journal of Systems and Software,* 12: 255-261.

Pinker, S. (1997). *How the Mind Works.* W.W. Norton & Co.

Pirsing, Robert M. (1976). *Zen and the Art of Motorcycle Maintenance: An Inquiry Into Values.* Bantam.

Provost, F. and Fawcett, T. (2013). *Data Science for Business: What you need to know about data mining and data-analytic thinking.* O'Reilly Media, Inc.

Pryor, Francis (2010). *The Making of the British Landscape.* Penguin.

Pylyshyn, Z.W. (1987). *The Robot's Dilemma: The frame problem in artificial intelligence.* Ablex Publishing Company.

Pyzdek, T. (2003). *The Six Sigma Handbook: A complete guide for green belts black belts, and managers at all levels.* McGraw-Hill.

Ragsdale, C.T. (2004). *Spreadsheet Modeling & Decision Analysis, 4th Ed.* South-Western.

Raut, S. (2017). *18 Big Data tools you need to know!* Posted May 27 to the Data Science Central group of LinkedIn.

Redman, T.C. (1992). *Data Quality: management and technology.* Bantam Books.

Redman, T.C. (2001). *Data Quality: The Field Guide.* Digital Press.

Redman, T.C. (2013). *Data Driven: Profiting from Your Most Important Business Asset.* Harvard Business Press, Dec. 30.

Redman, T.C. (2015). *Can Your Data be Trusted?* Harvard Business Review. October 29.

Redman, T.C. (2016a). *Data Quality Should Be Everyone's Job.* Harvard Business Review. May 20.

Redman, T.C. (2016b). *Assess Whether You Have a Data Quality Problem.* Harvard Business Review. July 28.

Redman, T.C. (2016c). *Bad Data Costs the US $3 Trillion Per Year.* Harvard Business Review. September 22.

Render, B. et al. (2015). *Quantitative Analysis for Management.* Pearson Education, Inc.

Rieuf, E. (2017). *6 Easy Steps to Learn Naïve Bayes Algorithm (with code in Python).* In Data Science Central group of LinkedIn, September 3, 2017.

Rockwell Automation. (2004). *Arena© Basic User's Guide.* Rockwell Software, Inc.

Ross, J.W. and Feeny, D.R. (1999). *The Evolving Role of the CIO,* Center for Information Systems Research, Sloan School of Management, MIT, CISR WP No. 308, Sloan WP No. 4089.

Salsburg, D. (2001). *The Lady Tasting Tea: How statistics revolutionized science in the twentieth century.* Owl Books; Henry Holt & Co.

Seife, C. (2010). *Proofiness: The Dark Arts of Mathematical Deception.* Viking Press.

Seila, A.F. (2004). *Spreadsheet Simulation.* Proceedings of the 2004 Winter Simulation Conference. Terry College of Business, University of Georgia.

Senge, P. (1990). *The Fifth Discipline: The Art & Practice of The Learning Organization.* Currency Doubleday.

Shewhart, W.A. (1931). *Economic Control of Quality of Manufactured Product.* New York, D. Van Nostrand Company, Inc. 50th Anniversary Commemorative Issue, Milwaukee, WI, Quality Press, 1981.

Simon, H.A. (1977). *The New Science of Management Decision.* Revised Edition. Prentice-Hall, Inc.

Smith, A. (2009). *Wall Street Revalued: Imperfect markets and inept central bankers,* Wiley.

Smith, P.G. (2003). A Portrait of Risk. *PM Network,* April.

Smith, P.G. and G.M. Merritt (2002). *Proactive Risk Management: Controlling Uncertainty in Product Development.* Productivity Press, New York.

Sumanth, D.J. (1997). *Total Productivity Management (TPmgt): A Systemic and Quantitative Approach to Compete in Quality, Price and Time.* CRC Press.

Tapscott, Don and Tapscott, Alex (2016). *The Blockchain Revolution: How the Technology Behind Bitcoin is Changing, Money, Business, and the World.* Portfolio.

Taleb, N.N. (2004). *Fooled by Randomness: The Hidden role of chance in life and in the market.* Random House, 2nd edition.

Taylor, F.W. (1912). *The Principles of Scientific Management*, Dover Publications.

Thaler, R.H. and Sunstein, C.R. (2003). Libertarian Paternalism. *The American Economic Review*, Vol. 93, no. 2, pp. 175-179.

Tian, J. (2005). *Software Quality Engineering: Testing Quality Assurance and Quantifiable Improvement*. IEEE/Wiley Interscience.

Tufte, E.R. (1983). *The Visual Display of Quantitative Information*, Graphics Press.

Tufte, E.R. (1990). *Envisioning Information*. Graphics Press.

Tufte, E.R. (1997). *Visual Explanations*, Graphics Press.

Tufte, E.R. (2006). *Beautiful Evidence*, Graphics Press.

Tukey, J.W. (1977). *Exploratory Data Analysis*, Addison-Wesley Publishing Co.

Tukey, J.W. and Wilk, M.B. (1970). "Data Analysis and Statistics: Techniques and Approaches," In E.R. Tufte, *The Quantitative Analysis of Social Problems*, Addison-Wesley Publishing Co., pp. 370-390.

Turkle, S. (2015). *Reclaiming Conversation*. Penguin Press, NY.

Vallianatos, F. (2009). *A Non-Extensive Approach to Risk Assessment*. Natural Hazards and Earth System Sciences. Copernicus Publications on behalf of the European Geosciences Union.

Walkenbach, John (2010). *Microsoft Excel 2010 Bible*. John Wiley and Sons.

Wand, Y., and Wang, R.Y. (1996). Anchoring data quality dimensions in ontological foundations. *Communications of the ACM*. 39(11): 86-95.

Warren and Brandeis. (1890). *The Right to Privacy*. Harvard Law Review, Vol. IV, No. 5.

Weaver, W. (1963). *Lady Luck: The Theory of probability*. Dover Publications.

Weelan, Charles (2013). *Naked Statistics: Stripping the Dread from the Data*. W.W. Norton and Company.

Weinberger, D. (2007). *Everything is Miscellaneous: The Power of the new digital disorder*, Times Books, Henry Holt and Company.

Western Electric Co. (1956). *Statistical Quality Control Handbook.* 2nd edition, 11th printing (1985). Charlotte, NC, Delmar Printing Co.

Wheeler, D.J. (2003). *Making Sense of Data*, SPC Press.

Wimmer, Andreas (2008). The Making and Unmaking of Ethnic Boundaries: A multilevel Process Theory. *Am. J. of Sociology* 113:970-1022.

Working, H. and Hotelling, H. (1929). Application of the Theory of Error to the Interpretation of Trends. *Journal of the American Statistical Association.* Vol. 24, no. 165, pp. 73-85.

Yau, N. (2013). *Data Points: Visualization that means something.* John Wiley & Sons, Inc.

Index

www.ingramcontent.com/pod-product-compliance
Lightning Source LLC
Chambersburg PA
CBHW081055220326
41598CB00038B/7110